KS002612068
B: Ks
DE01510496X

D1725029

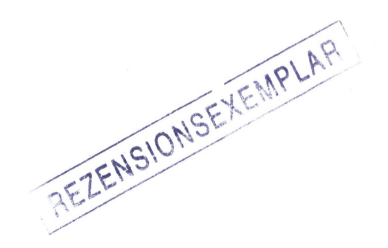

Springer Series on
ATOMIC, OPTICAL, AND PLASMA PHYSICS   32

Springer
Berlin
Heidelberg
New York
Barcelona
Hong Kong
London
Milan
Paris
Singapore
Tokyo

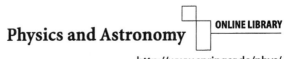

Springer Series on
ATOMIC, OPTICAL, AND PLASMA PHYSICS

The Springer Series on Atomic, Optical, and Plasma Physics covers in a comprehensive manner theory and experiment in the entire field of atoms and molecules and their interaction with electromagnetic radiation. Books in the series provide a rich source of new ideas and techniques with wide applications in fields such as chemistry, materials science, astrophysics, surface science, plasma technology, advanced optics, aeronomy, and engineering. Laser physics is a particular connecting theme that has provided much of the continuing impetus for new developments in the field. The purpose of the series is to cover the gap between standard undergraduate textbooks and the research literature with emphasis on the fundamental ideas, methods, techniques, and results in the field.

27 **Quantum Squeezing**
By P.D. Drumond and Z. Spicek

28 **Atom, Molecule, and Cluster Beams I**
Basic Theory, Production and Detection of Thermal Energy Beams
By H. Pauly

29 **Polarization, Alignment and Orientation in Atomic Collisions**
By N. Andersen and K. Bartschat

30 **Physics of Solid-State Laser Physics**
By R.C. Powell
(Published in the former Series on Atomic, Molecular, and Optical Physics)

31 **Plasma Kinetics in Atmospheric Gases**
By M. Capitelli, C.M. Ferreira, B.F. Gordiets, A.I. Osipov

32 **Atom, Molecule, and Cluster Beams II**
Cluster Beams, Fast and Slow Beams, Accessory Equipment and Applications
By H. Pauly

Series homepage – http://www.springer.de/phys/books/ssaop/

Vols. 1–26 of the former Springer Series on Atoms and Plasmas are listed at the end of the book

Hans Pauly

Atom, Molecule, and Cluster Beams II

Cluster Beams, Fast and Slow Beams,
Accessory Equipment and Applications

With 165 Figures

 Springer

Professor Dr. Hans Pauly
MPI für Strömungsforschung
Bunsenstrasse 10, 37073 Göttingen, Germany

ISSN 1615-5653

ISBN 3-540-67673-2 Springer-Verlag Berlin Heidelberg New York

Library of Congress Cataloging-in-Publication Data: Pauly, Hans, 1928–. Atom, molecule, and cluster beams/ Hans Pauly. p. cm.– (Springer series on atomic, optical, and plasma physics, ISSN 1615-5653 ; 28) Includes bibliographical references and index. Contents: 1. Basic theory, production, and detection of thermal energy beams. ISBN 3540669450 (v.1 : acid-free paper). 1. Particle beams. I. Title. II. Series. QC 793.3.B4 P38 2000 539.7'3–dc21   00-038829

This work is subject to copyright. All rights are reserved, whether the whole or part of the material is concerned, specifically the rights of translation, reprinting, reuse of illustrations, recitation, broadcasting, reproduction on microfilm or in any other way, and storage in data banks. Duplication of this publication or parts thereof is permitted only under the provisions of the German Copyright Law of September 9, 1965, in its current version, and permission for use must always be obtained from Springer-Verlag. Violations are liable for prosecution under the German Copyright Law.

Springer-Verlag Berlin Heidelberg New York
a member of BertelsmannSpringer Science+Business Media GmbH

© Springer-Verlag Berlin Heidelberg 2000
Printed in Germany

The use of general descriptive names, registered names, trademarks, etc. in this publication does not imply, even in the absence of a specific statement, that such names are exempt from the relevant protective laws and regulations and therefore free for general use.

Typesetting: Camera-ready copies by the author
Cover concept by eStudio Calmar Steinen
Cover design: *design & production* GmbH, Heidelberg

Printed on acid-free paper   SPIN: 10771865   57/3141/ba - 5 4 3 2 1 0

Preface

Volume 1 contains a description of the gas kinetic and gas dynamic foundations of molecular beam formation as well as a detailed survey of the methods of production and detection of thermal energy beams together with the underlying physical principles and examples of typical applications.

The second volume completes these methods with a description of the production and detection of fast molecular beams as well as of the production and analysis of cluster beams (Chaps. 1 and 2). The latter are not only an important means to investigate the properties of this "new phase of matter" intermediate between single atoms and molecules and the solid-state bulk material, but also offer many technical applications in lithography, epitaxy, catalysis, chemical engineering, and material science. The subsequent chapters, 3 and 4, deal with the accessory equipment of molecular beam work such as selectors and analyzers for velocity and quantum state. Finally, the reader is introduced to the fascinating new field of atom manipulation, slow and cold atoms, atom optics, and atom interferometry (Chap. 5). This new topic of molecular beam work, which has experienced an impetuous growth during recent years, has already led to great scientific success and holds important perspectives for the future.

As in Vol. 1, a special effort is made to outline the physical foundations of the various experimental methods and to explain and demonstrate them by numerous scientific applications in various research areas, paying special attention to recent developments. Numerous references allow readers interested in further details easy access to supplementary literature. Thoughout Vol. 2, references to Vol. 1 are marked by the Roman numeral I in front of the chapter or section number.

Thus, although the choice of the material reflects the tastes and inclinations of the author, it is hoped that this book will be both a useful reference for scientists engaged in research with molecular beams and a textbook for advanced undergraduates and graduate students in order to give them a solid background in these techniques. Scientists and engineers interested in different areas where molecular beams represent an alternative to other techniques, may also be interested in reading this book so that they can estimate the usefulness of molecular beams in their work.

Many people whose names have already been mentioned in the preface of Vol. 1, have helped me while writing this book. Again, their valuable aid is gratefully acknowledged. In particular, I am indebted to my long-standing colleague and friend Prof. R.K.B. Helbing from the University of Windsor for his valuable help while preparing the English version of this book and for numerous suggestions and improvements.

Finally, my thanks are also due to Springer Verlag for the pleasant cooperation and the expert help during the course of the whole project.

Göttingen *H. Pauly*
April 2000

Contents

1. Fast Beams, Production and Detection 1
  1.1 Charge Exchange 4
    1.1.1 Ground-State Particles 6
    1.1.2 Metastable Particles 13
    1.1.3 Rydberg Atoms 17
  1.2 Neutralization of Negative Ions by Electron Detachment 18
    1.2.1 Photodetachment 18
    1.2.2 Collisional Detachment 21
    1.2.3 Autodetachment 21
  1.3 Neutralization of Ions by Collisions with Metal Surfaces 22
  1.4 Aerodynamic Acceleration 24
  1.5 Plasma and Gas Discharge Sources 29
    1.5.1 Arc-Heated Jet Sources 29
    1.5.2 Radiofrequency Discharges 34
    1.5.3 Laser-Sustained Plasmas 34
    1.5.4 Hollow Cathode Discharges 37
    1.5.5 Corona Discharges 37
    1.5.6 Glow Discharges 37
  1.6 Laser Ablation 39
  1.7 Sputtering 42
  1.8 Photolysis 48
    1.8.1 Beam Experiments 48
    1.8.2 Gas-Phase Measurements 49
    1.8.3 Oriented Reactants 49
  1.9 Electron-Stimulated Desorption 50
  1.10 Other Methods 51
    1.10.1 Mechanical Acceleration 51
    1.10.2 Shock Waves 52
    1.10.3 Recoil Nuclei of Radioactive Decays 52
  1.11 Some Examples of Fast Beam Applications 53
    1.11.1 Coaxial Laser Spectroscopy with Fast Beams 53
    1.11.2 Photofragment Translational Spectroscopy 56
    1.11.3 Precision Measurements of Lifetimes 57
    1.11.4 Merged Beams 59

1.11.5 Injection of Fast Beams into Fusion Plasmas 62
　　　1.11.6 Fast Beam Methods for Plasma Diagnostics 63
　1.12 Fast Beam Detection 65
　　　1.12.1 Surface Ionization (Langmuir–Taylor Detector) 65
　　　1.12.2 Bolometer and Pyroelectric Detectors 67
　　　1.12.3 Laser-Induced Fluorescence 68
　　　1.12.4 Secondary Electron Emission 68
　　　1.12.5 Collision-Induced Fluorescence 69

2. Production and Diagnostics of Cluster Beams 71
　2.1 Survey of Methods for Cluster Formation 75
　2.2 Supersonic Jets 77
　　　2.2.1 Influence of Nozzle Shape 78
　　　2.2.2 Influence of Carrier Gases 80
　　　2.2.3 Technical Realization of Cluster Sources 81
　2.3 Gas Aggregation 82
　2.4 Surface Erosion 85
　　　2.4.1 Sputtering 86
　　　2.4.2 Laser Ablation 87
　　　2.4.3 Pulsed Arc Discharges 89
　　　2.4.4 Liquid Metal Ion Sources (LMIS) 91
　2.5 Laser-Induced Pyrolysis 92
　　　2.5.1 Multiphoton Infrared Dissociation and Photosensitization 93
　　　2.5.2 Source Design and Applications 95
　2.6 Doping of Clusters and Production of Mixed Clusters 96
　　　2.6.1 Formation of Mixed Clusters by Coexpansion 97
　　　2.6.2 Gas Aggregation 97
　　　2.6.3 Particle Capture ("Pick–up" Sources) 97
　　　2.6.4 Cluster Aggregation 99
　　　2.6.5 Laser Ablation 100
　　　2.6.6 Pulsed Arc Discharges 101
　2.7 Generation of Excited Clusters 101
　2.8 Determination of Size Distributions in Cluster Beams 102
　　　2.8.1 Electron Diffraction 102
　　　2.8.2 Light Scattering 105
　　　2.8.3 Gas Scattering 107
　　　2.8.4 Atom Diffraction 120
　　　2.8.5 Mass Spectrometric Methods 122
　　　2.8.6 Other Methods 125
　2.9 Some Cluster Applications 132
　　　2.9.1 Cluster Beam Deposition 132
　　　2.9.2 Cluster Impact Lithography 133
　　　2.9.3 Examples of Experimental Results 134
　　　2.9.4 Cluster Beams in High Energy Physics 136

3. Velocity Measurement and Selection ... 137
3.1 Mechanical Selectors ... 138
3.1.1 Slotted Disk Velocity Selector (Fizeau Principle) ... 139
3.1.2 Calibration ... 150
3.1.3 Special Designs ... 152
3.1.4 Slotted Cylinder Velocity Selector ... 153
3.1.5 Other Designs ... 154
3.2 Time-of-Flight Methods ... 156
3.2.1 Resolution and Methods of Deconvolution ... 157
3.2.2 Cross-Correlation Method ... 161
3.2.3 Experimental Details ... 164
3.2.4 Calibration ... 169
3.3 Doppler Shift Measurements ... 171
3.3.1 Experimental Technique and Resolution ... 172
3.3.2 Measurements of Differential Scattering Cross Sections ... 175
3.4 Deflection in Inhomogeneous Magnetic Fields ... 177
3.4.1 Two-Wire Field ... 178
3.4.2 Magnetic Hexapole Fields ... 184
3.5 Deflection by Gravity ... 191
3.6 Determination of the de Broglie Wavelength ... 193
3.6.1 Diffraction from Crystal Surfaces ... 193
3.6.2 Diffraction from Transmission Gratings ... 195
3.7 Beam Deflection by Photon Recoil ... 195

4. State Selection ... 197
4.1 Potentials of Cylinder-Symmetric and Planar Fields ... 199
4.1.1 Monopole Field ... 200
4.1.2 Multipole Fields ... 201
4.1.3 Two-Dimensional, Periodic Fields ... 202
4.2 Deflection in Inhomogeneous Magnetic Fields ... 204
4.2.1 Effective Magnetic Dipole Moment of Atoms ... 204
4.2.2 Two-Wire Field (Rabi Field) ... 206
4.2.3 Quadrupole Sector Field ... 211
4.2.4 Two-Pole Field ... 213
4.2.5 Multipole Fields ... 213
4.3 Deflection in Inhomogeneous Electrostatic Fields ... 220
4.3.1 Effective Electric Dipole Moment of Molecules ... 221
4.3.2 Rabi and Two-Pole Fields ... 222
4.3.3 Electrostatic Multipole Fields ... 224
4.3.4 Other Fields ... 230
4.4 Magnetostatic and Electrostatic Traps ... 230
4.4.1 Three-Dimensional Quadrupole Fields ... 232
4.4.2 Ioffe Trap ... 233
4.5 Nonadiabatic (Majorana) Transitions ... 234

X   Contents

    4.6 Technical Details 235
    4.7 Applications of State Selection by Inhomogeneous Fields 240
        4.7.1 Molecular Beam Magnetic and Electric Resonance Method 240
        4.7.2 Cesium Frequency and Time Standard 242
        4.7.3 Atomic and Molecular Collisions 244
        4.7.4 Cluster Investigations 247
        4.7.5 Atomic Polarizabilities 248
        4.7.6 State Analysis 249
        4.7.7 Gas–Surface Interaction 249
        4.7.8 Miscellaneous Applications 250
    4.8 Optical Methods for State Selection 251
        4.8.1 Optical Pumping 252
        4.8.2 Selective State Depopulation 252
        4.8.3 Selective Population of an Atomic State 254
        4.8.4 Selective Population of a Molecular State 255
        4.8.5 Two-Photon Processes 257
        4.8.6 Photodissociation 260
        4.8.7 State Selection in Excited States 260
        4.8.8 Stimulated Raman Adiabatic Passage (STIRAP) 263

5. Slow Atom Beams, Traps, and Atom Optics 267
    5.1 Radiation Pressure Forces 268
        5.1.1 Photon Recoil Force 269
        5.1.2 Optical Dipole Force 271
        5.1.3 Optical Beam Slowing by Photon Recoil 273
        5.1.4 Atomic Beam Deflection by Photon Recoil 278
    5.2 Trapping and Cooling of Atoms 282
        5.2.1 Optical Molasses and Cooling Mechanisms 282
        5.2.2 Atom Traps 284
        5.2.3 Atom Traps as Sources for Slow and Cold Atoms 287
        5.2.4 Methods of Beam Compression 289
    5.3 Examples of Applications 291
        5.3.1 Precision Spectroscopy and Frequency Standards 291
        5.3.2 Atomic Collision Processes 292
        5.3.3 Bose–Einstein Condensation 293
        5.3.4 Photoassociative Spectroscopy 295
        5.3.5 Atom Lithography 297
    5.4 Atom Optics 298
        5.4.1 Atom-Optical Elements 299
        5.4.2 Lenses 300
        5.4.3 Mirrors 303
        5.4.4 Atom Waveguides 308
        5.4.5 Diffraction Gratings 309
        5.4.6 Prisms 315

5.5 Atom Interferometry 316
    5.5.1 A Mach–Zehnder Interferometer 317
    5.5.2 Some Experimental Results 318

References 323

Subject Index 369

Conversion Factors for Pressure and Energy Units 375

Contents (Volume 1)

1. The Role of Molecular Beams in the 20th Century
  1.1 Historical Development
  1.2 Main Applications of Molecular Beams
  1.3 Thermal Energy Molecular Beam Applications in other Fields.
  1.4 Fast Beam Applications
  1.5 Examples of Molecular Beam Machines

2. Fundamentals of Kinetic Gas Theory
  2.1 Ideal Gases in Thermodynamic Equilibrium
  2.2 Quantum Statistics
  2.3 Molecular Flow Through an Ideal Aperture
  2.4 Molecular Flow Through Channels

3. Fundamental Principles of Gas Dynamics
  3.1 Some Fundamentals of Thermodynamics
  3.2 Governing Equations of Steady Flow
  3.3 One-Dimensional Flow
  3.4 Two-Dimensional Flow
  3.5 Free-Jet Expansion
  3.6 The Transition to Nonequilibrium Conditions
  3.7 Internal Energy Relaxation
  3.8 Binary Gas Mixtures
  3.9 Condensation and Cluster Formation

4. Thermal Energy Molecular Beam Sources
  4.1 Experimental Requirements
  4.2 Gas Sources (4–600 K)
  4.3 Ovens for Gases and Solids
  4.4 Laser Ablation
  4.5 Sputtering Sources
  4.6 Recirculating Sources and Sources for Special Applications
  4.7 Sources for Beams of Radicals

4.8 Production of Metastable Particles
4.9 Rydberg Atoms
4.10 Pulsed Beam Sources
4.11 Sources of Slow and Cold Atoms

5. Detection Methods
5.1 Accumulation Detectors
5.2 Momentum Detectors
5.3 Special Vacuum Gauges
5.4 Surface Ionization (Langmuir–Taylor Detector)
5.5 Field Ionization
5.6 Universal Molecular Beam Detector
5.7 Thermal Detectors
5.8 Detection of Metastable Particles
5.9 Spectroscopic Detection Methods

References

Subject Index

1. Fast Beams, Production and Detection

Thermal energy molecular beams have energies of between 0.3 and 300 meV, corresponding to source temperatures of between 4 K and 3500 K. As will be discussed in this chapter, many applications in physics and chemistry, however, require neutral particles with much higher energies. For example, energies of between 1 and 20 eV are of particular interest to chemistry, since the energies of all chemical bonds and the activation energies of most reactions lie in this energy region. The same energy range is also of great significance for the investigation of atom–surface and molecule–surface impact phenomena, since the thresholds for various ionization processes, secondary electron emission, and sputtering are found here [Amirav (1990)]. Moreover, this energy interval is also very important for technological applications. The velocities of satellites and space missiles and their interaction with atoms, molecules, and larger aggregates of the upper atmosphere correspond to this energy range. Furthermore, a considerable part of the material problems encountered in controlled fusion research is caused by neutral particles with hyperthermal energies.

The low hyperthermal energy range, however, is a rather difficult region to work in, and many principles have been proposed and investigated to produce beams with energies below 20 eV. Among these are methods of mechanical acceleration (see Sect. 1.10.1) [Bull and Moon (1954)], acceleration of molecules in time-dependent inhomogeneous electric fields [Auerbach et al. (1966)], shock tube sources, in which a reflected shock wave is used to heat the source gas to very high temperatures (see Sect. 1.10.2) [Skinner and Moyzis (1965)], pulsed beam formation by gun acceleration of a small, gas-filled container [Fenn (1967)], sputtering (see Sect. 1.7) [Wehner (1958)], aerodynamic acceleration (see Sect. 1.4) [Becker and Henkes (1956)], plasma-heated sources (see Sect. 1.5.1) [Kessler and Koglin (1966), Winicur and Knuth (1967)], recoil atoms from radioactive decays (see Sect. 1.10.3) [Wolfgang (1965)], laser ablation (see Sect. 1.6) [Levine et al. (1968)], and the neutralization of low-energy ions at metal surfaces (see Sect. 1.3) [Cuthbertson et al. (1992)] as well as electron-stimulated desorption (see Sect. 1.9) [Hoflund and Weaver (1994), Wolan et al. (1997)]. Only a few of these methods prevailed and have been developed to standard methods in present-day experiments.

The production of beams with energies above 20 eV is much easier. A universal method, which has been extensively used, is based on resonant (or near resonant) charge exchange between fast positive ions and thermal energy neutral particles. Ions are accelerated to the desired energy and then neutralized in a gas or vapor target (see Sect. 1.1). In principle, no upper energy limit exists for this method. Near-resonant charge transfer may be used to produce fast beams of metastable species or highly excited Rydberg states. Negative ions can also be used to produce fast neutral beams. Their negative charge is removed after acceleration to the desired energy by photodetachment, autodetachment, or collisional detachment of the electron (see Sect. 1.2).

Neutral beams with energies below 20 eV have been predominantly used to investigate atomic and molecular collisions as well as atom–surface and molecule–surface impact phenomena. In the first case, elastic, inelastic (vibrational and electronic excitation, ionization processes), and reactive collisions have been investigated. In the latter case, chemical reactions, ionization, fragmentation, and collision-induced desorption processes have been studied.

The energy range up to a few thousand volts is also of great interest for atomic collision experiments. While elastic collisions at thermal energies are essentially determined by the attractive part and the well region of the interaction potential, its repulsive part can be investigated with increasing collision energy down to very small internuclear distances. This has been done by measurements of incomplete total cross sections, a method that was introduced by Amdur at the end of the 1930s and that has been applied to numerous collision pairs [Amdur (1968)]. More recent investigations of this type use measurements of differential scattering cross sections [Newman et al. (1986), Nitz et al. (1987), Gao et al. (1987), (1989), Johnson et al. (1988), Chitnis et al. (1988)]. Other scattering experiments in this energy range have been used to study inelastic processes (e.g. vibrational and electronic excitation, ionization, dissociation) [Pauly (1973), Fluendy and Lunt (1983), Madeheim et al. (1990), Kuzel et al. (1994), Siegmann et al. (1994)].

Another application of fast beams in the keV energy region is investigations of low-energy ion–atom and ion–molecule reactions using the "merged beams" method [Neynaber (1968), (1969), Newman et al. (1982), Havener et al. (1989), (1995)]. This technique consists of two beams (an ion beam and a fast neutral beam, each with an energy in the keV region), traveling in the same direction along a common axis, having a small adjustable energy difference, which determines the collision energy between ions and neutrals (see Sect. 1.11.4). The same method has also been used with two fast beams of metastable particles in order to study the associative ionization in collisions between electronically excited reactants [Neynaber and Magnuson (1975)].

Fast neutral beams with energies of a few keV are also used in surface physics and technology in cases where ion beams cannot be used due to electrical surface charging, for instance, for sputtering, etching, and analysis of insulating surfaces [Kuwano and Nagai (1985), Shimokawa (1992), Shimokawa and Kuwano (1994), Giapis et al. (1995), Churkin et al. (1998)]. Another application is the implantation of foreign atoms into solid surfaces [Volosov and Churkin (1997)].

Beams with energies between 10 and 100 keV find increasing application in fast beam coaxial laser spectroscopy [Neugart (1987), Demtröder (1992)] (see

Sect. 1.11.1). This method combines excellent resolution with high sensitivity and is, therefore, particularly useful in high resolution studies of radioactive isotopes. Further applications are investigations of fast dynamical processes in excited atoms and molecules as well as detailed measurements of photofragmentation processes in molecules [van der Zande et al. (1988), (1992) Cosby and Helm (1989)] (see Sect. 1.11.2).

Another important application of fast beams in the energy region mentioned above is precision measurements of atomic lifetimes [Tanner (1995), Rafac et al. (1999)]. A fast atomic beam is excited by a laser beam at a well defined location, and the exponential decay in time of the excited state is observed by monitoring the fluorescence precisely as a function of the position along the beam line (see Sect. 1.11.3).

In plasma physics, fast beams of light atoms (H, He, Li) with energies between 10 and 100 keV are used for diagnostic purposes in order to determine electron and ion densities, ion temperature and the density of impurity atoms within the plasma. The beam energy determines the penetration depth of the beam and thus the thickness of the plasma layer which can be investigated [Fiedler (1995), Marmar et al. (1997), Sarkissian et al. (1997)].

High energy neutral beams (with energies up to 10 MeV) have also been used to study the ionization of highly excited atoms in electric and magnetic fields [Riviere (1968), Aseyev et al. (1996)], a phenomenon which is of special interest for applications in plasma physics and astrophysics.

In nuclear physics, fast beams with energies up to 1 MeV have been used for injection into tandem accelerators to extend the energy range accessible to electrostatic accelerators [Wittkower et al. (1964)]. Energies up to 10 MeV have been employed to investigate the interaction of fast beams with matter [Ogawa et al. (1996)].

Finally, the production of fast beams by charge exchange of positive ions or collisional detachment of negative ions has been realized in large-scale technical plants for controlled fusion (neutral beam injection). H^+ and D^+ ion beams of 50 to 100 A are produced in large-area ion sources, accelerated to energies up to 1 MeV and neutralized by charge exchange. Similar designs using negative ions and collisional detachment are under investigation. Without being influenced by the magnetic field confining the plasma, the high energy neutral atoms are injected into the fusion reactor, ionized, and trapped in the magnetic field, contributing considerably to plasma heating. They may also be used for fueling and controlling the reaction rate (see Sects. 1.11.5 and 1.11.6) [Gabovich et al. (1989), Okumura et al. (1996), Ohara (1998)].

The production of fast molecular beams has been described in many books and review articles, and only a selection of references is given here [Anderson et al. (1965), Fenn (1967), Amdur (1968), Hasted (1972), Leonas (1972), Jordan et al. (1972), Fluendy and Lawley (1973), Massey et al. (1974), Vályi (1977), Kempter (1975), Pauly (1988), McDaniel et al. (1993), Chutjian and Orient (1996)].

1.1 Charge Exchange

In the process of single charge exchange between an ion X^+ (projectile) and a neutral particle Y (target), an electron (and negligible kinetic energy) is transferred from the target particle to the projectile ion according to the reaction

$$X^+ + Y \to X + Y^+ + \Delta E. \tag{1.1}$$

The reaction is a glancing one, taking place at comparatively large impact parameters. Consequently, the trajectory of the projectile remains practically unchanged, while the product ion is scattered nearly perpendicular to the projectile path. The energy defect ΔE of the process is equal to the difference of the ionization potentials of the two particles involved. In the case of identical particles according to

$$X^+ + X \to X + X^+, \tag{1.2}$$

the reaction is called "symmetrical" or "resonant" charge transfer. Its cross section may be as large as, or larger than, gas kinetic cross sections.

In discussing charge transfer processes, two regions characterized by $v/u \ll 1$ and $v/u \gg 1$ must be distinguished, where v is the relative velocity of the collision partners and u the velocity of their outer shell electrons (of the order of 10^8 cm/s). In the high velocity region ($v/u \gg 1$), Born's approximation may be employed yielding a cross section that falls off rapidly with increasing velocity v. In the low velocity region ($v/u \ll 1$), the electrons move fast compared to the nuclei. They have enough time, therefore, to re-adjust to the slowly changing conditions during a collision. Consequently, electron and nuclei motion can be separated (Born–Oppenheimer approximation) and the collision may be described in an adiabatic approach. In the simplest case only the two states must be considered, which belong asymptotically to the separated particles.

The theoretical treatment of resonant charge transfer (1.2) was given many years ago [Massey and Smith (1933), Bates et al. (1953), Gurnee and Magee (1957)], while the first quantum mechanical calculations of the asymmetric charge exchange date back to Rapp and Francis (1962). They solved the time-dependent Schrödinger equation using hydrogen-like 1s electronic wave functions to calculate charge exchange cross sections as a function of the relative velocity v of the particles and their ionization potentials. These calculations have been improved and complemented in further investigations [Mapleton (1972), Olson (1972), Dewangan (1973), Sakabe and Izawa (1991), (1992), Bransden and McDowell (1992), Croft and Dickinson (1996)].

A comparison of these calculations with numerous experimental data show that these rather simple theoretical considerations are able to describe the general velocity dependence of the cross section quite well, if details such as interference structures are neglected. Correspondingly, the cross section for charge exchange as a function of relative velocity has qualitatively the following behavior: In the case of resonant charge transfer, the cross section $\sigma(v)$ decreases monotonically with increasing relative velocity v. Disregarding the transition elements, the cross section may be written as

$$\sigma(v) = (A - B \ln v)(I_0/I)^{3/2}, \tag{1.3}$$

where I is the ionization potential of the atom considered and I_0 is the ionization potential of the hydrogen atom (13.56 eV). The constants A and B can be numerically represented by $A = 1.81 \times 10^{-14}$, $B = 9.21 \times 10^{-16}$, and v is the relative velocity in cm/s. Equation (1.3) is a simple empirical formula, and the coefficients A and B have been determined by the analysis of a large volume of experimental data [Sakabe and Izawa (1992)]. It is valid for relative velocities $v < 10^8$ cm/s and yields, in general, a good estimate of the cross section.

For nonresonant charge exchange ($\Delta E \neq 0$) the cross section is very small at the lowest relative velocities v, rises to a rather flat maximum of the order of the gas kinetic cross section, and then falls off in the manner of the symmetrical resonance process. The velocity of the maximum can be estimated from the "adiabatic" criterion (Massey criterion), which may be written in the form

$$\frac{a \Delta E}{h v} \approx 1, \tag{1.4}$$

where a is the range of interaction between the atoms (adiabatic parameter). In its maximum, the cross section for nonresonant charge exchange is of the same order of magnitude as for resonant charge exchange.

The Massey criterion follows from simple and very general considerations [Massey (1949)]. During the collision, the partners exert a time-dependent force F(t) upon each other, which can be thought to be expanded into a Fourier integral. It is only the components of this expansion which have frequencies between ν and $\nu + d\nu$ that will produce any appreciable forced oscillation on the interacting system. In order that these components of F(t) should be strong, it is necessary that the time τ of collision should be comparable to the natural period of the oscillator ν. For $\tau \nu \approx 1$, therefore, a strong excitation of the oscillator may be expected. With $\tau = a/v$, and replacing the frequency ν by $\Delta E/h$ according to quantum mechanics, where ΔE is the internal energy change involved in the electronic transition, (1.4) is immediately obtained.

If, at a fixed velocity v, the charge exchange cross section σ is plotted against the energy defect ΔE, a resonance curve is obtained, as is shown in Fig. 1.1, where measured values of σ for charge exchange between Li^+, Na^+, K^+, Rb^+, and Cs^+ ions and Rb atoms are plotted as a function of ΔE ($v = 10^7$ cm/s) [Perel and Daley (1971)].

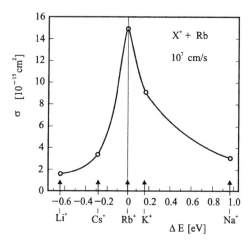

Fig. 1.1. Experimentally determined cross sections (in 10^{-15} cm^2) for charge exchange between Li$^+$, Na$^+$, K$^+$, Rb$^+$, and Cs$^+$ ions and Rb atoms as a function of the energy defect ΔE at a relative velocity $v = 10^7$ cm/s [Perel and Daley (1971)]

1.1.1 Ground-State Particles

The principle of fast beam production by resonant charge exchange is illustrated by Fig. 1.2, which shows schematically the essential elements of a neutralized-ion-beam system. From the ion source (1), in which ions may be produced by electron bombardment, surface ionization, or by a gas discharge, an ion beam is extracted by a suitable electrostatic lens system (2) and focused onto the entrance aperture of a charge-exchange chamber (3). This chamber usually contains the same gas (or vapor) which has been ionized in the ion source. It may also be replaced by a target beam. Since the resonant charge exchange cross sections are much larger than the cross sections for energy or momentum transfer, a large fraction of the ions is neutralized without changing their direction or energy [Massey et al. (1974)]. Consequently, a beam of neutral particles, having the energy and direction of the ion beam, emanates from the charge exchange chamber. Ions which have not been neutralized are removed from the beam by an appropriate electric field (4) at the exit of the neutralization chamber. The energy spread of the fast beam is determined by the corresponding energy inhomogeneity of the ion beam and depends on the ion source used. The smallest energy spread is obtained in the case of surface ionization (determined by the temperature of the surface), but this method is not universally applicable (see Sect. I.5.4). It can be kept rather small in electron impact ion sources (some tenths of an eV), and it is largest in gas discharge ion

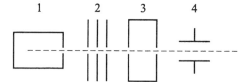

Fig. 1.2. Schematic drawing of a setup for fast beam production. (1) ion source, (2) accelerating electrodes, (3) charge-exchange chamber, (4) deflection electrodes

sources (up to a few eV). But also in this case it is possible to achieve energy spreads below 1 eV, if the potentials in the ion source and the discharge pressure are chosen appropriately [Gaus et al. (1994)].

At a fixed number density n in the neutralization chamber and assuming single collision conditions, the intensity of a fast molecular beam produced by charge exchange is determined by the charge exchange cross section σ and the scattering cross sections σ_i and σ_0 for ions and neutrals, respectively. It may be readily calculated from the following rate equations:

$$\frac{dN_0(x)}{dx} = N_+(x) n\sigma - N_0(x) n\sigma_0, \tag{1.5}$$

$$\frac{dN_+(x)}{dx} = -N_+(x) n\sigma_i - N_+(x) n\sigma, \tag{1.6}$$

where $N_+(x)$ and $N_0(x)$ are the numbers of ions and neutrals, respectively, in the beam at a given location x in the neutralization chamber. The integration of these equations over the total length L of the neutralization chamber yields

$$N_0(L) = N_+(0) \frac{\sigma}{\sigma_i - \sigma_0 + \sigma} \exp(-n\sigma_0 L) \left[1 - \exp(-nL(\sigma + \sigma_i - \sigma_0))\right], \tag{1.7}$$

where $N_+(0)$ is the ion flux entering the neutralization chamber. In (1.7), the expression within the brackets causes a fast rise of the neutral beam intensity N_0 with increasing number density n in the neutralization chamber. After passing through a maximum, N_0 decreases with a further increase of n due to the exponential factor in front of the brackets. The density n_{max} of maximum neutralization yield is given by

$$n_{max} = \frac{1}{L(\sigma_i + \sigma - \sigma_0)} \left[\ln \frac{\sigma_i + \sigma}{\sigma_0} \right]. \tag{1.8}$$

Correspondingly, the maximum neutralization yield is a function of the cross sections only:

$$\frac{N_0^{(max)}(L)}{N_+(0)} = \frac{\sigma}{\sigma_i + \sigma} \left(\frac{\sigma_0}{\sigma_i + \sigma} \right)^{\sigma_0/(\sigma_i - \sigma_0 + \sigma)}. \tag{1.9}$$

Due to the very steep increase of the ion–atom and atom–atom potential at small internuclear distances, the scattering cross sections σ_i and σ_0 are nearly energy independent. Therefore, the energy dependence on the maximum neutralization yield is determined by the energy dependence of the charge exchange cross section σ, which is also rather weak (see (1.3)) within the adiabatic collision range. As a consequence, the maximum neutralization yield shows a weak energy dependence too. At higher energies, however, σ decreases rapidly, giving rise to a strong decrease of the neutralization efficiency.

As already mentioned, the neutralization of ions by charge exchange does not necessarily require a resonance process. For example, Benoit and Gauyacq (1976) produced a hydrogen atom beam by charge exchange between protons and argon atoms ($\Delta E = 2.14$ eV). In general, however, the energy defect ΔE between the ionization potentials of the particles involved should be as small as possible (near resonant charge exchange) since the cross section increases with decreasing ΔE (see Fig. 1.1). Apart from fast ground state atoms, nonresonant charge exchange usually yields a broad distribution of excited states (metastable particles and Rydberg states). This is often used in experiments with fast excited particles. An example is the neutralization of helium ions in charge exchange collisions with argon atoms [Claytor et al. (1995)], which has been used to investigate the fine structure of highly excited Rydberg states (see Sect. 1.1.3).

In nonresonant charge exchange, the energy defect ΔE can be reduced by optical excitation of the particles in the neutralization chamber, resulting in a higher neutralization efficiency. Kushawaha (1983), for example, neutralized a proton beam by directing it into a sodium cell. Through a lateral window of the cell, Lyman-α radiation emitted by excited hydrogen atoms was used as a measure of the charge-transfer cross section of the reaction

$$H^+ + Na \rightarrow H^*(n=2) + Na^+. \tag{1.10}$$

In the case of ground state sodium atoms, the energy defect is $\Delta E = 1.74$ eV, and the charge exchange cross section is rather small. When the sodium atoms were excited to the 3P state using a dye-laser collinear with the ion beam, an enhancement of the Lyman-α radiation of up to 90% ($\Delta E = -0.36$ eV) was observed. The ratio of the cross sections with and without laser excitation of the target as a function of the proton velocity was found to be in qualitative agreement with the adiabatic approximation mentioned before. This has been confirmed by further measurements of similar charge exchange reactions [Finck et al. (1988), Royer et al. (1988), MacAdam et al. (1990), Gieler et al (1993)]. For the production of a fast beam of sulfur atoms in highly excited Rydberg states, Deck et al. (1993) use charge exchange collisions with a thermal Rb beam, which is excited by three lasers to the 8F or 10F Rydberg states [Fisher et al. (1997)] (see Chap. 4). Charge exchange of molecules using an electronically excited target have also been investigated [Bruckmeier et al. (1995)]. Beams of ArH^+, ArD^+, and D_3^+ have been crossed with an atomic beam of cesium atoms in the ground state and in the excited P state. In this case, the emitted rovibronic band spectra from the excited neutral molecules was used to indicate the charge exchange efficiency. An enhancement of the neutralization yield of more than an order of magnitude was observed.

As has been pointed out in Sect. I.5.6.2, electron impact ionization of molecules in the ion source leads not only to the formation of the parent ions, but also to fragment ions (for example, the electron impact ionization of H_2O molecules yields the ions H_2O^+, OH^+, and H^+). In order to obtain a fast beam of a single species, a mass spectrometer or another suitable mass selector must be added between the ion source and the neutralization chamber [Savola et al. (1973), Fleischmann et al. (1974), Eriksen et al. (1975), Smith et al. (1996)]. A Wien filter is often used. In this manner, it is possible to produce fast beams of reactive atomic species such as H, N, and O beams or of the corresponding stable molecules H_2, N_2, and O_2 [Foreman et al. (1976), Nitz et al. (1987), Quintana and Pollak (1996)].

Fast beams of radicals such as CF_3, CF_2, and CF with energies of several keV have been produced in a similar way. The single ionized ions of these radicals are formed in a discharge source containing CF_4 and after mass selection are neutralized in collisions with xenon atoms. Nearly all neutral radicals are in their ground state with vibrational energies below 0.5 eV [Tarnovsky and Becker (1993)].

Molecular ions, which cannot be obtained from a single parent molecule, may also be produced in a gas discharge source, using appropriate gas mixtures and a subsequent ion–molecule reaction (e.g. ArH^+ from a discharge in a H_2–Ar mixture) [Bruckmeier et al. (1995)].

It should be mentioned finally, that mass-selected neutral clusters have also been produced by charge exchange of cluster ions. An example has been described by Abshagen et al. (1991). Pb_N^+ ions (with $3 \leq N \leq 12$) have been neutralized by charge exchange with sodium vapor. This process is nearly resonant, since the ionization energies of small Pb clusters differ only by a few tenths of an eV from the ionization energy of sodium (5.14 eV). This not only has the advantage of a large charge exchange cross section, but also prevents the clusters from fragmentation, since the neutralization produces predominantly clusters in their ground state, which minimizes the transfer of internal energy. This has been experimentally confirmed by measurements of the angular distribution of the neutral clusters, which is practically identical with that of the cluster ions.

The intensity of a beam produced by charge exchange depends strongly on its energy. The reason for this is the space charge of the ions, which puts an upper limit on the current density that can be extracted from an ion source at a particular energy. This upper limit is described by Langmuir's space charge law. According to this law, the maximum current density, j, may be written as

$$j = c \frac{U^{3/2}}{\sqrt{M}}, \qquad (1.11)$$

where U is the acceleration voltage and M the ion mass number. The factor c depends only on the geometry of the extraction region of the ion source. Consequently, since both the charge exchange and the scattering cross sections show a rather weak dependence on energy, the neutral beam intensity shows a similar energy dependence, as was first pointed out by Fenn and co-workers [Anderson et al.

(1965), Fenn (1967)]. This can also be seen from Fig. 1.3, where beam intensities reported from various authors are plotted as a function of energy. The straight line drawn through the measurements has a slope of 3/2. Although there is a large spread in the data, because of the various system geometries and masses that have been used (methods to compensate space charge effects have also often been employed in the low energy region), the straight line according to (1.11) describes the general behavior quite well over six decades of energy. Thus, the flux of a charge exchange beam may be estimated by

$$j_N = 2 \times 10^9 \frac{U^{3/2}}{\sqrt{M}} \quad \text{particles s}^{-1}\text{cm}^{-2}, \qquad (1.12)$$

where the acceleration potential U is measured in V. M is the molecular weight of the particles. For an energy of 10 eV, this estimate (1.12) yields a beam intensity (for M = 40) of about 10^{10} particles/s cm^2, which is of the same order of magnitude as the flux of an effusive beam with mechanical velocity selection. In spite of these low intensities, experiments with fast beams produced by charge exchange have been carried out down to energies of a few eV [Utterback and Miller (1961), Utterback (1966), Amme and Hangsjaa (1969), Neumann and Pauly (1967), Helbing and Rothe (1968), Kwan (1974)].

The space charge limit estimated by (1.12) can be overcome if the space charge spreading of the ion beam at low energies is compensated by the injection of free electrons into the beam [Boring and Humphries (1969), Doughty and

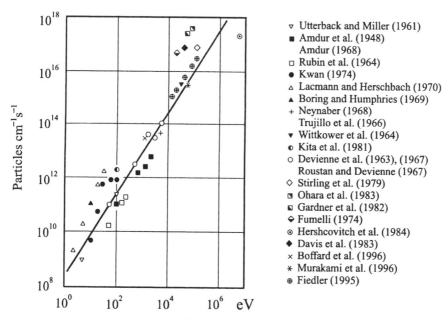

Fig. 1.3. Intensities (in particles/s cm^2) of charge exchange beams as a function of energy. The straight line corresponds to (1.12) with M = 40

Schätzle (1969)]. The electrons are emitted from a circular filament surrounding the ion beam. In this manner, the beam intensities at low energies can be increased by one or two orders of magnitude. Space charge compensation is also used in neutral beam injectors employed in plasma physics in order to transport intensive ion beams over long paths [Sherman et al. (1998)].

As has been already mentioned, in principle there is no energy limitation at high energies. Fast beams have been used up to energies in the MeV region [Pedersen (1974), Davis et al. (1983), Luhmann and Peebles (1984), Murakami et al. (1996)].

Since ion beams can be easily pulsed, focused, or deflected, the same is possible for charge exchange beams before neutralization [Gersing et al. (1973), Eccles et al. (1986), Quintana and Pollak (1996)].

Figure 1.4 shows an example of a fast beam source that has been used to produce intense beams of barium atoms for plasma diagnostics [Murakami et al. (1996)]. It consists of an ion beam source based on surface ionization and a Li vapor cell (neutralization chamber).

Barium atoms effuse from an oven (1) and hit a diffuser (7), so that they are deflected to the wall of a hot rhenium-foil cylinder (4) (5 mm ∅, length 21 mm) in front of the oven aperture, where they are surface-ionized. The rhenium-foil cylinder is heated to about 3000 K by radiation from the surrounding filament (5) and by electron bombardment. Several tungsten sheet radiation shields (6) are used to protect the surroundings from overheating. The Ba^+ ions are accelerated by a lens system (8) and pass through a neutralization chamber (9) filled with lithium vapor (near-resonant charge exchange $|\Delta E| = 0.16$ eV). About 75% of the ions are

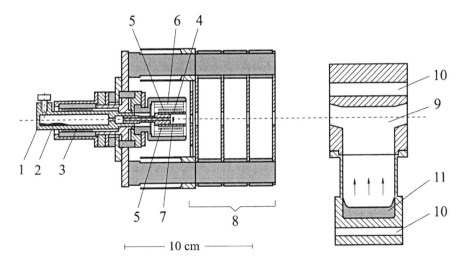

Fig. 1.4. Source for fast Ba beams. (1) Ba oven, (2) material sample, (3) heating filament, (4) rhenium tube for surface ionization, (5) tungsten filament to heat the rhenium tube, (6) radiation shields, (7) diffuser sheet, (8) lens system, (9) neutralization chamber (heat pipe principle), (10) bore holes for heating wires, (11) Li supply. Insulation materials are gray-colored [Murakami et al. (1996)]

neutralized here. With an acceleration voltage of 40 keV, neutral flux densities of 3.2×10^{15} particles/s cm^2 are obtained. The charge exchange chamber makes use of the "heat pipe" principle, which is discussed in Sect. 1.1.2.

The production of fast neutral atom beams is particularly simple for alkali metals [Hollstein and Pauly (1966), Neumann and Pauly (1967)], since alkali metals can be ionized by surface ionization at moderate temperatures. Fig. 1.5 shows a source which has been used to produce fast alkali atom beams with energies between 7 and 4000 eV. The alkali metal is evaporated from the oven chamber (1) into the charge exchange chamber (2), which has two slits (3) and (4) on opposite sides. Alkali atoms effusing from slit (3) strike a heated tungsten filament (5), where they are ionized and accelerated back into the neutralization chamber by a voltage which is applied between the filament and the neutralization chamber. A fraction of the ion beam is neutralized, and the beam leaving the exit slit (4) consists of three components: Alkali ions, fast neutrals, and thermal energy atoms. The ions are deflected out of the beam by a suitable electric field, as is indicated in Fig. 1.2. The thermal energy atoms are distinguished from the fast ones by using a cold surface ionization detector (see Sect. 1.12.1), which is sensitive only to fast alkali atoms. Another possibility to separate thermal energy and fast atoms employs an inhomogeneous magnetic field (Rabi field), which deflects the thermal component out of the beam, while the deflection of the fast atoms is negligible [Hubers et al. (1976)].

In lifetime measurements with fast beams (see Sect. 1.11.3), the thermal energy beam component gives rise to an additional background of scattered light. Tanner et al. (1995), therefore, produce a fast lithium beam by charge exchange in rubidium vapor.

Devices for fast alkali atom beams similar to the one shown in Fig. 1.5, sometimes using a target beam for charge exchange instead of a neutralization cell, have often been used in scattering experiments [Rubin et al. (1964), Anderson et al. (1969), Lacmann and Herschbach (1970), Düren et al. (1974), Kempter et al. (1974), Aten and Los (1975), Aten et al. (1977), Mochizuki and Lacmann (1977), Young et al. (1981)].

For higher energies, a modification of this source, in which ionization and neutralization are independent of each other, rather similar to the one presented in Fig. 1.4, is more adequate [Fluendy et al. (1975), Andresen et al. (1983)]. By separating the accelerating electric field from the charge exchange chamber, electrical discharges, which may occur at high energies, can be avoided.

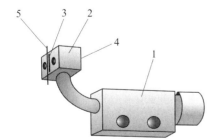

Fig. 1.5. Fast alkali atom source. (1) oven, (2) neutralization chamber, (3) ion entrance slit, (4) fast beam exit slit, (5) ionizing filament. The bore holes provide for heating elements [Neumann and Pauly (1967)]

The lowest energies have been achieved with a source in which surface ionization, acceleration and neutralization take place in a single chamber filled with alkali metal vapor [Helbing and Rothe (1968), Kwan (1974), Hubers et al. (1976)]. The ions are produced by ionization at a tungsten wire and are then accelerated by an appropriate potential toward a fine-meshed grid extremely close to the wire. A fraction of the ions is neutralized between the grid and the exit aperture of the chamber, leaving it as a neutral fast beam. This device has been used to produce cesium beams with energies between 2 and 200 eV. Operating the source at energies above 200 eV leads to destruction of the fine-meshed grid due to sputtering.

In certain applications, the linear arrangement of the ion source and charge exchange chamber (schematically shown in Fig. 1.2), leads to undesired stray light, for example in measurements of differential scattering cross sections utilizing particle–photon coincidence. In these cases, the ion beam may be deflected by an appropriate magnetic or electric sector field before neutralization [Kita et al. (1981)].

1.1.2 Metastable Particles

An example for a near-resonant charge exchange to produce fast metastable particles is the reaction

$$H^+ + Cs \rightarrow \{H(2S), H(2P)\} + Cs^+ - 0.49 \text{ eV}, \tag{1.13}$$

which was first proposed by Donnally et al. (1964). Vu Ngoc Tuan et al. (1974) have investigated this process in detail in the energy range between 400 and 3000 eV. The proton beam is produced in a duoplasmatron ion source, extracted by an electric potential of 10 kV, focused by an einzel lens, analyzed in a 90° magnetic sector field, and then decelerated to the desired energy. The protons then cross a cesium nozzle beam for charge exchange. A "thin" target is required to have only single collisions, since the cross section for destruction of H(2S) in cesium is very large (about 50 Å^2). After leaving the charge exchange region, the beam passes through a drift region, where the remaining ions can be removed with a weak transverse electric field without appreciable quenching of the metastable atoms. Similar sources, utilizing neutralization cells instead of a target beam, are described in the literature [Spiess et al. (1972), Pradel et al. (1974), Dose et al. (1975)].

The rate of material loss through the entrance and exit aperture of a neutralization cell can be reduced by either using long channels with a low transmission for a diffuse gas or vapor (see Chap. I.2) (for example, of 4 mm ⌀ and 40 mm length [Spiess et al. (1972)]), or using the "heat pipe" principle for the neutralization chamber [Bacal and Reichelt (1974), Brisson et al. (1980), Boffard et al. (1996), Lagus et al. (1996)].

The "heat pipe" principle was first used by Grover et al. (1964) and has found widespread applications, particularly in spectroscopy. It has been repeatedly described in text books and review articles [Dunn and Reay (1976), Ivanovskii et al. (1982), Vidal (1996)]. A simple, open system consists essentially of a tube with a mesh structure on its inner surface which acts as a wick. It is heated in its middle (heat source) and has a lower temperature at its ends (heat sink) and contains the material sample to be evaporated.

In such a device the vapor is driven from the heat source to the heat sinks at the tube ends, where it condenses and returns as a liquid back to the source due to capillary action of the wick. A uniform wetting of the mesh structure of the wick is an important requirement for proper operation.

As an example, Fig. 1.6 shows a neutralization chamber for cesium vapor based on this principle [Boffard et al. (1996)]. The bottom of the cesium reservoir (1) is maintained at a much higher temperature than the charge exchange chamber (6). This temperature gradient results in a cesium vapor pressure gradient. It causes a beam of Cs atoms to travel up through the connecting tube (4) (the throat, which has a high flow resistance) into the charge exchange cell. The cesium throughput, and hence its number density, is determined by the reservoir temperature and the conductance of the throat. After several wall bounces, the Cs atoms stick to the sloped inner surface of the neutralization chamber, gather into liquid droplets, and trickle back down into the reservoir. This recirculation is further facilitated by a fine stainless-steel mesh covering the wall of the throat, that acts as a wick. The angle of the slope in the charge exchange region is a critical parameter for proper recirculation, which depends on the material used.

In the case of cesium, a slope of 7° is sufficiently steep. The whole device is constructed of stainless steel. The temperature gradient is maintained by heating the reservoir and cooling the upper part of the throat using a water-cooled copper block (5), fixed to its top.

The main advantage of this design is that the cesium has a high density only within the beam that emerges from the throat. Keeping the Cs vapor pressure in the rest of the neutralization chamber low allows one to have relatively large apertures for the passage of the fast beam, enabling an easy adjustment of the arrangement.

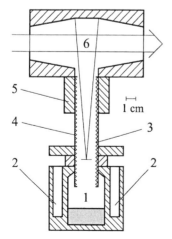

Fig. 1.6. Charge transfer cell utilizing the "heat pipe" principle. (1) cesium reservoir, (2) boreholes for cartridge heaters, (3) stainless-steel mesh, (4) throat, (5) water-cooled copper block, (6) charge exchange region [Boffard et al. (1996)]

Charge exchange collisions between He$^+$ ions and alkali metal atoms also have often been used to produce fast metastable helium beams [Hollstein et al. (1969), Boffard et al. (1996)]. The sources used are rather similar to those described above. After charge exchange, the neutralized atoms are in the 2 ^{1}S, 2 ^{1}P, 2 ^{3}S, and 2 ^{3}P states. Since the P states decay immediately to the ground state, the final fast beam has three components: the two metastable species (2 ^{1}S, 2 ^{3}S) and ground-state He atoms. The ratio of singlet to triplet states in the beam has been investigated by Reynaud et al. (1979) in the energy range from 100 to 1500 eV. They were able to show that charge exchange with sodium atoms yields an almost pure beam (> 97%) of He(2^{1}S) atoms.

Charge exchange with all other alkali metals leads to a mixture of both metastable states. While for rubidium at energies below 500 eV more than 50% of the helium atoms are found to be in the singlet state, charge exchange with cesium shows the same result only at energies below 150 eV. The largest fraction of metastable atoms in the triplet state (70 to 80%) is obtained when neutralizing helium ions in potassium vapor. Evidence for the higher abundance of the triplet metastables in this energy range has already been reported [Coleman et al. (1973), Morgenstern et al. (1973)].

Independent of these fractions a pure beam of helium atoms in the 2^{3}S state can always be produced, since the singlet state can be quenched by irradiating the beam containing both states with light of the 2^{1}S–2^{1}P transition (2.06 μm). The 2 ^{1}P state decays immediately into the ground state. At high velocities, however, the quench lamp must be sufficiently long, since the excitation probability strongly decreases with increasing beam velocity (proportional to 1/v at high velocities v). One reason for this is the decreasing residence time of the atoms within the irradiated region, the other reason is the Doppler shift, which reduces the number of photons which can be utilized for excitation. An example of a simple quench lamp is presented in Fig. 1.7.

In a low-pressure discharge, the width of the line corresponding to the 2^{1}S–2^{1}P transition, emitted by the quench lamp, is determined by the Doppler effect. This, in turn, is given by the temperature and the mass of the emitting helium atoms according to:

$$\frac{\Delta v}{v_0} = 2\sqrt{\ln 2}\frac{v_w}{c} = \frac{2}{c}\sqrt{\frac{2RT\ln 2}{M}} = 7.16 \times 10^{-7}\sqrt{\frac{T}{M}}. \qquad (1.14)$$

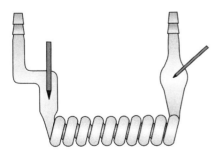

Fig. 1.7. Example of a helium quench lamp. Helium at a pressure between 5 and 50 mbar flows through the glass helix. A self-sustained electric discharge between the tungsten electrodes produces the resonance light which excites the He(2^{1}S) atoms passing along the axis of the helix into He(2^{1}P) states, which decay into the ground state

v_0 is the frequency of the line, Δv is its halfwidth, v_w is the most probable atom speed in the discharge, c is the velocity of light, T the absolute temperature, R the gas constant and M the molecular weight of the emitting atoms [Kessler and Crosswhite (1967)]. Due to this linewidth, the beam particles are not only excited by photons which move perpendicular to the beam, but also by photons which cross the particle beam under angles ϑ (smaller than $\pi/2$), as long as the necessary frequency v is within the Doppler-broadened linewidth. The frequency v required for excitation is given by

$$v = v_0\left(1+(v_b/c)\cos\vartheta\right) = v_0\left(1+(v_b/c)\sin\Theta\right), \qquad (1.15)$$

where v_b is the velocity of the beam particles. The angles ϑ and Θ are defined in Fig. 1.8. If the halfwidth of the Doppler profile is assumed to be a measure for this frequency range, the angle of acceptance Θ may be written as

$$\Theta = \arcsin\left(\sqrt{\ln 2}\,(v_w/v_b)\right). \qquad (1.16)$$

Correspondingly, the angle of acceptance and thus the number of useful photons decreases for high beam velocities v_b according to $1/v_b$.

Using charge-transfer collisions between helium atoms and alkali metal atoms both low-energy metastable helium atoms (5 to 500 eV) (see previous references) and metastable atoms in the keV region have been produced [Neynaber and Magnuson (1975), (1977), Neynaber and Tang (1979), Aseyev et al. (1996)].

The same technique has also been applied to generate metastable neon atoms at low energies (9 to 91 eV) [Coleman et al. (1973)]), medium energies (30 to 1000 eV) [Alvarino et al. (1984)], and high energies (30 to 300 keV). The latter experiment utilized charge exchange in sodium vapor to perform two-photon spectroscopy in a fast metastable neon beam [Poulsen and Winstrup (1981)]. Metastable argon beams have also been produced by charge exchange with potassium atoms [Neynaber and Magnuson (1977)]. Finally, metastable helium beams produced with this technique have been used as target beams in electron-impact experiments [Boffard et al. (1996)]. Another application uses metastable $He(2^3S)$ beams to produce fast beams of highly excited Rydberg atoms (n > 18) (see Sect. 1.1.3). The metastable atoms are excited in two steps by superimposing two coaxial dye-laser beams, according to the transitions $2\,^3S \to 3\,^3P \to n\,^3S$ or $n\,^3D$, respectively [Aseyev et al. (1995), (1996)].

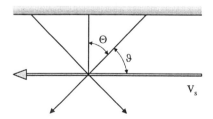

Fig. 1.8. Definition of the angles used in (1.13) and (1.14). The atom beam is represented by the arrow, the radiation emerges from the top

1.1.3 Rydberg Atoms

The production of Rydberg Atoms by charge exchange according to the reaction $A^+ + B \rightarrow A_{n,\ell} + B^+$ has occasionally been mentioned in the previous sections. Since the cross section for this process scales as $1/n^3$ (n = principle quantum number), the fraction of ions that is neutralized into a specific Rydberg state is rather small (similar to the case of electron impact excitation (see Sect. I.4.9.1)). Typical values are $< 10^{-3}$ for n = 10 and $< 10^{-4}$ for n = 30. States with low angular momentum quantum number ℓ are predominantly populated, but subsequent collisions in the neutralization chamber may result in a redistribution of the m,ℓ states, because the cross sections for transitions between these states are extremely large [Kachru et al. (1983)]. Consequently, higher ℓ and m states are also populated. With the cross sections being smaller than those for transitions between m and ℓ states, a possible collisional redistribution of the n states does not change the $1/n^3$ dependence of the population in practice, since all states are populated by the charge exchange process.

A possibility to influence the distribution of excited states obtained by charge exchange, has already been mentioned in Sect. 1.1.1. Using charge exchange collisions between sulfur ions and a thermal energy rubidium beam, which is excited by three lasers to the 8F or 10F Rydberg states, respectively, those Rydberg states of the sulfur atoms that are nearly resonant with the excited rubidium states are populated predominantly [Deck et al (1993), Fisher et al. (1997)].

In general, however, experiments with fast atoms in a specific Rydberg state require additional measures to narrow the state distribution which is produced by charge exchange. A simple method uses field ionization [Bayfield and Koch (1974)]. A beam of Rydberg atoms passes through an electric field, which can be alternately switched between two field strengths E_1 and E_2. The ionization of a Rydberg state with principle quantum number n requires a field strength proportional to $1/n^4$ (see (5.10)). The field strength E_1 ionizes all atoms with $n > n_1$. Similarly, at the field strength E_2 (with $E_2 < E_1$) atoms with $n > n_2$ are ionized. Thus, the signal difference of a measurement with E_1 and a measurement with E_2 is due to atoms with n values in the interval $n_1 \leq n \leq n_2$. With an electric field that has been changed between 28.5 and 41.0 V/cm, Bayfield and Koch were able to work with selected hydrogen atoms in Rydberg states with ($63 \leq n \leq 69$). A higher selectivity requires optical methods [Stebbings et al. (1975), Palfrey and Lundeen (1984)]. Usually, continuous lasers operating at a fixed frequency are used in these cases, since the fine-tuning of the excitation frequency can be obtained via the Doppler effect (Doppler tuning). Doppler tuning is achieved by either changing the angle between the laser beam and the fast atom beam or by changing the velocity of the atom beam at a fixed angle between laser and atom beam.

An early example of this procedure is the excitation of a fast metastable H(2s) beam into the np Rydberg states with $40 < n < 55$, using the UV-line of an argon ion laser [Bayfield et al. (1977)].

Another possibility makes use of a CO_2 laser. From the broad distribution of Rydberg states produced by charge exchange all states with n > 10 are removed by field ionization. Then, a CO_2 laser tuned to the transition n = 10 → n = 30 is used to selectively populate the state with n = 30 [Koch and Mariani (1980)].

1.2 Neutralization of Negative Ions by Electron Detachment

Another method of producing fast neutral beams involves the generation of a beam of negative ions, ion acceleration to the desired energy, and finally ion neutralization by an electron detachment process. Either photodetachment or collisional detachment can be used. In favorable cases, autodetachment is also possible. These methods are described in more detail in the following sections.

1.2.1 Photodetachment

Photodetachment has become possible with the advent of high power lasers [van Zyl et al. (1976), (1978), Havener et al. (1989)]. As an example, we describe in the following a device which has been used by Stephen et al. (1996) to produce neutral beams of oxygen atoms with energies between 4 and 1000 eV. Negative ions are extracted from a low-voltage gas-discharge source (operated with N_2O which was found to be the optimum species for O^- generation), accelerated to an energy of 1000 eV, and mass selected by a Wien filter. The selected O^- ions are then decelerated to the desired energy and focused by an electrostatic ion optics and finally neutralized by electron detachment using intracavity laser radiation. Having obtained their final energy, the O^- ions are electrostatically deflected through an angle of about 9° by deflectors (1) shown in Fig. 1.9.

This deflection is essential to keep more energetic O atoms, produced upstream from this point by O^- collisions with the background gases (before the deceleration process), from contaminating the final atomic oxygen beam. The ions then enter a three-element einzel lens assembly (2), which renders the O^- trajectories nearly parallel in the detachment region (3). This einzel lens is important for

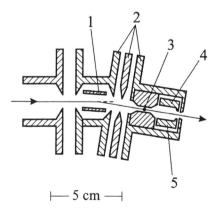

Fig. 1.9. Details of the photodetachment region [Stephen et al. (1996)]. (1) deflection plates, (2) einzel lens, (3) detachment laser beam (perpendicular to the drawing plane), (4) ion repeller, (5) ion collector

energies below 50 eV to counteract the space-charge spread of the ion beam. Subsequently, electron detachment occurs within the cavity of a 25 W argon ion laser. The intracavity laser axis crosses the plane of Fig. 1.9 perpendicularly. Depending on the laser configuration used, neutralization efficiencies (at an energy of 5 eV) between 5 and 25% are observed. At this energy, typical beam intensities are of the order of 10^{12} atoms/s cm^2.

Independent of the beam energy, the energy spread is about 1.5 eV. The remaining O$^-$ ions in the beam are finally forced to the ion collector (5) by the ion repeller (4). Here, the difference in the arriving O$^-$ currents with the laser on or off can be accurately measured yielding the neutralization efficiency. Since the probability of electron detachment is proportional to the residence time of the ions in the laser field, the neutralization efficiency decreases with increasing particle velocity v as 1/v. To increase the efficiency of neutralization at higher beam velocities, mirror devices (multipass cells) are usually used, in which the laser beam crosses the ion beam many times [Anheier et al. (1989)].

A source for groundstate O atoms, developed for generating beams with energies between 2 and 100 eV and which employs a multipass cell, is schematically shown in Fig. 1.10 [Orient et al. (1990), (1992)]. Electrons emitted from a hot tungsten filament (1) are extracted at high energies and then decelerated by an electron optics (2) to a final energy of 8 eV at a gas nozzle (3). NO molecules effusing from this nozzle dissociatively attach to form O$^-$(^{2}P) ions. Electrons and ions are confined by a strong (6 T), uniform solenoidal magnetic field (11) and the latter are accelerated to the desired final energy. Using a trochoidal deflector (4), the faster electrons are separated from the slower ions. The electrons are trapped in a Faraday cup (5). With a 20 W argon ion laser (7) (using all visible lines) and a multiple pass mirror assembly (6) the electrons are detached from the O$^-$ ions to form exclusively O(^{3}P) atoms. At an atom energy of 5 eV, the neutralization efficiency is about 15%. Due to space-charge effects, the divergence of the neutral beam is rather high (about 20°). Beam fluxes are of the order of 10^{12} particles/s.

The device shown in Fig. 1.10 has been used to investigate recombination reactions of O atoms with NO and CO molecules adsorbed on a MgF$_2$ target (9). The reactions are detected by the luminescence emitted from the NO$_2$ or CO$_2$ molecules formed in the reaction. Negative oxygen atoms cannot reach the target due to an appropriate potential at the mirror (8) and at a grid in front of the mirror (not shown in Fig. 1.10). The light emanating from the target is imaged by the mirror and a lens (10) onto the entrance slit of an optical spectrometer [see also Orient et al. (1994), (1995)].

Similar devices have also been used to produce beams of H and D atoms in the energy from 10 eV to 10 keV [van Zyl et al. (1976), (1978), Aberle et al. (1980), Havener et al. (1989), (1995), Folkerts et al. (1995), Pieksma and Havener (1998)]. With increased intracavity laser power in Nd:YAG lasers, neutralization efficiencies of 5% at 1000 eV and 0.5% at 10 keV can be achieved. In this energy range the attainable intensities are comparable to the intensities of beams produced by charge exchange of positive ions. At lower energies, however, the neutral beam

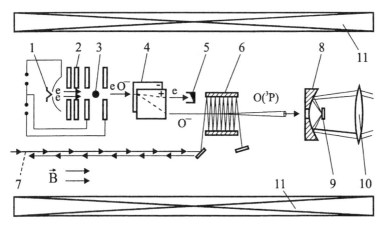

Fig. 1.10. Source for groundstate O atoms [Orient et al. (1990)]. (1) tungsten filament, (2) electron optics, (3) gas target (NO), (4) electric field separating electrons from ions, (5) Faraday cup for electrons, (6) multipass assembly for photodetachment, (7) argon ion laser beam, (8) mirror for luminescence light, (9) target, (10) focusing lens (11) solenoid

intensities obtainable with photodetachment are higher than those achievable with charge exchange, due to the v^{-1} dependence of the neutralization yield.

Fast beam production by photodetachment of negative ions has several advantages. It can be used in ultrahigh vacuum systems without any problems, which is important when using fast beams for surface investigations. Another advantage over the charge exchange technique is that photodetachment with light of an appropriate wavelength guarantees the generation of only groundstate atoms. This may be illustrated by an example: The electron affinity of atomic oxygen is 1.46 eV, and the lowest excited state (^{1}D) of oxygen lies 1.97 eV above the ground state. Using photons with energies smaller than the sum of these two energies (E < 3.43 eV or λ > 361 nm) results in the formation of only groundstate atoms. Furthermore, the current of detached electrons provides an excellent measure of the beam intensity. This makes an absolute calibration of the detector possible and allows monitoring of the beam (at high beam energies usually by secondary electron emission from a surface) [Stephen and Peko (2000)].

The method described above has also been used to produce pulsed fast beams of radicals such as N_3, NCO, and CH_2NO_2. An appropriate parent molecule is expanded with a carrier gas from a pulsed source. Immediately outside the nozzle aperture, the supersonic expansion is crossed with a continuous 1 keV electron beam. Low energy secondary electrons formed by electron impact ionization result in the production of the desired negative ions via dissociative attachment. These negative ions undergo rotational and vibrational cooling during the remainder of the expansion. After acceleration of the ions to the desired energy and mass selection by a time-of-flight spectrometer, the anions are photodetached by a tunable, pulsed, excimer-pumped-dye laser system, timed to intersect the mass of interest. The detachment wavelength is chosen to avoid the production of vibrationally excited neutral radicals [Continetti et al. (1993), Cyr et al. (1992), (1993), Osborne et al. (1995), Leahy et al. (1995)]. These pulsed beams have been used to investigate the dynamics of photodissociation.

Finally, negative cluster ions have been used to produce fast beams of mass selected neutral clusters by photodetachment [Zheng et al. (1985), Suzuki et al. (1991)] (see Chap. 2).

1.2.2 Collisional Detachment

In the adiabatic region, characterized by v/u << 1, where v is the relative velocity of the collision partners and u the velocity of their outer shell electrons (of the order of 10^8 cm/s), the resonant or near-resonant charge exchange of positive ions in a gas- or vapor-filled chamber is a very efficient process, which depends only very weakly on the energy. This behavior, however, changes at higher velocities (v > 10^8 cm/s), where the neutralization efficiency decreases drastically (see Sect. 1.1). In the case of hydrogen atoms, for example, the neutralization efficiency decreases at energies below 50 keV only from 90% to 80% [Wiesemann (1981)]. Beyond the adiabatic region (E > 50 keV), the neutralization efficiency decreases to about 20% at 100 keV and less then 10% at energies above 200 keV.

At these energies, the neutralization of negative ions by collisions with gas molecules, in which the electron is removed by a stripping reaction, is a very efficient process with neutralization yields as high as 60% [Grisham et al. (1982)]. Even higher neutralization yields (about 85%) can be obtained using collisional detachment of negative ions passing through an argon or hydrogen plasma [Ivanov and Rosyakov (1980), Hershcovitch et al. (1984)]. Present investigations concerning neutral beam injection for fusion reactors, therefore, concentrate on the development of high intensity sources for negative ions and high density plasma devices for neutralization [Tanaka et al. (1998)].

1.2.3 Autodetachment

Large molecules, which may form long-lived (from a few to several hundred microseconds) temporary negative ions by electron attachment, can be rather easily used to produce fast molecular beams [Greene et al. (1975)]. Electrons emitted from a hot filament and confined by an axial magnetic field are accelerated into a Faraday cage, where they form negative ions in attachment collisions with the molecules admitted to the cage. Electrons undergoing no attachment reaction are collected outside the Faraday cup. The ion production is maximized by adjusting the electron energy. The negative ions are extracted from the cage perpendicular to the direction of the electron beam and focused into a drift tube. Here they travel for a sufficient length of time until the negative ions decay back to neutral molecules. This source has been used to generate fast beams of SF_6 molecules for scattering experiments. It is, however, capable of producing beams of a variety of other large molecules. The beam intensities reported are rather low (about 4×10^4 molecules/s cm^2 at an energy of 1200 eV).

1.3 Neutralization of Ions by Collisions with Metal Surfaces

The neutralization of ions at metal surfaces was already observed in the first half of the 20th century [Oliphant (1929), Oliphant and Moon (1930)]. The first systematic investigations, however, began in the middle of the 1970s, when methods utilizing the scattering of low-energy ions were developed to analyze and characterize surfaces under ultrahigh vacuum conditions [Hagstrum (1977), Brako and Newns (1981), Sulston et al. (1988), Nordlander (1990), Kimmel and Cooper (1993), Davison et al. (1995), Okada and Murata (1997)].

Electron transfer between a singly charged positive ion and a metal surface is a very important and much studied process in surface science. Various mechanisms can be involved. In many cases, the dominating process is resonant charge transfer, in which an electron from the conduction band of the metal is transferred to a valence state of suitable energy of the ion (see (2) in Fig. 1.11) [Borisov et al. (1996)]. If more than one electron plays an active role in the neutralization process, the charge transfer is classified as an Auger transition [Hagstrum (1977), Almulhem (1997)]. In this case, a metal electron is captured into a low-lying state of the incident ion and the energy associated with this capture is used to excite a metal electron beyond the Fermi limit (see (3) in Fig. 1.11). Further channels for the charge exchange process are the excitation of collective modes or the excitation of plasmons during the electron-capture process [Lorente and Monreal (1997)], as well as radiative neutralization (see (1) in Fig. 1.11), which, however, can be neglected in general. Which mechanism prevails in a given case depends strongly on the energy levels of the ion in the vicinity of the surface. The neutralization yield depends also on the lifetimes of the ion states, and on the ion velocity

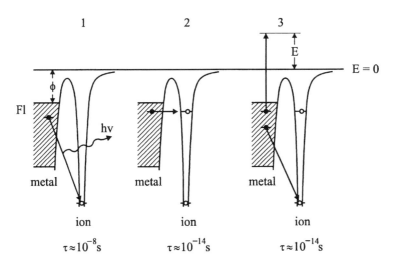

Fig. 1.11. Electron transfer mechanisms in ion–surface collisions: (1) radiative neutralization, (2) resonance neutralization, (3) Auger neutralization. Fl = Fermi level, ϕ = work function of the metal, E = kinetic energy of the emitted Auger electron, τ = lifetime of the initial state

v, which determines the time available for the electron transfer process [Shao et al. (1995)].

For the experimental investigation of the electron transfer process in ion–surface collisions extremely clean single crystal surfaces under ultrahigh vacuum conditions are usually used. Varying the ion energy and the angles of ion incidence and reflection, and analyzing the reflected particles according to their energy and charge state yields a detailed picture of the dynamics of the charge exchange between ions and metal surfaces [Kimmel and Cooper (1993), Okada and Murata (1997)]. The results of these investigations, however, cannot be applied directly to the production of fast beams by charge exchange with metal surfaces, since polycrystalline materials under little-known surface conditions are usually employed in such situations.

Figure 1.12 shows a cross-sectional view of a hyperthermal neutral beam source utilizing the neutralization of ions impacting upon a metals surface [Cuthbertson et al. (1992)]. It has three basic sections: a plasma chamber containing the coaxial plasma source (3), a neutralization plate (4) and, perpendicular to this, a differentially pumped intermediate section leading to the beam utilization chamber (6). The ions (with energies E < 100 eV) are generated in a coaxial plasma source driven by an electrodeless radiofrequency discharge (frequency 2.45 GHz, power 1–2 kW) which is confined by an axial magnetic field (3.4–4 kG) to a column of 1 cm diameter [Motley et al. (1979)]. This plasma column hits

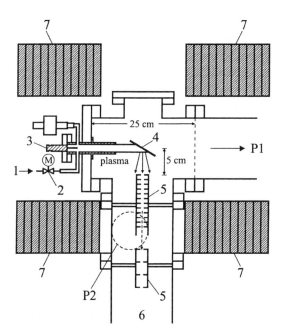

Fig. 1.12. Cross-sectional view of a low-energy (10 to 40 eV) neutral beam source [Cuthbertson et al. (1992)]. (1) gas supply with piezoelectric valve (2), (3) radiofrequency plasma source, (4) neutralization target, (5) baffle, (6) beam utilization chamber, (7) water-cooled coils for the magnetic field, P1 and P2 vacuum pumps

the neutralization surface, which consists of a thin plate of the chosen reflecting metal mounted on a water-cooled copper block. It can also be heated by means of an internal heating element (up to 300 °C) for purposes of outgassing. The reflecting material used is usually polycrystalline molybdenum or tantalum, chosen for their high atomic mass, resistance to reaction with oxygen, and low sputtering rates. The reflecting plate is kept on a negative potential (with respect to the plasma) and this potential determines the impact energy (10–70 eV) of the ions. Because the plasma is quasineutral, there is no space charge limitation of the ion current, nor any electrostatically caused divergence. Thus, a high-intensity ion current (about 4 A/cm^2) hits the target, which is inclined by an angle between 45° and 55° against the direction of the plasma beam. The ions are neutralized very efficiently near the surface by picking up an electron tunneling from the metal (in an Auger or resonant transition). If the metal surface atoms are much heavier than the incident particles, the latter are reflected back from the surface, forming a beam of hyperthermal neutrals.

The system is typically operated in a pulsed mode with pulse lengths of a few milliseconds and duty cycles of 5% to 10%, in order to avoid a thermal overload of the central electrode, which is used to excite the radiofrequency discharge. The gas supply can also be pulsed using a piezoelectric valve. Optimal results are obtained by applying the radiofrequency power just after the end of the gas pulse.

Perpendicular to the plasma beam, the reflected neutral beam is directed across the magnetic field, and thus the field keeps charged particles away from the neutral beam direction. The atomic beam passes through a system of apertures (baffle) through the differential pumping stage and finally enters the beam utilization chamber. The neutralization efficiency depends upon the exact species of the incident particles and surface atoms, their relative masses and the strength of the interatomic forces acting between them, and upon the condition of the surface with respect to roughness and impurities or heterogeneity. Detailed experimental investigations show that the energy E_n of the neutrals is proportional to the energy E_i of the incident ions ($E_n = 0.4E_i + 2.4$ V, if E_i is measured in V). The halfwidth (FWHM) of the energy distribution (measured with an argon beam) is at energies up to 20 eV constant (about 8 eV) and then increases with the energy of the neutral particles (14 eV at $E_n = 30$ eV). The absolute intensity of the neutral beam is very high. Using a duty cycle of 7%, the time-averaged intensity at a distance of 9 cm from the neutralizing target is 5×10^{15} atoms/cm^2 s. High-intensity beams of oxygen, nitrogen and rare gas atoms have been produced with this source.

1.4 Aerodynamic Acceleration

The seeded beam technique, which is commonly used in modern molecular beam work, has various applications. In this section, however, we discuss only the technique's ability to produce fast neutral beams in the energy range from 1 to 20 eV. The basic concept of this method has already been outlined in Sect. I.3.8.1.

During the expansion of a dilute solution of a heavy species in a light carrier gas, the heavy particles obtain approximately (if slip effects are neglected) the velocity of the light particles due to the permanent collisional interaction between heavy and light particles. In the case of gases, an appropriate gas mixture is used, whereas in the case of vapors the carrier gas is simply admitted to the oven. If E_h is the energy which would be achieved by expanding the pure heavy species, the energy E obtained with the carrier gas is (under the above assumptions) given by

$$E = \frac{M}{m} E_h. \qquad (1.17)$$

M is the mass of the heavy particles, m the mass of the carrier gas. Consequently, for large ratios M/m a considerable energy amplification can be achieved [Becker et al. (1955), Becker and Henkes (1956)].

As an example, Fig. 1.13 shows measured energies of various atoms and molecules with different masses (hydrocarbons, CO, N_2, CO_2, Na, K, and Xe), expanded at different temperatures with He and H_2 as carrier gases, plotted as a function of the heavy particle mass [Fenn (1967), Hüwel et al. (1981), (1982)]. It can be clearly seen from this plot, that seeded beams, together with source heating, can be easily employed to produce beam energies between 1 and 20 eV. There is good agreement between the experimental data for helium and calculations that assume isentropic expansion to infinite Mach number and no contribution to the specific heat (except translational energy) from the heavy species. All the data at higher source pressures indicate the absence of appreciable slip effects. In the case of hydrogen as carrier gas, the observed velocities of the heavy component are

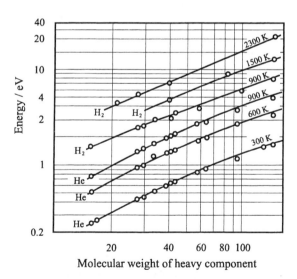

Fig. 1.13. Measured energies of seeded beams as a function of the molecular weight of the heavy species (1% of the carrier gas) at various source temperatures. The carrier gases are He and H_2.

smaller (about 20%) than under the previously mentioned assumptions. This may be a result both of velocity slip effects and of incomplete rotational relaxation during the expansion (see Sect. I.3.8).

The low concentration of the heavy species required to achieve high acceleration means a loss in intensity. The intensity gains, however, that nozzle sources provide compared with effusive sources are so large that even if the source gas contains only a 1% fraction of heavy particles, the net beam intensity of the heavy component is still larger than the total intensity of an effusive source. Moreover, as has been shown in Sect. I.3.8, the heavy species has a higher perpendicular speed ratio than the light one. Accordingly, the lateral beam spreading is larger for the light gas component than for the heavy species, resulting in a preferential concentration of the latter on the beam centerline, increasing the ratio of heavy to light species in the final beam considerably.

As has been shown by Varentsov and Hanseanov (1992), this ratio can be further increased by direct injection of the heavy gas (or vapor) into the cold expanding supersonic carrier gas jet, instead of expanding a homogeneous binary mixture. Their device is shown schematically in Fig. 1.14. The heavy species is injected through an inner tube (4) on the axis of a convergent–divergent nozzle (5), which terminates within the divergent part of the nozzle in the low-pressure region of the supersonic carrier gas expansion. Thus, the carrier gas jet acts as a pump, effectively evacuating the heavy gas species from the injection tube. Due to the interaction with the already cooled carrier gas molecules, the heavy species is effectively cooled, and since the streamlines of the carrier gas flow in the vicinity of the expansion centerline are parallel to the axis, the main portion of the heavy species remains in the vicinity of the axis. Measurements with PbI molecules, injecting the vapor at temperatures between 800 and 900 K into a nitrogen expansion, yield a terminal translational temperature of 20 K and a PbI beam intensity of 1.4×10^{19} molecules/cm^2 sr. Compared to an expansion of a binary gas mixture, a much higher fraction of heavy particles can be transferred through the skimmer, resulting in a substantial increase in intensity. This makes the direct injection of the heavy species into a carrier gas expansion attractive for experiments

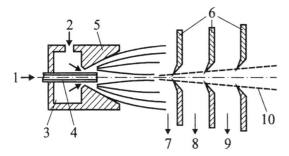

Fig. 1.14. Schematic view of a nozzle with direct injection of a heavy species into a carrier-gas expansion. (1) heavy species supply, (2) carrier-gas supply, (3) carrier gas stagnation chamber, (4) injection tube, (5) convergent–divergent quartz nozzle, (6) skimmer separating the differential pumping stages (7), (8) and (9), (10) molecular beam.

which require intense beams of expensive gases or vapors [Varentsov et al. (1995)].

Using aerodynamic acceleration, essentially two possibilities exist to vary the beam energy. First, the concentration of the heavy species in the carrier gas may be changed while the nozzle temperature is kept constant. In the second method, a mixture of two carrier gases (He and H_2, for instance) may be used, together with a variation of the source temperature. Additionally, the mixture ratio of the two carrier gases may also be changed.

Both methods are frequently applied. Sheen et al. (1978), for example, who investigated the collision-induced ion pair formation of CsCl and Cs_2Cl_2 interacting with argon, krypton, and xenon atoms, varied the energy from 5.2 to 8.9 (for CsCl) and from 6.2 to 13 eV (for Cs_2Cl_2), by reducing the concentration of the heavy species from 2.3% to 0.3%. To produce rare gas beams, they also changed the temperature of the source from about 1000 to 1500 K. The corresponding beam energies were 4.8 to 6.8 eV for argon, 9.1 to 12.1 eV for krypton, and 12.9 to 18.3 eV for xenon. Similar energies have been achieved by Parks et al. (1982), (1984).

The high-temperature source shown in Fig. 4.10 has been used to produce a fast Xe beam by using gas dynamic acceleration in hydrogen (with 0.23% Xe) to investigate the UV emission in collisions between Xe atoms in the vicinity of the threshold. With a maximum source temperature of 2900 K, a Xe beam energy of 30 eV has been achieved [Buck et al. (1980)]. Similar Xe beam energies have been reported by Campargue and co-workers. Using pressures up to 60 atm in a high-temperature source, a maximum Xe beam energy of 37 eV could be obtained [Campargue (1984)].

The device presented in Fig. 4.10 has also been used to produce N_2 and CO_2 beams seeded in H_2 in order to investigate vibrational to electronic energy transfer in collisions with alkali metal atoms [Buck et al. (1980)]. At a source temperature of 2300 K, an energy of 5.4 eV was achieved for N_2, and of 5.1 eV (at a source temperature of 2000 K) for CO_2. Since the beam energy can be varied either by changing the gas mixture or by varying the source temperature (which changes the distribution of vibrational states of the beam molecules), the electronic excitation at a given collision energy can be measured with different vibrational state distributions, allowing one to distinguish between vibrational and translational to electronic energy transfer.

Also, the technique of aerodynamic acceleration has often been used to produce hyperthermal energy alkali atoms for scattering experiments [Larsen et al. (1974), Hüwel et al. (1980), (1981), (1982), Lackschewitz et al. (1986), Brencher et al. (1988)]. The two-chamber oven shown in the lower half of Fig. 4.8, for example, has been used for this purpose. Applying a temperature of 2300 K (using H_2 as carrier gas) sodium beams with energies up to 4.6 eV and potassium beams with energies up to 7 eV have been produced. Using metal vapors requires a heated skimmer (for alkali metals at a temperature of about 500 K) to avoid plugging up its aperture.

Another application of hyperthermal energy beams generated by gas dynamic acceleration is the investigation of molecule–surface reactions. Jones et al. (1996), for example, have studied the reactions of GeH_4 and Ge_2H_6 molecules with Ge(100) and Ge(111) surfaces at impact energies between 1 and 3 eV. Further investigations of molecule–surface impact phenomena are studies of the ionization processes of I_2 molecules, using the expansions of I_2 in He and H_2 carrier gases together with a variation of the source temperature. I_2 beam energies between 2 and 10 eV have been achieved [Kolodney and Amirav (1983), (1984)].

Large polyatomic molecules have also been accelerated using the seeded beam technique. In this case, however, the nozzle temperature usually cannot exceed temperatures of 400 °C, since most of these molecules are thermally unstable or chemically reactive in hot helium or hydrogen beams [Parks et al. (1982), Budrevich et al. (1995)]. An example is SF_6, which has been accelerated up to 15 eV to investigate dissociation processes in molecule–molecule collisions [Parks et al. (1984)]. Other examples are large organic molecules such as cholesterol ($C_{27}H_{45}OH$), which has been accelerated (in H_2–Ar mixtures) to energies between 5.3 eV and 22 eV [Dagan et al. (1992)], or nicotine ($C_{10}H_{14}N_2$) (3 to 11 eV), as well as caffeine ($C_8H_{10}N_4O_2$) (6 to 13 eV). The ionization and fragmentation of these molecules impinging on an oxidized rhenium surface has been investigated as a function of the impact energy [Dagan et al. (1995)]. To vary the beam energy, the concentration of argon is changed in the H_2–Ar carrier-gas mixture.

Budrevich et al. (1995) applied the gas dynamic acceleration to fullerenes, in particular to C_{60}, to investigate their interaction with solid surfaces. In this case, the seeded beam technique offers the unique situation of high fluxes ($\geq 10^{15}$ molecules/s sr) of neutral C_{60} molecules in their electronic ground state, low vibrational temperature due to cooling in the expansion and well controlled and variable kinetic energy in the extremely large range from 1 to 80 eV. To achieve such high energies, temperatures up to 2000 K are required. This has been achieved by using a two-chamber oven similar to the one shown in Fig. 4.8, but made completely from ceramics [Danon and Amirav (1987), Kolodney et al. (1994), Budrevich et al. (1997)]. The first stage is the C_{60} evaporation chamber, temperature-controlled to better than 1% and typically maintained in the region of 950 to 990 K (corresponding to vapor pressures between 0.1 and 0.2 torr). The molecular vapor pressure regulates the flow through the second stage (nozzle chamber), which is a narrow ceramic tubing resistively heated to 2000 K with an extremely small volume (10^{-3} cm^3), to keep a possible reaction of the C_{60} molecules with the hot carrier gas as low as possible. A ruby watch bearing is used as a nozzle, to avoid a catalytic decomposition of the fullerene molecules at hot metal surfaces. The C_{60} molecules are seeded in helium or hydrogen at pressures between 200 and 2400 torr, the highest dilution ratio used is about 2.5×10^4. With this source, using He as carrier gas, the maximum energy achieved was 56 eV, while H_2 as carrier gas results in a maximum energy of 73 eV. The energy distribution in the fast fullerene beam has been measured with an electrostatic energy selector after ionization of the C_{60} molecules yielding a halfwidth of the energy spread $\Delta E/E$ of about 11%.

1.5 Plasma and Gas Discharge Sources

Using gas dynamic acceleration, the energy gap between thermal energy beams and beams produced by ion neutralization can by closed for many species of sufficiently high molecular weight, especially if the acceleration is combined with high temperature sources. The temperature, however, is limited by the oven material to about 3000 K. Therefore, higher temperatures require methods to heat the gas (or vapor) without heating the surrounding oven material.

1.5.1 Arc-Heated Jet Sources

This problem is solved by arc discharges, which can be operated at pressures sufficiently high to permit a gas dynamic expansion. The first devices of this kind were already used during the 1960s. They have been considerably improved since then and employed for various gases [Clausnitzer (1961), Knuth (1964), Kessler and Koglin (1966), Winicur and Knuth (1967), Young et al. (1969)].

The basic principle is the following: The beam gas inside the source is heated to temperatures of the order of 10 000 to 30 000 K by an arc discharge between a cylindrical tungsten rod as anode and a water-cooled anode containing the nozzle aperture. An axial magnetic field confines the arc discharge to the vicinity of the axis between cathode and anode. The hot plasma expands through the nozzle and its core passes as a molecular beam through a skimmer and further collimating apertures, which separate the subsequent differential pumping stages. The beam energy depends on the plasma temperature, which in turn is determined by the discharge current and the gas pressure. The beam flux is proportional to the pressure. Using pressures between 2 and 3 atm and discharge currents of 30 to 250 A, argon and krypton beam intensities of about 10^{19} particles/s sr and energies between 0.7 and 5.2 eV have been achieved [Nagata et al. (1991)]. Combining arc heating and the seeded beam technique, argon beams with energies up to 21 eV have been reported [Liu et al. (1974)].

Plasma-heated sources are also often used to dissociate molecules in order to generate intense hyperthermal beams of atoms. Of special interest are nitrogen atom beams, since N_2 molecules cannot be dissociated by pyrolysis (see Sect. I.2.1.7). Supersonic beams of N atoms with intensities above 10^{17} atoms/s sr have been produced in arc-heated nozzle sources using Ar/N_2 mixtures. With moderate plasma temperatures of about 6000 K energies between 1 and 1.5 eV have been achieved [Bickes et al. (1976), Love et al. (1977)]. Plasma temperatures up to 18 000 K and nitrogen atom translational energies up to 2.5 eV have been obtained with a similar device by Shobatake and Tabayashi (1982) [see also Tabayashi et al. (1984)]. Xu et al. (1997) describe a design which can be operated in pure nitrogen producing N atom beams with intensities of 10^{19} atoms/s sr and energies between 0.5 and 5 eV. Intense beams of H and D atoms have also been generated with similar sources [Young et al. (1969), Way et al. (1976), Götting (1985), Götting et al. (1986), Samano et al. (1993)].

30   1. Fast Beams, Production and Detection

The use of these sources with oxygen is problematic because of the rapid corrosion of the metal electrodes. To overcome this problem, devices have been used in which a rare gas (helium, argon or a mixture of both) is heated in a commercially available plasma torch. The heated gas flows through a rather long tube and the oxygen is injected at the tube end just before the expansion. The distance between the arc region and the nozzle has the advantage that almost any metastable and ionized species is quenched before beam formation takes place. On the other hand, however, this channel causes a considerable drop of the initial plasma temperature and thus reduces the final beam energy. A high dissociation efficiency (close to 100%), high beam intensities (3×10^{17} atoms/s sr), and energies up to 1.4 eV characterize this device [Silver et al. (1982)].

An improved design which has been developed on the basis of the design of Götting et al. (1986) is presented in Fig. 1.15 [Ferkel et al. (1991), Feltgen al. (1993), Pikorz (1994)]. It can be used to produce both fast groundstate atom beams (for example of H and D atoms) and metastable atoms.

Similar to the plasma sources described above, the discharge burns between a tungsten cathode (1) and the anode (2) containing the nozzle (3). The source pressure can be varied between 0.1 and 2 bar and the discharge currents range from 30 to 100 A. The cathode is a thoriated (4%) tungsten rod (commercial welding electrode with 3.2 mm ∅), which is spherically ground at the tip, in order to keep its burning off as low as possible. The circular anode with the concentric nozzle is made of tungsten and brazed into a water-cooled copper disk. The distance

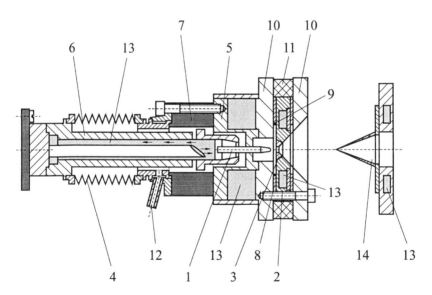

Fig. 1.15. Cross-sectional view of an arc-heated nozzle source [Götting et al. (1986), Ferkel et al. (1991)]. (1) tungsten cathode, (2) copper anode (3) with tungsten nozzle (3), (4) bellows tubing, (5) stainless steel plasma chamber (with water-cooling (13)), (6) cathode holder, (7) polyamide insulation, (8) Teflon gasket and insulation, (9) alumina washer, (10) disk-shaped yokes of soft iron, (11) 16 permanent magnets, (12) gas supply, (13) water cooling, (14) tungsten skimmer

between cathode and anode, which is critical for achieving high temperatures, is adjustable from outside the vacuum chamber by means of a bellows tubing (4). Typical distances during operation lie between 1 and 2 mm. Because of the high plasma temperatures, the copper shaft (6) supporting the cathode as well as the anode and the walls of the stainless steel plasma chamber (5) are cooled by water (13). To achieve optimum cooling, the surfaces in contact with water are mechanically roughened; this suppresses the formation of an insulating steam layer (Leidenfrost phenomenon), which may reduce the heat conductivity drastically. The plasma chamber is electrically insulated against the cathode by a polyamide block (7) and against the anode by a Teflon gasket (8). The latter is protected against heat radiation from the plasma by a felt-like alumina (Al_2O_3) washer (9). The axial magnetic field is produced between two cylindrical soft iron disk-shaped yokes (10), which are magnetized by 16 small permanent magnets (11) between these yokes around the outer circumference. This magnetic field of about 0.1 T confines the plasma to the vicinity of the axis. The gas is admitted to the plasma chamber by four tangentially arranged channels (12). The skimmer (14), located downstream of the nozzle is water-cooled and made from tungsten. Its distance from the nozzle is adjustable from outside the vacuum chamber in the range from 10 to 60 mm.

A commercial welding transformer supplies the necessary dc current up to 130 A at voltages between 18 and 30 V. Its no-load voltage is 110 V, its current is continuously variable between 0 and 180 A with a maximum power of 12.5 kW.

To start operation with helium, the arc is ignited by a short ignition pulse with a voltage of 500 V and a current up to 60 A at a He pressure between 60 and 150 mbar. Under these conditions a glow discharge is obtained which turns into the desired arc discharge with increasing pressure. In the case of hydrogen gas, the discharge is commonly started in pure argon, since it proved to be rather difficult to ignite a discharge in pure hydrogen. Then the feed gas is changed to an argon/hydrogen mixture with increasing fraction of hydrogen, and after a few minutes, the gas mixture is changed to pure hydrogen [Way et al. (1976), Götting (1985)]. Meanwhile it has been demonstrated, however, that a hydrogen discharge can also be directly ignited in pure hydrogen [Samano et al. (1993)], starting with a glow discharge at low pressures.

As an example, Fig. 1.16 shows two velocity distributions, obtained under different operation conditions of the arc-heated helium source. The low-velocity distribution has been measured with a low pressure (50 mbar) in the source and an arc current of 40 A, the fast distribution has been obtained with a pressure of 800 mbar and an arc current of 60 A. The curves drawn through the measurements are calculated for a nozzle beam speed distribution (drifting Maxwellian distribution) (for a flux detector) according to (3.128), given by

$$f(x) = C\left(\frac{v}{v_w}\right)^3 \exp\left(-\frac{(v-w)^2}{v_w^2}\right) = C x^3 \exp\left(-(x-S)^2\right), \tag{1.18}$$

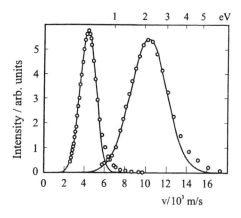

Fig. 1.16. Measured speed distributions of an arc-heated nozzle source for different operation conditions. The full curves are calculated according to (1.18)

where v_w is the most probable velocity corresponding to the terminal parallel temperature T in the beam, w is the flow velocity, and C is a normalization constant. In the second part of this equation we have made the substitutions

$$x = v/v_w, \quad v_w = (2kT/m)^{1/2}, \quad \text{and} \quad S = w/v_w. \tag{1.19}$$

Fitting the distribution (1.18) to the experimental data, the speed ratio S and the most probable speed v_w in the beam and thus the terminal parallel beam temperature T and the flow velocity w can be determined. Assuming an isentropic expansion of an ideal gas, a constant specific heat ratio κ, and gas dynamic flow, the stagnation temperature T_0 in the source can be calculated using the relation (3.37), which (by introducing the speed ratio instead of the Mach number) can be written in the form

$$T_0/T = \left(1 + S^2(\kappa - 1)/\kappa\right). \tag{1.20}$$

Table 1.1 shows the results for the speed distributions shown in Fig. 1.16 together with the operating conditions of the source. The maximum of the distribution (1.18) is given by

$$v_{max} = \frac{v_w}{2}\left(S + \sqrt{S^2 + 6}\right). \tag{1.21}$$

A further increase of the gas pressure (up to 2 bar) shifts the maximum of the distribution to higher velocities, so that by utilizing the rather broad distribution

Table 1.1 Operation conditions and beam data for the distributions of Fig. 1.16

P [mbar]	I [A]	S	v_w [m/s]	w [m/s]	v_{max} [m/s]	T [K]	T_0 [K]
50	40	3.2	1203	3851	4350	348	1773
800	60	3.5	2637	9231	10250	1672	9864

sufficient beam intensities are available up to 20 000 m/s. This corresponds to an energy of more than 8 eV.

Similar distribution functions have been measured with hydrogen and deuterium beams, covering the energy range up to 10 eV [Way et al. (1976), Götting et al. (1986)]. Beam intensities are of the order of 10^{19} particle/s sr.

The material stress of such a source is rather high, especially when operated at high plasma temperatures. The efficiency of cooling the electrodes is decisive, as well as the durability of the insulation material between anode and plasma chamber, which is simultaneously used as a gasket. The device shown in Fig. 1.15 has proved itself in this respect, permitting undisturbed operation in helium for about 200 h with arc currents up to 60 A. This time reduces to about 50 h when using a hydrogen discharge with arc currents of 100 A. After this time, the electrodes are burnt down and must be replaced.

Apart from groundstate atoms, this source produces also metastable helium atoms, but almost only $He(2\ ^3S)$ states. Due to the high pressure in the source, the $He(2\ ^1S)$ states are effectively quenched by collisions with helium atoms and slow electrons. The fraction of ions produced in this source can be estimated using Saha's equation, which describes the thermal ionization of gases and vapors. Assuming a quasineutral thermal plasma, in which electron and ion temperatures are equal to the gas temperature (an assumption which is justified at the high pressures usually used in the source), the degree of ionization ξ, defined as the ratio of ion density to the total plasma density, is given by

$$\xi = \sqrt{\frac{F(T,P,U_i)}{1+F(T,P,U_i)}} \quad \text{with} \tag{1.22}$$

$$F(T,P,U_i) = \frac{(2\pi m_e)^{3/2}(kT)^{5/2}}{Ph^3}\exp(-eU_i/kT)$$

$$= 3.33\times 10^{-7}\frac{T^{5/2}}{P}\exp\left(-11.6\times 10^3(U_i/T)\right). \tag{1.23}$$

m_e is the electron mass, k is Boltzmann's constant, T is the temperature, h is Planck's constant, P is the pressure, and U_i is the ionization potential of the gas. In the numerical part of (1.23), P is measured in bar, U_i in V, and T in kelvin. For some gases and vapors, Fig. 1.17 shows the degree of ionization ξ as a function of the temperature T at a pressure of 1 bar. Up to temperatures of the order of 10^4 K, the degree of ionization ξ is rather small, especially for helium. With increasing temperature, ξ increases rapidly and finally approaches its asymptotic value $\xi = 1$.

The fraction of metastable atoms, especially of $He(2\ ^1S)$ atoms, can be considerably increased by employing an additional discharge between nozzle and skimmer. This discharge is operated at a voltage between 50 and 150 V and currents from 1 to 10 A. The metastable singlet atoms formed in this region are not collisionally quenched due to the low density between nozzle and skimmer. In this

34   1. Fast Beams, Production and Detection

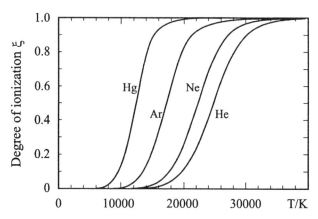

Fig. 1.17. Degree of ionization ξ of some gases and vapors as a function of temperature according to Saha's equation (1.22) and (1.23)

manner, metastable He beams with intensities of some 10^{15} atoms/s sr can be produced; about 30% of these atoms are in the He(2^1S) state [Ferkel et al. (1991)].

1.5.2 Radiofrequency Discharges

The production of high-temperature plasmas (up to 20 000 K) by high-power radiofrequency discharges was also investigated during the 1960s [Reed (1961), Eckert et al. (1968)], but these plasmas have been applied only occasionally to hyperthermal beam production. An example is a quartz nozzle source, in which the carrier gas is heated using a 100 kW radiofrequency discharge (0.485 MHz) [Madson and Theby (1983)]. The discharge is ignited in argon at low pressures (10^{-1} to 10^{-2} torr). Then the pressure is continuously increased up to 25 torr, and finally the gas is gradually changed to helium using an appropriate mixing device. The gas (CO_2) to be accelerated by the helium plasma is modulated by a rotary valve gas pulser and injected into the expansion through three holes (100 μm ⌀) in the nozzle wall, spaced at equal angles. The CO_2 pulse adapts quickly to the flow velocity of the carrier gas during the remaining expansion, resulting in a fast modulated supersonic CO_2 beam with rather narrow velocity spread. Accordingly, only the desired beam species is modulated, while the carrier gas beam and the photon emission from the plasma yield a continuous background signal at the detector. Hence the lock-in technique can be applied to separate the modulated beam signal from the background signal.

1.5.3 Laser-Sustained Plasmas

Continuous CO_2 Lasers. Another possibility to generate a high-temperature plasma in a gas dynamic beam source was first employed by Cross and Cremers (1986). They developed a laser-sustained plasma technique for producing high-intensity beams in the 1 to 5 eV energy region. Using a 70 W CW CO_2 laser, plasma temperatures of 8000 to 9000 K were obtained in xenon, resulting in a mean beam energy of about 2 eV after gas dynamic expansion. The beam energy can be easily increased by increasing the laser power. This has been demonstrated by Campargue and co-workers, who used a CW CO_2 laser with a maximum power of 380 W, which is focused to a point in the immediate vicinity of the nozzle entrance. Thus, a continuous plasma (in argon) with temperatures up to 20 000 K can be sustained at pressures between 2.5 and 7 bar [Girard et al. (1993), (1994), Lebéhot and Campargue (1996)]. With plasma temperatures of 12 500 K, the argon atom energy at the maximum of the energy distribution is 2.6 eV.

Laser-sustained plasmas have the advantage that no electrodes are used which evaporate during operation and thus may give rise to beam impurities. For the same reason there is no limit of the operation time due to consumption of electrodes. On the other hand, the use of a high-power CO_2 laser to operate a beam source is rather expensive.

As in all discharge sources, aside from hyperthermal energy particles in the electronic ground state, the beam also contains electronically excited particles. In essence, metastable and highly excited Rydberg states survive in the beam. Spectroscopic investigations of the light emitted from the beam as a function of the distance from the nozzle yields information on the population of excited states and their relaxation during the expansion [Beulens et al. (1994)].

Pulsed Lasers. Pulsed CO_2 lasers combined with pulsed sources have also been used to produce plasma-heated hyperthermal energy beams. This method has been introduced by Caledonia et al. (1987), who generated pulsed O atom beams with energy distributions ranging from 2 to 14 eV [Oakes et al. (1995)]. A similar device has been used to produce beams of F atoms in the same energy range [Giapis et al. (1995), Minton et al. (1997)].

The essential parts of such a device are schematically shown in Fig. 1.18. The radiation (5) of a pulsed CO_2 laser (with pulse energies up to 10 J at $\lambda = 10.6$ μm and pulse lengths of 2.5 μs) is focused by a gold-plated mirror (6) (50 cm apart

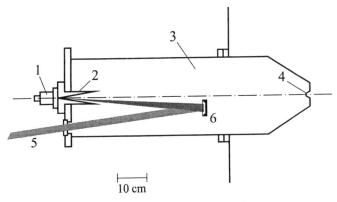

Fig. 1.18. Schematic drawing of a laser detonation source [Giapis et al. (1995)]. (1) pulsed valve, (2) nozzle, (3) source chamber, (4) skimmer, (5) laser beam, (6) gold-plated mirror

from the nozzle) to the nozzle throat of a pulsed piezoelectric valve (1). The water-cooled aluminum nozzle (2) is 10 cm long and has a cone angle of 20°. The maximum power density in the laser focus, which is located just in front of the nozzle throat (with 1 mm ∅), is 2×10^9 W/cm^2. The duration of the gas pulses can be varied between 100 and 200 µs, stagnation pressures range from 3 to 6 bar. The laser pulse is fired 200 µs after opening the valve. It causes an electrical breakdown in the expanding gas, resulting in a detonation wave which heats the gas pulse to yield a completely dissociated high-temperature plasma.

Therefore, these sources are often called "laser detonation sources". The hot plasma expands through the nozzle; the nozzle dimensions allow electron–ion recombination but not atom–atom recombination. Hence, an intense beam of neutral O atoms with a flux of 10^{18} atoms/pulse is obtained.

SF$_6$ has been used to produce a pulsed, hyperthermal energy beam of fluorine atoms with a laser detonation source [Giapis et al. (1995), Minton et al. (1997)]. They use a water-cooled copper nozzle of the same geometry as described above and a stagnation pressure of 9.6 bar. The repetition rate of the laser lies between 1 and 2 Hz. The intensity of F atoms per pulse obtained (30 cm downstream the nozzle) is about 2×10^{15} atoms/s cm^2; the degree of dissociation is 96%.

A measured energy distribution of the beam is presented in Fig. 1.19. It can be influenced by the stagnation pressure in the source and the delay time between gas and laser pulse in such a manner that the maximum of the distribution can be shifted between 3 and 7 eV. The full curve drawn through the measurements shown in Fig. 1.19 is calculated for a drifting Maxwellian distribution for a density detector ($\nu = 2$ in (3.128)). The fit to the measured distribution yields a beam temperature T = 2698 K, a speed ratio S = 4.2, and a stagnation temperature of $T_0 = 21\,500$ K.

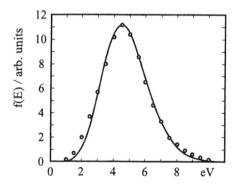

Fig. 1.19. Measured energy distribution f(E) (circles) in a F atom beam produced in a laser detonation source [Giapis et al. (1995)]. The full curve is calculated (see text)

Kinoshita et al. (1998) used a Q-switched YAG laser (200 mJ/pulse) to operate a laser detonation source for O atoms. The electron density in the plasma is considerably enhanced by photoelectrons released by the laser from a tungsten surface located close to the laser focus. The O atoms have a mean energy 4.7 eV, the halfwidth of the energy distribution is 5.5 eV.

1.5.4 Hollow Cathode Discharges

To generate metastable atom beams in the energy range of 1 to 10 eV, Verster and co-workers have developed a reliable source which is based on a hollow cathode arc discharge, in which metastables are produced by both electron impact and heavy particle collisions [Theuws et al. (1977), (1982)]. Intensities of 10^{13} and 10^{14} atoms/s sr (for krypton and argon, respectively) have been measured. The broad velocity distribution of the emerging metastable atoms (1–10 eV) makes velocity selection necessary. The lifetime of the cathode tubes is limited by the high operation temperatures to about 40 h for argon and krypton and to about 15 h for neon. In addition to the metastable beam, this source also produces an intense beam of groundstate atoms [Theuws et al. (1982)]. Intensities up to 10^{18} atoms/s sr have been observed, with a broad energy distribution ranging from about 1 to 10 eV.

1.5.5 Corona Discharges

In Sects. I.4.7 and I.4.8 sources are described which utilize a corona discharge to produce radicals or metastable species of thermal energy. As was first observed by Leasure et al. (1975), these sources not only produce thermal energy beams, but also have a fast beam component if the needle-shaped electrode is used as anode and the skimmer as cathode. The fraction of fast particles increases with increasing pressure in the source. At pressures between 20 and 90 torr (dependent on the gas used), intensities are of the order of 10^{14} atoms/s sr. Under these conditions, the speed distribution of the fast particles is rather narrow, and the energies obtained are 5 eV for helium, 30 eV for neon, and 74 eV for argon. At higher (300 torr) and lower pressures (< 20 torr), the speed distributions become broader.

A similar source has been described by Fahey et al. (1978). Under certain operation conditions they were able to produce fast neutral helium atoms with intensities of 10^{15} atoms/s sr and a mean energy of 800 eV. Pulsed corona discharges, which yield a thermal energy and a fast beam (helium atoms of 190 eV) component, have also been investigated [Lo et al. (1997)]. The high energy, the polarity of the electrodes, and the pressure dependence of the fast beam signal indicate that charge exchange processes play an important role in the fast neutral particle production.

1.5.6 Glow Discharges

Figure 1.20 shows a schematic drawing of a glow discharge source, which has been developed to investigate and to etch insulating solid surfaces [Kuwano and Nagai (1985), Shimokawa (1992), Shimokawa and Kuwano (1994)]. It can be used for gaseous substances and consists essentially of a tubular cathode made of graphite (70 mm long, 40 mm ⌀), closed at its rear end by a graphite plate ending in a gas supply tube. The front side containing the exit aperture of (30 mm ⌀) is also made of graphite, and the exit aperture is covered with a multiaperture grid. The ring-shaped anode (also made of graphite) is inside the tubular cathode and equidistant from each end. The complete device is surrounded by a solenoid, which can produce an axial magnetic field up to 900 G.

When a voltage of several kV is applied between anode and cathode, a glow discharge is ignited at pressures between 10^{-5} and 10^{-3} torr, which is confined to the region of the tube axis by the axial magnetic field similar to a Penning discharge. Electrons of the glow discharge oscillate back and forth through the ring anode, thereby increasing the electron path considerably. This increases the number of ions produced. Some of these ions are accelerated towards the end of the cathode, and most of them are converted into fast atoms by neutralization in charge-exchange collisions with gas molecules or by recombination with low-energy electrons. Consequently, a fast neutral beam exits the source aperture. Ions are removed from the beam by a suitable electric field applied to a pair of deflection electrodes. Under optimum conditions of gas pressure and magnetic field strength, using a discharge voltage of 3 kV, the measured flux density of the neutral beam is of the order of 6×10^{15} particles/s cm^2. High neutralization efficiencies (up to 95%) are achievable. The energy distribution f(E) of the neutral particles (see Fig. 1.21) has its maximum in the vicinity of the discharge voltage with a relative halfwidth $\Delta E/E$ of 8 to 10%. About 60% of the neutral particles have energies in this interval. The rest — corresponding to the different places of neutralization — have a broad energy distribution containing all energies lower than the maximum energy given by the discharge voltage.

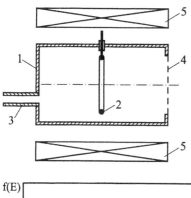

Fig. 1.20. Schematic view of a glow discharge source [Shimokawa and Kuwano (1994)]. (1) graphite tube (cathode), (2) ring-shaped anode, (3) gas supply, (4) exit aperture with grid, (5) solenoid

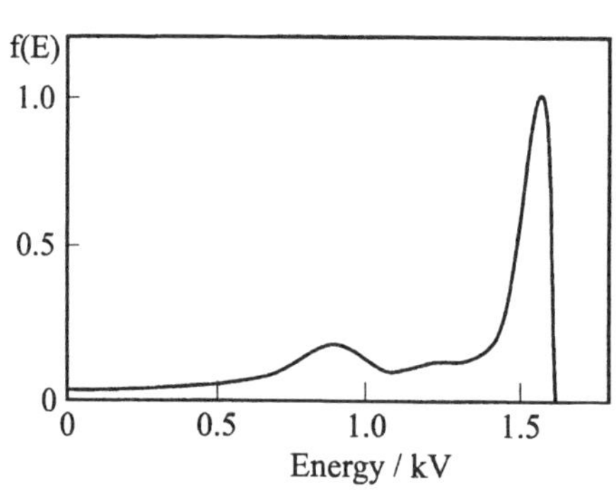

Fig. 1.21. Energy distribution of neutral atoms at a discharge voltage of 1.6 kV and a pressure of 6×10^{-4} torr

1.6 Laser Ablation

Levine et al. (1968) were the first to observe hyperthermal energy particles evaporating from massive gold targets when irradiated with the light of a Q-switched ruby laser. Utilizing this effect, Früchtenicht and co-workers have developed a pulsed beam source for atoms in the energy range from 1 to 10 eV [Friichtenicht (1974)]. The laser light (0.5 J/pulse) is focused on a target consisting of a thin film deposited on a glass microscope slide. A collimating aperture in front of the target forms a beam from the expanding vapor. This aperture is followed by a pair of electrostatic deflection plates to remove charged particles from the beam. Time-of-flight measurements show that the energy distribution of the emitted particles ranges from about 1 to 10 eV, having a maximum between 5 and 7 eV (dependent on the target material and its structure). Uranium, aluminum, and fluorine beams have been generated in this way. The total number of atoms contained in one beam pulse, measured at a distance of 60 cm from the target, is of the order of 2×10^{13} particles/cm^2 pulse. A similar device has been used by Tang et al. (1981) to produce carbon beams with energies between 0.1 and 10 eV. Thin films of graphite deposited on transparent microscope slides are irradiated by the light of a Q-switched ruby laser. The authors observed not only C atoms, but also dimers (C_2) and trimers (C_3). The relative intensities of these beam components can be varied by changing the laser focusing as well as the film thickness.

The velocity distribution of beams produced by laser ablation can be described by a drifting Maxwellian distribution according to (3.128) (see, for example, Figs. 1.16, 1.19, and 1.24). From this fact it may be concluded that the laser pulse produces a hot plasma, which subsequently experiences a weak expansion.

To obtain the same target conditions for all laser pulses, two techniques for the operation of laser ablation sources have been developed, essentially. In the first method, the laser beam is focused to a point (typical diameter 0.5 mm) on a solid target, which is mechanically moved in such a manner that each laser pulse finds a new, unused part of the surface. The movement of the target can be realized in different ways. A cylindrical rod of the material to be ablated is often used as the target, which is kept in a precise helical motion during the laser irradiation (combined rotary and longitudinal motion) by a suitable motor (see Sect. I.4.4) [Reid et al. (1992)]. In other devices, a circular, disk-shaped target is used, which rotates while the center of rotation makes a longitudinal motion, so that the laser focus follows a spiral on the target [Maruyama et al. (1990)]. Finally, rectangular targets are also used, which are moved by step motors in two dimensions, so that the laser focus describes a zigzagged or square-wave line (see Sect. 2.4.2) [Sugai and Shinhara (1997), Lievens et al. (1997)].

In the second method (see Fig. 1.22), the laser is focused onto a thin film of the beam material, which is continuously deposited onto a transparent target which is in close contact with a cold trap (back-illumination technique). The beam material effuses from the aperture of a simple source close to the target. F atom beams, for example, have been produced by this method with energies up to 4 eV utilizing a frequency quadrupled Nd:YAG laser at λ = 266 nm and thin films of XeF$_2$ [Levis et al. (1990)]. The kinetic energy of the ablated particles can be changed by varying the thickness of the deposited film. Using a constant deposition rate, the kinetic energy of the F atoms increases with decreasing repetition rate of the laser. Apart from F atoms, Xe atoms and F$_2$, XeF, and XeF$_2$ molecules are also observed in the beam. The heavier particles are slower than the F atoms, but their most probable energy is larger than the most probable energy of the light atoms. From this observation it may be concluded that the light particles accelerate the heavier ones in a kind of gas dynamic expansion.

This method has been used to produce hyperthermal energy beams of various molecules, such as CH$_2$I$_2$ [Domen and Chuang (1989)], H$_2$S [Harrison et al. (1988)], Cl$_2$ [Cousins and Leone (1988), (1989)], NO [Natzle et al. (1988), Cousins et al. (1989)], HBr [Cho et al. (1988)], as well as of metal atom beams [Knowles and Leone (1997), Ching et al. (1997)]. Target materials consisting of two components (Cl$_2$ and Xe, for example) have also been used [Campos et al. (1993)].

As an example, Fig. 1.22 shows schematically a setup which has been used to produce and investigate NO beams [Cousins et al. (1989)]. A MgF$_2$ target (3) (2.54 cm ∅, thickness 0.2 cm), in direct contact to a helium cryostat (2) and having a temperature between 20 and 30 K, forms the substrate surface for the NO layer. A continuous beam of NO effusing from the small gas supply tube made from copper (4.3 mm ∅) (4) is used for film deposition. Its aperture is close to the target (0.7 cm). The flow rate can be varied in the range of 1×10^{17} to 2.5×10^{18} molecules/s in order to produce layers of different thickness. The transparent target is irradiated by the light (1) of a weakly focused excimer laser (20 mJ/cm^2, 193 nm) with a repetition rate between 3 and 5 Hz. Deflection plates (6) remove

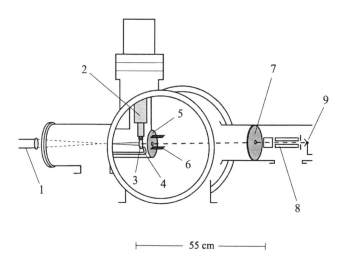

Fig. 1.22. Schematic of a laser ablation source [Cousins et al. (1989)]. (1) laser beam, (2) cryostat, (3) target (MgF$_2$), (4) target deposition source, (5) aperture, (6) condenser plates, (7) aperture, (8) mass filter with ion source, (9) ion collector

ions from the beam, which is finally ionized in the ion source of a mass filter (8) and detected by a channeltron (9). Applying time-of-flight measurements, the translational energy of the ablated particles can be determined, while multiphoton ionization with a second laser can be used to determine the distribution of rotational states in the beam. The measured translational energies lie between 0.14 and 0.71 eV, the rotational temperatures are of the order of 100 to 200 K.

The number of NO molecules per laser pulse is of the order of 10^{15}, with an ion fraction of 10^{-7}. Under the operational conditions used, the fraction of clusters in the beam is about 1%. In general, however, the fraction of clusters in laser-ablated beams is higher, particularly under appropriate operation conditions. Therefore, laser ablation is often used in combination with a gas dynamic carrier gas expansion to produce beams of clusters [Powers et al. (1981), Bondybey et al. (1983), Hopkins et al. (1983), Morse and Smalley (1984), Parks et al. (1988), Weiller et al. (1989), Parks and Riley (1990), Parks et al. (1996)]. These sources are described in Sect. 2.4.2.

A two-stage laser-ablation source based on the principle shown in Fig. 1.22, which also permits a focusing of the neutral particles, is illustrated in Fig. 1.23. A rotating disk contains the material (Ba) to be evaporated and a hole on opposite sides. In the first step the ablation laser is fired when the solid barium slab is situated below the transparent substrate (at a vertical distance of 2–5 mm) (see left half of Fig. 1.23). In this position, material is ablated from the slab and deposited onto the substrate. In the second step, the laser is fired again when the disk has rotated by an angle of 180°. Now the hole is located below the substrate (see right half of Fig. 1.23). Accordingly, barium from the substrate is ablated and forms a pulsed beam which passes through the hole. Beam focusing is achieved by using a cylindrically or spherically curved glass substrate. The beam is detected by a condensation target detector, using an imaging technique (see Sect. I.5.1.1) to obtain a three-dimensional plot of the density distribution in the beam [Kadar-Kallen and Bonin (1989)]. Hence, a pulsed, focused beam with half the repetition rate of the laser is obtained. A modification of this source for highly refractory metals has been used to measure the polarizability of uranium [Kadar-Kallen and Bonin (1994)].

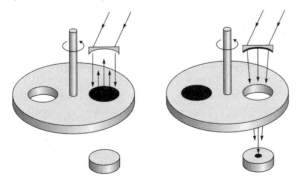

1.23. Two-stage laser ablation [Kadar-Kallen and Bonin (1989)]. The atom beam is produced in two steps (see text)

42   1. Fast Beams, Production and Detection

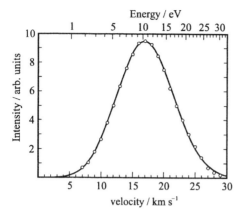

Fig. 1.24. Measured velocity distribution (circles) of a Li beam produced by laser ablation [Abel et al. (1997)]. The full line is calculated according to (3.128) for a temperature T = 20 000 K and a speed ratio S = 2.05

Li atom beams produced by laser ablation are used in plasma physics to obtain the radial edge profile of the electron density of fusion plasmas. Abel et al. (1997) use a movable target holder with three target slides. The targets can be moved in two directions by a motion feedthrough using a computer-controlled positioner. The LIF-C targets are made of a thin 50 nm film deposited over a 500 nm layer of carbon evaporated onto a glass substrate having a predeposited 5-nm thin film of chromium. A 1.5 J, 30 ns pulse halfwidth Q-switched ruby laser is used to ablate the target. The laser beam is focused to a 2 mm spot on the rear side of the film through the glass substrate of the target.

The velocity distribution of the beam has been determined by time-of-flight measurements. The mean beam energy is about 10 eV, the velocity distribution (see Fig. 1.24) can be described by a drifting Maxwellian distribution according to (3.128). The density distribution of the electrons in the edge plasma is reconstructed from the emissivity of the Li ($2p\ ^2P-2s\ ^2S$) transition at 670.8 nm (lithium beam spectroscopy).

Ching et al. (1997) produce a pulsed sodium beam by laser ablation from a target consisting of a fused-silica glass slide with a 700 Å thick vanadium film followed by a 1000 to 3000 Å thick NaAu film. The purpose of the vanadium layer is to increase absorption of the 1.06 μm Nd:YAG laser light at the metal–glass interface. The Na beam is used for spectroscopic diagnosis of a high-power ion diode and is injected into the diode acceleration gap to measure electric and magnetic fields from Stark and Zeeman effects through the observation of the fluorescence using two-dimensional space- and time-resolved spectroscopy.

1.7 Sputtering

Although sputtering of solid surfaces by ion bombardment was discovered in the middle of the 19th century, it was about a hundred years before detailed investigations began to reveal the underlying fundamental physical processes [Behrisch (1964), Kaminsky (1965), Carter and Colligon (1968)]. The interest in sputtering phenomena at that time was stimulated by its bearing on the behavior of materials in outer space and in various plasma devices. Wehner (1955) was the first to observe preferential ejection of atoms from single crystals under low-energy (about 2 to 10 keV) ion bombardment, and a few years later he measured the mean kinetic energy of the emitted particles and found values of the order of 10–20 eV [Wehner (1958), (1959), (1960)]. On the basis of these results, Wehner suggested that the directionality and energy range of sputtered atoms might be utilized to produce beams of hyperthermal neutral particles.

In the following years the results of Wehner were confirmed and complemented by various authors. Most experiments dealt with measurements of the sputtering yield versus energy for many ion–target combinations, but a significant amount of work was also done in investigating the angular and energy distributions of sputtered particles. Single crystals, polycrystals, amorphous and liquid targets have been studied. The results obtained led to a detailed understanding of the main features of sputtering in terms of a series of quasi-elastic collision processes induced by the bombarding ions [Nelson (1968), Sigmund (1969), Gabovich et al. (1989)]. In the simplest case of a random target, sputtering may be described by the following qualitative picture: An incident ion undergoes a series of collisions in the target, and recoil atoms with sufficient energy undergo secondary collisions, thereby creating a second generation of recoiling atoms (collision cascade). This process continues to yield higher-order generations of recoil atoms. Both the ion itself and energetic recoil atoms may get scattered back to the surface by a series of collisions from a depth that depends on their energy. These particles account for most of the sputtered energy but constitute only a minor portion of the number of ejected particles. The slowing-down paths of both the ion and all energetic recoil atoms are surrounded by clouds of higher-order recoil atoms with low energy. These atoms have small ranges and therefore only get ejected from the solid if they are located close to the surface. But there are so many of them that they account for the major portion of the sputtering yield. For this reason, the energy distribution of sputtered atoms peaks at small energies (see Fig. 1.26).

In the case of single crystals this picture must be modified in order to account for effects of the regular lattice structure, which in the case of polycrystals are mostly averaged out.

In the theoretical formulation of the above picture, an integro-differential equation for the sputtering process has been developed from the general Boltzmann transport equation. Input quantities are the cross sections for ion–atom and atom–atom collisions and the atomic binding energies of the solid. The solutions of this equation yield the characteristic quantities describing the sputtering process [Sigmund (1969)], in particular, the sputtering yield S and the sputtering rate v_S. The yield is defined as the number of atoms knocked out of the solid by one incoming ion, the latter is the layer thickness x of the target material removed by sputtering per unit time:

$$v_s = \frac{dx}{dt} = \frac{1}{\rho A}\frac{dm}{dt}. \qquad (1.24)$$

ρ is the density of the solid, A the bombarded area, and dm the mass removed during the time dt. Sputtering yield and sputtering rate are connected by the relation

$$v_s = \frac{m_a S}{e\rho} j, \qquad (1.25)$$

where m_a is the mass of the target atoms, j the ion current density, and e the elementary charge. Sputtering starts at a threshold energy E_t of the ions, which depends on the mass of the ions and of the target atoms as well as on the binding energy E_b of the atoms in the solid (sublimation heat of the material). Already the simple assumption of elastic collisions permits a rather realistic estimate of the threshold energy E_t. The maximum energy, which can be transferred from an ion of energy E_i and mass m_i to a target atom at rest of mass m_a (assuming head-on collisions), follows from the conservation of energy and momentum:

$$E_a = \frac{4 m_i m_a}{(m_i + m_a)^2} E_i. \qquad (1.26)$$

If we assume that at the threshold this energy E_a is equal to the binding energy in the solid E_b, ($E_i = E_t$), the threshold energy is given by

$$E_t = \frac{(m_i + m_a)^2}{4 m_i m_a} E_b. \qquad (1.27)$$

Empirically determined threshold energies have values between some eV (4 eV for Hg^+–Au, for example) and a few hundred eV (400 eV for H^+–W).

From the threshold energy, the sputtering yield increases with increasing ion energy, passes through a broad maximum and then decreases again with a further increase in ion energy. At a fixed ion energy, the dependence of the sputtering yield on the angle of ion incidence Θ, referred to the surface normal ($\Theta = 0$), also passes through a rather pronounced maximum. Contrary to the energy dependence of the sputtering yield, this maximum is rather independent of the ion–target combination and is experimentally observed at an angle of about $\Theta = 70°$.

Hyperthermal atom beam sources utilizing sputtering of solids were developed at the end of the 1960s both in Amsterdam [Politiek et al. (1968)] and Freiburg [Hulpke and Kempter (1966)]. They have been used in numerous scattering experiments.

Figure 1.25 shows a schematic view of the Amsterdam design. The Ar^+ ions produced in an unoplasmatron source (1) are focused onto a rectangular spot on

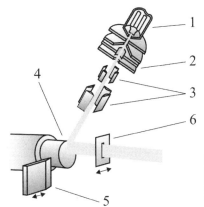

Fig. 1.25. Sputtering source [Politiek et al. (1968)]. (1) ion source, (2) lenses, (3) quadrupole lenses, (4) alkali metal, (5) turning knife, (6) collimating slit

the target (4) (8×0.2 mm^2) by an einzel lens (2) and two quadrupole lenses (3). Typical ion currents are of the order of 500 μA at an acceleration voltage of 6 kV. The angle of incidence is about 70° to obtain the maximum sputtering yield. Because the angular distribution of the particles sputtered from a polycrystalline target is a cosine or cosine squared distribution, the particles ejected perpendicular to the surface are collimated by a slit (6) to form the atomic beam. There then follows a slotted disk velocity selector (see Chap. 3), not shown in Fig. 1.25. The target (alkali metal) is cooled to low temperatures and rotates at 120 rpm to achieve uniform removal of material. The oxide skin on the target can be removed mechanically by a small turning knife.

Figure 1.26 shows a velocity distribution of sputtered potassium atoms. The broad distribution has its maximum at about 2200 m/s and extends to velocities of more than 14 000 m/s. It can be described by

$$f(v)dv = \frac{v^2}{\left(v^2 + v_B^2\right)^{n+1}} dv, \tag{1.28}$$

with $1 \leq n \leq 2$ (full curve in Fig. 1.26). This follows from the theory of linear collision cascades [Thompson (1968), Sigmund (1977)] and has been confirmed by numerous measurements [Pellin et al. (1981), Husinsky et al. (1983)]. v_b is the velocity which corresponds to the binding energy $E_b = mv_b^2/2$ at the surface. For low velocities, the distribution can be approximated by a Maxwellian distribution corresponding to a temperature of 11 630 K (dashed curve in Fig. 1.26). In the case discussed here, the total intensity in the forward direction is $1.9 \times A \times 10^{15}$ atoms/s sr, where A is the bombarded target area in mm^2.

A much simpler sputtering source for alkali-metal atoms has been described by Hulpke and Kempter (1966). They use a Penning discharge in an axial magnetic field, which is sustained between two cathodes and a ring-shaped anode (similar to a Penning gauge) for both ion production and sputtering. One cathode contains the solid alkali metal, the other one has an aperture, through which the

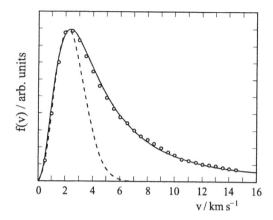

Fig. 1.26. Measured velocity distribution of sputtered potassium atoms (circles) [Politiek et al. (1968)]. The full line is calculated according to (1.28) (with n = 1.15, v_b = 2.55 km/s), the dashed line corresponds to a Maxwellian distribution with a temperature of 11 630 K

sputtered atoms may leave the source. The discharge is pulsed at 1000 Hz, yielding a pulsed beam of sputtered atoms, which can easily be velocity selected by means of a rotating slotted disk phase-correlated to the discharge pulse.

The energy distribution of this source has its maximum at about 1 eV, with a broad wing to high energies (up to 20 eV) similar to the distribution shown in Fig. 1.26. The intensity of the beam depends on the operation conditions of the discharge. Optimum results were obtained with a low-pressure (2×10^{-3} torr) discharge in argon, applying about 2000 V to the electrodes, which resulted in a 100 mA discharge current. Under these conditions, the beam intensity at an energy of 1 eV (using an energy resolution of 25%) is 5×10^3 atoms/pulse. This gives an average intensity of 2×10^9 atoms/sr s. This is an order of magnitude less than that achievable with a source utilizing separated ion and neutral production.

The low beam intensity of these sources is the reason why they have been supplanted by sources based on other principles, which have been developed later (such as carrier gas acceleration with plasma heating or laser ablation). An exception are refractory elements, for which sputtering is a real alternative to electron bombardment evaporation or laser ablation.

As an example, Fig. 1.27 shows a sputtering device which has been used for laser-spectroscopic investigations [Wakasugi et al. (1993), (1996)]. Argon ions, produced in an electron impact source (1), are accelerated to an energy of 8 keV (0.4 A) and focused by means of an einzel lens (2) onto a target (3) (1 mm focus diameter) of a refractory metal (such as Hf, Ta, and W, for example). From the sputtered atoms a beam (5) (about 3.5×10^{14} atoms/s sr) is collimated by an aperture (4) (1 mm ∅) which is crossed under right angles by the light (6) of a dye laser. This reduces the Doppler broadening substantially, so that the hyperthermal beam velocity and the broad velocity distribution plays no essential role.

The resonant fluorescence photons are collected by a spherical mirror (7) and detected by a photomultiplier (8). The fluorescence spectrum and the transmitted

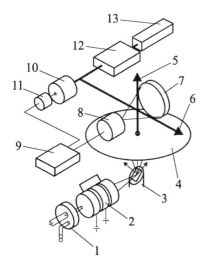

Fig. 1.27 Atomic beams of refractory metals produced by sputtering [Wakasugi et al. (1993)]. (1) ion source, (2) einzel lens, (3) target, (4) collimator. (5) laser beam, (6) atom beam, (7) spherical mirror, (8) photomultiplier, (9) computer, (10) Fabry–Perot interferometer, (11) photodiode, (12) dye laser, (13) Ar ion pump laser

signal of a Fabry–Perot interferometer (10) are recorded simultaneously with a computer (9). This setup has been used to determine the magnetic dipole moment and the electric quadrupole moment of the radioactive tantalum isotope ^{179}Ta from high-resolution laser spectroscopic investigations of the hyperfine structure [Wakasugi et al. (1996)].

Sputtering sources have found widespread application in molecular beam epitaxy for accelerated atom-beam-assisted deposition, which improves the adhesion properties of thin films and for the implantation of foreign atoms [Greene and Lee (1997)]. Contrary to ion beams, the atom flux has no space charge limitation and produces no surface charges on dielectric materials. With sintered targets consisting of several materials, multicomponent neutral beams can be produced and used for the deposition of corresponding films [Endo et al. (1997)].

A source for accelerated atoms for the above mentioned applications, utilizing both sputtering and charge-exchange collisions for neutral beam production, has been developed by Churkin et al. (1998). They use a cylindrically symmetric, self-sustained discharge in a radial electric and an axial magnetic field. The discharge, confined by the magnetic field, burns in an auxiliary gas (hydrogen or nitrogen). The main plasma components, however, are heavy metal atoms formed in the plasma by ionization of sputtered atoms from the cathode. A fraction of the neutral particles, produced by sputtering of the cathode material, can leave the source through a lateral aperture (in the radial direction). These neutrals form the low-energy beam component ($E < 100$ eV). Another fraction of the sputtered particles which have been ionized and accelerated in the electric field, may undergo charge-exchange collisions resulting in a second beam component with a broad energy distribution (between 0.1 U and U, where U is the voltage applied between cathode and anode), since they have experienced different accelerations between ionization and neutralization. The beam species is determined by the cathode material, the total beam intensity depends on the discharge conditions. Typical discharge voltages U range from 1 to 10 kV and discharge currents between 30 and 150 A.

48   1. Fast Beams, Production and Detection

The magnetic field strength is of the order of 0.2 to 0.6 T. Achievable beam intensities are about 10^{18} particles cm^{-2} s^{-1}.

Apart from fast atoms, sputtering of metals also produces neutral [Baede et al. (1971), Staudenmaier (1972), Können et al. (1974)] and charged clusters. Accordingly, sources similar to the one shown in Fig. 1.25 are frequently used in cluster research to produce cluster ions [Fayet and Wöste (1986), Lindsay et al. (1989), Harbich et al. (1990), Hu et al. (1991), Wang et al. (1995)]. The cluster ions are mass selected (e.g. using a Wien filter), and then deposited on a target (cooled by liquid helium) together with rare gas atoms of thermal energy and electrons (for neutralization). The mass-selected clusters embedded into the rare gas matrix can be used for spectroscopic investigations, e.g. to obtain information on the cluster structure (see Chap. 2).

1.8 Photolysis

If a diatomic molecule is dissociated by a photon having an energy larger than the binding energy of the molecule, the released, well defined energy (up to a few eV) is divided between the two atoms according to the ratio of the corresponding atom mass to the total mass (due to conservation of energy and momentum). If the molecule consists of two atoms of very different mass, the light atom obtains practically the total released energy. Accordingly, the photolysis of HBr, HCl, and HI produces hydrogen atoms with energies of more than 2 eV, which can be used to investigate endothermic chemical reactions. This method has been used in beam experiments, in time-resolved gas-phase measurements, and in the expansion of a nozzle beam. This is illustrated in the following sections by selected examples.

1.8.1 Beam Experiments

An example of a beam experiment has already been described in detail in Sect. I.5.9.3 [Schnieder et al. (1991)]. A hydrogen atom beam, produced by photolysis of a pulsed HI beam, is crossed with a pulsed D_2 beam in order to study the reaction H + D_2 → HD + D by measuring the velocity and angular distribution of the D atom product. The photolysis wavelength is 266 nm, generated by the fourth harmonic of a Nd:YAG laser. At this wavelength the HI dissociates according to the reaction HI → H + I($^2P_{3/2}$, $^2P_{1/2}$), yielding hydrogen atoms with kinetic energies of E_{kin}(H) = 1.59 and 0.66 eV, respectively, associated with the lower and upper spin-orbit states of the iodine atom.

Another possibility to produce fast D atom beams utilizes the photolysis of a pulsed D_2S nozzle beam [Buntin et al. (1987)] at 193 nm using the radiation of an ArF laser (250 mJ). Photolysis takes place at the end of the expansion zone, where no further collisions occur to avoid a relaxation of the translational energy. Thus, the energy of the D atoms is simply the sum of the translational energy of the beam and the energy released in the dissociation. If the velocity of the dissociated

D atoms is larger than the beam velocity, only those atoms contribute to the beam which dissociate in the beam direction. This fraction is about 2×10^{-4} of the total number of dissociated atoms. In this manner, a D atom beam with an energy of 2.2 eV (and a reciprocal velocity resolution $\Delta v/v = 11\%$) has been produced to investigate the $D + H_2$ reaction. Using the photolysis of a pulsed HI beam at 248 nm and 193 nm, hydrogen atom beams with energies of 1 eV and 2.9 eV, respectively, have been generated by the same technique. These beams have been employed to study the desorption of D atoms adsorbed on a Si(100) surface due to fast H atom impact [Buntin (1996)].

It has also been suggested to apply this technique to the production of photofragments with extremely low velocities [Pang et al. (1986)], utilizing those particles that dissociate in a direction opposite to the beam direction. Their energy is the difference of beam energy and energy released in the dissociation reaction. By an appropriate choice of the molecules to be dissociated, the beam velocity, and the photon energy, dissociated atom beams with any desired low velocity should be obtainable.

1.8.2 Gas-Phase Measurements

Time-resolved measurements of chemical reactions in the gas phase with fast atoms produced by photolysis are the main application of this technique. To illustrate the method we briefly discuss the reaction $H + CO_2 \rightarrow OH + CO$ [Quick and Tiee (1983), Buelow et al. (1985)]. Many other reactions have also been studied by this method [Kleinermanns and Wolfrum (1984), de Juan et al. (1988)]. In a mixture of HBr and CO_2 the HBr molecules are dissociated by a pulsed ArF laser (193 nm) with a quantum yield 1, according to the reaction scheme:

$$HBr + h\nu \rightarrow H + Br\left(^2P_{3/2}\right) + 2.69\,eV, \quad HBr + h\nu \rightarrow H + Br\left(^2P_{1/2}\right) + 2.23\,eV. \quad (1.29)$$

85% of the photolysis follows the first channel and 15% the second channel. Due to the large mass difference between H and Br, 98.76% of the released energy is transferred to the hydrogen atom. After a well defined delay time, which is short enough so that the energetic H atoms can only undergo one collision with a CO_2 molecule, a second pulsed laser (for instance, a dye laser pumped by an N_2 laser) is used for state-selected detection of the OH reaction product by laser-induced fluorescence. Absolute values of reaction cross sections and the distribution of the reaction energy among the various decrees of freedom (translation, rotation, and vibration) can be directly determined by this method.

1.8.3 Oriented Reactants

Wittig and co-workers have refined the above described method to obtain additional information on the influence of the spatial orientation of the reactants upon the reaction [Buelow et al. (1985)]. As has been demonstrated by molecular beam studies in which reactants have been oriented in external electric fields (see Chap. 4), the spatial orientation of reactants in bimolecular collisions affects both the reaction probability and the product state distribution.

The method involves van der Waals complexes having a well defined equilibrium configuration, in which the constituents are oriented relative to one another. The degree of orientation is limited only by the zero-point fluctuations associated with the vibrational modes. Thus, the direct photodissociation of one component of the van der Waals complex produces a reagent whose orientation and velocity are well defined with respect to the other component.

As an example we consider the reaction (1.29). The precursor van der Waals complexes $CO_2 \cdot HBr$ are produced in a pulsed gas dynamic expansion using a mixture of 0.5% HBr and 3% CO_2 with helium (pressure 1300–1400 torr, 0.5 mm nozzle ∅) as carrier gas. The photolysis of the HBr molecules which are weakly bound to the linear complex OCOHBr takes place in the expansion about 50 nozzle diameters downstream of the nozzle using a pulsed ArF laser (193 nm). Hence the initial H atom velocity is directed along the HBr axis and because HBr forms a complex with CO_2, the impact parameters and the orientations of the H + CO_2 collisions are well defined. The nascent reaction product $OH(X\,^2\Pi)$ is detected by laser-induced fluorescence of the $A\,^2\Sigma \leftarrow X\,^2\Pi$ system using a frequency doubled, pulsed dye laser which is collinear to the photolysis laser beam. Comparing the results with measurements performed under bulk conditions yields the desired information about the influence of the orientation of the reaction partners on the reaction [Radhakrishnan et al. (1986), Buelow et al. (1987), Häusler et al. (1987), Rice et al. (1988)].

1.9 Electron-Stimulated Desorption

A new principle to generate hyperthermal O atom beams (mean energy 5 eV) has been developed by Hoflund and Weaver (1994). It is based on the dissociative adsorption of O atoms, their permeation through a metal membrane, and their forced desorption by electron bombardment. The source consists of a metal tube terminated by a metal alloy membrane (Ag with 0.5% Zr, 0.1 cm thick, 1.5 cm ∅), which separates the source from the ultrahigh vacuum apparatus. Filled with high-purity oxygen (200 torr), the oxygen molecules undergo dissociative adsorption on the surface of the membrane, which is heated to 550 °C. At this temperature, the oxygen atoms permeate through the membrane and form a layer of adsorbed O atoms on its outer surface, which acts as the source "aperture", since the atoms desorb with hyperthermal energies when bombarded by electrons (with an energy of 1650 eV). The effective aperture area of the source is determined by the size of the focus of the electron beam on the surface. With a focus area of 1 mm^2 and an electron current density of 1.5 mA/cm^2, the emitted atom flux is some 10^{10} atoms/s. This is comparable to the intensity of an effusive beam. The fraction of O^+ ions in the beam is about 2.5×10^{-7}. It can be eliminated by a suitable electric field. The beam intensity increases linearly from 0 to 4.5×10^{12} atoms s^{-1} cm^{-2} as the electron current density increases from 0 to 1.6 mA/cm^2, indicating that the intensity is not limited by the permeation rate through the membrane.

Because many applications in surface physics require beam sources with large emitting areas, Hoflund and Weaver use a ring-shaped filament in front of the membrane to produce the electrons. Together with a cylindrical reflector and an appropriate choice of the potentials this device yields a homogeneous electron density across the membrane surface. Hence, the entire surface area can be utilized for atom emission. The same principle has also been used to produce H atom beams [Wolan et al. (1997)]. Diffusion though the membrane occurs already at room temperature, beam intensities of 10^{14} atoms s^{-1} cm^{-2} with a mean energy of 1 eV have been measured. Only groundstate atoms have been observed, the beam contains neither excited atoms nor ions. Thus, the source is well suited for applications in ultrahigh vacuum and in surface science, where beams of extreme purity are required.

1.10 Other Methods

For completeness, we mention in the following sections several methods which have been successfully used to produce neutral particles with hyperthermal energies. Due to inherent disadvantages, however, they have been supplanted by the methods described in the foregoing sections.

1.10.1 Mechanical Acceleration

The first attempts to produce neutral molecules with hyperthermal energies of 1 eV were made by mechanical acceleration, utilizing atoms and molecules thrown off the tip of a fast rotating metal vane [Marshall et al. (1948), Marshall and Bull (1951), Moon (1953), Bull and Moon (1954)]. Placing potassium on the rotor tip, a pulsed atom beam was generated. A pulsed beam of CCl_4 molecules was formed by simply rotating the vane in a chamber filled with CCl_4 vapor at low pressures, yielding a continuous thermal energy beam and a superimposed pulsed fast beam. The velocity of the latter is the sum of the peripheral speed of the rotor tip and the thermal speed of the molecules leaving the tip. The maximum attainable peripheral speed depends on the shape and the material of the rotor, not on its size. With the best nickel-chromium-molybdenum steel available at that time, and proper shaping of the rotor (see Sect. 3.2.3), a rotor tip speed of about 10^5 cm/s was achieved. For CCl_4 molecules, this corresponds to a translational energy of about 0.8 eV. During the 1970s, when new materials with higher tearing strengths became available, this development resumed, leading to the construction of new rotor sources [Moon et al. (1974), Nutt et al. (1977)]. These utilize a magnetically suspended rotor having two carbon fiber arms, which rotate in a horizontal plane and at one point intercept a beam of sample gas molecules from a simple effusive source inclined at an angle of 45° to the plane of rotation. The sample molecules which strike the rotor and then re-evaporate normal to the rotor surface can pass along a molecular beam axis defined by two apertures. Due to the increased tearing strength of the material, the peripheral rotor speeds were about a factor of two larger than those of the early designs, corresponding to a factor of four in the energy of the accelerated particles. Heavy atoms or molecules, therefore, can be accelerated up to energies of the order of 3 to 4 eV.

Compared to gas dynamic acceleration, mechanical acceleration involves a rather complicated technique and produces beams with very low intensities. Consequently, apart from early attempts to study chemical reactions [Bull and Moon (1954)], rotor sources have never been applied in molecular beam experiments.

1.10.2 Shock Waves

In a shock wave a gas sample may be heated from room temperature to 10^4 K within nanoseconds, and pressures up to 10^3 atm can be obtained. If the high-temperature gas is expanded through a nozzle into a high vacuum chamber, a pulse of more than 10^{18} particles with an energy of a few eV can be obtained and used for a given experiment [Skinner (1961), Treanor and Skinner (1961), Skinner and Moyzis (1965), Peng and Liquornik (1967)].

The main disadvantage of this hyperthermal energy beam source is the short running time ($\leq$ a few milliseconds) and the extremely low repetition rate. The latter is determined by the time required to prepare the shock tube for a new experiment (of the order of 10 to 15 min).

1.10.3 Recoil Nuclei of Radioactive Decays

Another method to produce hyperthermal energy neutral particles utilizes the recoil of a nucleus in a radioactive decay. The recoil atoms emerge from a thin film of the material and are detected by a second radioactive decay process. An example of a suitable decay is Cd^{107}. Most of these atoms decay through K capture into an excited state of Ag^{107}. The silver atoms, which have a recoil energy of 9 eV, can be detected by a subsequent γ decay [Frauenfelder (1950), Pauly (1961)].

Before being able to produce molecular beams of sufficient intensity in the range of 1 to 20 eV, the use of atoms produced by recoil processes in nuclear transformations played an important role in the investigation of those chemical reactions which occur above the threshold or activation energy ("hot atom chemistry") [Wolfgang (1965)]. Reactions of tritium atoms, halogen atoms, and carbon atoms have been studied predominantly. These atoms have been generated in situ by irradiating suitable parent atoms with neutrons or charged particles. Since the recoil energies released in nuclear reactions which lead to the emission of protons or α particles are usually very high (in the MeV range), the recoil atoms, which are generally produced as high energy, multiply charged ions, must be decelerated in a moderator material to energies of the order of 20 eV. During this deceleration process, numerous charge transfer processes take place, so that the average charge of the recoil atoms decreases with decreasing energy and finally neutral atoms may be obtained again and may eventually make a collision (with the desired energy of 10 to 20 eV) in which they become chemically combined. The products incorporating the hot atoms are then separated and analyzed utilizing the radioactivity of the hot atoms. The important factor of a well defined reaction energy, however, is lost in this deceleration process and neither the charge state nor the electronic state of the reacting particles can be predicted with any accuracy. Nevertheless, the recoil method combined with a quantitative kinetic treatment has yielded valuable information on chemical reactions at energies beyond their maximum efficiency.

1.11 Some Examples of Fast Beam Applications

1.11.1 Coaxial Laser Spectroscopy with Fast Beams

As already mentioned in the introduction, fast neutral beams with energies between 10 and 100 keV have found increasing applications for high-resolution studies using collinear laser spectroscopy (in particular for radioactive isotopes) [Demtröder (1992), Boudjarane et al. (1994), Ludin and Lehmann (1995), Keim et al. (1995), Young (1996), Müller et al. (1997)].

As an example, we mention here measurements of the optical isotope shifts of a given transition in the isotopes of krypton at $\lambda = 760$ nm (with the mass numbers A = 78, 80, 82, 84, 86, 88, and 90) [Schuessler et al. (1992)]. ^{88}Kr and ^{90}Kr are short-lived β emitters (half-lives $t_{1/2} = 2.84$ h and $t_{1/2} = 32.3$ s), which are produced online as fission products in a target of ^{235}U when bombarded with thermal neutrons (of a high flux reactor). The uranium target is located inside a high-temperature plasma ion source. Here the fission products are ionized, extracted from the source at an energy of 43 keV, magnetically mass separated, converted into a fast neutral beam by passing through a charge exchange cell containing alkali vapor, and finally injected into a collinear fast-beam laser-spectroscopy apparatus. The essential components of such a device are shown in Fig. 1.28. The exciting laser beam (1) is collinear with the fast neutral beam (2). Its frequency is kept fixed while the absorption frequency of the atoms in the beam is Doppler-tuned through the resonance. This is achieved by applying a variable potential to the charge exchange cell (5). The fluorescence photons are collected along a path of 20 cm by using cylindrical optics (6) and a light pipe (7). Collinear laser spectroscopy with fast beams provides a considerable narrowing of spectral lines due to "acceleration cooling" [Kaufman (1976)]. Very important is a low energy spread δE of the fast beam, which in turn is determined by the energy spread of the ion beam. Values of δE are typically of the order of 1 eV (see Sect. 1.1.1). Correspondingly, the fluctuations of the acceleration voltage U must be equal to or even lower than the

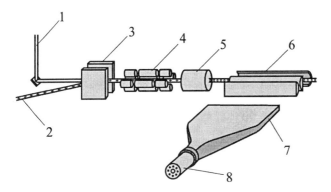

Fig. 1.28. Setup for collinear fast-beam laser spectroscopy [Neugart (1987)]. (1) Laser beam, (2) mass separated ion beam, (3) deflecting magnet, (4) quadrupole lenses, (5) charge exchange chamber, (6) cylindrical optics, (7) light pipe, (8) photomultiplier

energy spread δE. This requires a stabilization of the acceleration voltage to fluctuations and long-term drifts of about $\delta U/U \approx 10^{-5}$. From the kinetic energy

$$E = eU = mv^2/2, \tag{1.30}$$

where U is the acceleration voltage, the energy spread δE follows as

$$\delta E = mv\delta v \quad \text{or} \quad \delta v = \delta E/mv. \tag{1.31}$$

Since δE is conserved during the acceleration process, the velocity spread δv, which determines the Doppler broadening of the spectral lines, decreases with increasing beam velocity v. According to (1.30) and (1.31), the Doppler width δv_D (for a parallel beam) caused by the velocity spread δv of the fast beam, is given by

$$\delta v_D = v_0 \frac{\delta v}{c} = v_0 \frac{\delta E}{\sqrt{2eUmc^2}} = \frac{2.317 \times 10^{-5} \delta U}{\sqrt{MU}} v_0. \tag{1.32}$$

In the numerical part of (1.32) the unit for U and δU is V and M is the molecular weight. Assuming a fast atom beam of mass 100 with an energy spread $\delta E = 1$ eV at a total energy of 50 keV, and an optical frequency of 5×10^{14} Hz (600 nm), the Doppler width according to (1.32) is $\delta v_D = 5$ MHz. The natural linewidth of atomic transitions with lifetimes of a few 10^{-8} s is of the same order of magnitude. Consequently, all atoms are simultaneously excited when the frequency is tuned through the resonance, making this method particularly suited for applications in which only low intensity beams are available. Beam fluxes as low as 10^4 atoms/s have been used [Neugart (1987)].

The collinearity of laser and atom beam contributes also to the sensitivity of the method. Contrary to crossed beams, collinear beams offer a large photon–atom interaction region and thus an increased sensitivity if the fluorescence light is collected along the beam path by appropriate optics (see Fig. 1.28).

A beam divergence with an aperture angle α yields an additional line broadening of second order only, given by

$$\Delta v_D = \frac{1}{8} v_0 \frac{v}{c} \alpha^2. \tag{1.33}$$

In contrast to transverse excitation, this makes this method insensitive to small beam divergences and deviations from an exact parallel alignment of laser and fast atom beam.

The fact that fixed-frequency CW lasers can be used for collinear laser spectroscopy is a further advantage of this method. As already mentioned above, the atom frequency is Doppler-tuned through the resonance by simply changing the acceleration voltage U of the ions. A change ΔU of the acceleration voltage U yields a frequency change Δv, which is given by

$$\frac{\Delta v}{v} = \frac{\Delta v}{c} = \frac{e\Delta U}{c\sqrt{2emU}}. \tag{1.34}$$

While the resolved hyperfine structure yields information about nuclear spin, magnetic dipole and electric quadrupole moments, the isotope shift of heavy nuclei is essentially determined by the proton distribution within the nucleus, which depends on the number of neutrons in a nucleus of given proton number. Connected with this is a deformation of the nucleus and a change of the mean square nuclear charge radii $\langle r^2 \rangle$ within a sequence of isotopes. The latter can be determined as a function of the number of neutrons in the nucleus from the measured isotope shift. Such data, therefore, can be used to test models of the spatial distribution of protons and neutrons within the nucleus [Buchinger et al. (1992)]. The high resolution and the high-detection sensitivity are particularly important for measurements with radioactive isotopes, since in most cases only small amounts of online-produced short-living isotopes are available.

A fundamental experiment employing fast beam laser spectroscopy has been used to test the time dilation predicted by special relativity. According to this, the frequency of a clock in a moving frame is modified for a stationary observer by a factor $\gamma = (1 - v^2/c^2)^{-1/2}$, where v is the relative velocity with respect to the observer and c is the velocity of light. Since the transverse Doppler effect is a direct consequence of time dilation, it can be used to test this effect. Measuring the frequency difference between a two-photon transition in a fast neon beam and a thermal energy neon beam, the time dilation effect has been verified to an accuracy level of 2×10^{-6} [McGowan et al. (1993)]. This experiment represents a more than ten-fold improvement over earlier Doppler shift measurements, the most accurate one of these also used a fast beam [Kaivola et al. (1985)].

Fast beams are also used for precision measurements in radiofrequency and microwave spectroscopy [Lundeen (1991), Hessels et al. (1992), Young et al. (1993), Ward et al. (1996)].

1.11.2 Photofragment Translational Spectroscopy

Investigations of dynamical processes in molecules, induced by photon or electron impact excitation, represent another field in which fast neutral beams are frequently used (photofragment translational spectroscopy) [Continetti et al. (1993), Osborne et al. (1995), (1997), Leahy et al. (1995), Sherwood and Continetti (1996), Wouters et al. (1996)]. Such processes are, for example, dissociation, fragmentation, ionization, autoionization, ion pair production, or electron detachment. These studies make use of the fact that the center-of-mass velocity is conserved during the reaction, so that the reaction products have energies in the keV region and can therefore be detected with high efficiency (with a channeltron or a microchannel plate, for instance).

To illustrate this method, we describe photodissociation studies of H_2 molecules in their metastable c $^3\Pi_u^-$ state. A schematic drawing of the fast beam apparatus is shown in Fig. 1.29. H_2^+ ions from a low pressure, hot filament ion source are extracted at energies of several keV, mass selected in a Wien filter, and then neutralized in a charge exchange cell (1) filled with Cs vapor. Ions remaining in the beam are removed by deflection plates (2). The neutral beam passes through a collimating aperture (3) (1.5 mm ∅) and is crossed at right angles by the intracavity beam of a tunable CW dye laser, the polarization vector of the laser light being parallel to the fast beam direction. The metastable H_2 molecules are excited into the n = 3 dissociative states. The resulting photofragments spread along the parent beam direction in a cone within which the intensity distribution is determined by the angular distribution of fragments in the center-of-mass system. For a given initial molecular level, the maximum laboratory frame opening angle of this cone is determined by those fragments that dissociate at 90° with respect to the beam line.

After a path length of 150 cm the photofragments are detected by a position- and time-sensitive detector (6) (microchannel plate) [van der Zande et al. (1992)]. A V-shaped beam flag (5) prevents undissociated molecules from reaching the detector. The microchannel plate detector consists of two opposite sectors (opening angle 20°) which allow a separate detection of the two H atoms produced in a single dissociation event. The detector (and its electronics) permits the measurement of the spatial distance, R, of the two fragments in the plane of the detector (with a resolution of 70 µm), The flight time difference between the two H atoms can be determined with a resolution of 500 ps. Hence, space- and time-resolved spectra of photofragments can be recorded with the dye laser tuned to an absorption transition. If the dissociation occurs on a time scale short compared to the flight time between the dissociation region and the detector, the time and spatial coordinates of the fragment pair provides all the information required to determine

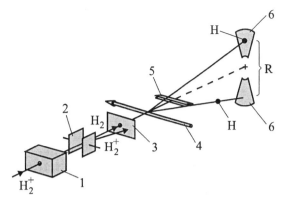

Fig. 1.29. Schematic view of a device to study photodissociation with fast beams [Helm and Cosby (1987)]. (1) Charge exchange chamber, (2) deflection plates to remove ions, (3) collimating aperture, (4) laser beam, (5) beam flag, (6) detector (microchannel plates)

the energy released in the photodissociation process (in the center-of-mass frame) and the angle of dissociation with respect to the laser polarization.

The same setup has also been used to investigate electron impact dissociation of diatomic molecules, using an electron beam instead of the intracavity laser beam [Cosby (1993)].

1.11.3 Precision Measurements of Lifetimes

Precision measurements of lifetimes of excited atom states (with an accuracy better than 1%) are another important application of fast beams with energies from 10 to 50 keV [Tanner (1995), O'Brien and Lawler (1996)]. This technique reduces the time measurement to a path length and a velocity measurement. The principle of the idea is rather old. Already in 1919 Wien used this method to determine lifetimes in fast beams of positive ions (canal rays) [Wien (1919)].

Accurate absolute values of lifetimes (or oscillator strengths) are of great importance for the interpretation of spectroscopic data both in astrophysics and in plasma physics [Ryan et al. (1996)]. Furthermore, precise lifetime measurements provide information important for further development of atomic and molecular theory and to test atomic structure calculations. For example, this is necessary for the interpretation of parity nonconservation experiments in atomic physics, which require accurate knowledge of transition matrix elements [Hunter et al. (1991), Stacey (1993), Fortson (1993), Chen et al. (1995), Argoustoglou et al. (1995), Warrington et al. (1995)].

Figure 1.30 shows a schematic diagram of an experiment which has been performed to determine the lifetimes of the $2p\ ^2P_{1/2,3/2}$ states of lithium [Tanner (1995)]. Neutral lithium atoms are prepared from a 30–50 keV beam of $^7Li^+$ ions (1) by charge exchange in a cell (2) filled with Rb vapor. A dye laser (3) at 670 nm, aligned perpendicular to the fast beam, is used to excite the atoms. Collisional excitation is also possible by passing the beam through a metal foil or a gas

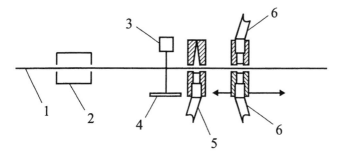

Fig. 1.30. Lifetime measurements utilizing fast beams [Tanner (1995)]. (1) Li$^+$ beam (30 to 50 keV), (2) charge exchange chamber (Rb), (3) dye laser, (4) mirror, (5) fiber bundle to monitor detector, (6) translating fiber bundle pair of the fluorescence detector

target. Laser excitation, however, yields the highest accuracy, since the excitation is efficient, state selective, and occurs at a well defined location. Both CW and pulsed lasers are used. After excitation, the fluorescence signal from the selected state decays in flight as the atoms move along the beam line. The fluorescence photons are precisely detected as a function of position using a translating fiber-optics light-collection system (6). A similar system (5) remains stationary; its signal is used to monitor the beam and to eliminate beam fluctuations.

The accuracy of the time scale is determined by the accuracy of position fixing, the spatial resolution, and the beam velocity. The latter can be adjusted to obtain optimum spatial resolution. A total accuracy of a few tenths of a percent is achievable, the main uncertainty results from the beam velocity, which is measured when the laser beam is aligned antiparallel to the beam using the Doppler shift of the atomic resonance. A careful calibration of the acceleration voltage of the ion beam has also been used in some cases.

As discussed in Sect. 1.2, excited states may also be produced in the charge exchange process which, by cascading fluorescence transitions, may feed the level whose lifetime is to be measured. Thus, the decay curve of the fluorescence can be changed, resulting in an erroneous lifetime determination. This problem, however, can be eliminated by measuring the signal difference with and without laser excitation [Demtröder (1992)].

The lithium $^2P_{1/2}$ state is an example of a lifetime that plays an important role in the interpretation of astrophysical data. In astrophysics, stellar surface abundances of lithium are determined relative to that of hydrogen by observing the emission spectra of stars. The data are interpreted using the oscillator strength for the lithium $2\,^2P \to 2\,^2S$ transition at 670 nm. Since the $2p\,^2P_{1/2,3/2}$ states have only a single decay mode, the determination of lithium abundances depends directly on the accepted value of the 2p lifetime. On the other hand, lithium abundances are of great importance in models of stellar evolution, in determining the primordial lithium abundance, and in understanding the origin of the universe [Ryan et al. (1996)]. The lithium 2p lifetime has been determined using the fast beam method to an accuracy of 0.15% [Gaupp et al. (1982)], a precision that has not been achieved so far with any other method. Another example is lifetime measurements with fast Cs beams, where the lifetimes of the 6p states have been measured with an accuracy of 0.23% (for the $^2P_{3/2}$ state) and 0.27% (for the $^2P_{1/2}$ state), respectively [Rafac et al. (1994), Rafac et al. (1999)]. These lifetimes can be used to test many-body perturbation theory approaches to the calculation of atomic structures in many-electron systems.

1.11.4 Merged Beams

A possibility to investigate low-energy ion–atom or ion–molecule reactions is provided by the merged beams method, which has already been mentioned in the introduction [Gentry et al. (1975), Havener et al. (1989), (1995), Havener (1997). Huq et al. (1989), McCollough (1995), Folkerts et al. (1995), Pieksma et al. (1996), (1997), Pieksma and Havener (1998), Bliek et al. (1997)]. This technique consists of two beams, each with a rather high laboratory energy (several keV), traveling in the same direction along a common axis with a low, adjustable relative energy (from thermal energy up to a few hundred eV). This method can also be used to investigate ion–ion collisions [Weiner et al. (1971), Plastridge et al. (1995)], electron–ion reactions [Bell et al. (1994), Olamba et al. (1996), Pedersen et al. (1999)], or ionizing collisions between metastable particles [Neynaber and Magnuson (1975)]. The basic principle is rather old. Already in 1929 beams of electrons and α particles were superimposed in order to study their interaction [Davis and Barnes (1929), (1931), Webster (1930)].

The relative velocity vector **g** is given by

$$\mathbf{g} = \mathbf{v}_2 - \mathbf{v}_1, \qquad (1.35)$$

where $\mathbf{v}_1$ and $\mathbf{v}_2$ are the velocities of the two beams. The energy W of relative motion, which is the interaction energy of the collision, is

$$W = \frac{\mu}{2}g^2 = \frac{\mu}{2}[\mathbf{v}_2 - \mathbf{v}_1]^2 = \mu\left(\frac{E_2}{m_2} + \frac{E_1}{m_1} - 2\sqrt{\frac{E_1 E_2}{m_1 m_2}}\cos\Theta\right), \qquad (1.36)$$

where μ is the reduced mass of the colliding particles, m_1 and m_2 are the masses, and E_1 and E_2 are the laboratory energies of the particles in each beam. Θ is the angle between the velocity vectors $\mathbf{v}_1$ and $\mathbf{v}_2$, which is zero under ideal experimental conditions. From (1.36) the relative energy spread δW can be expressed as

$$\frac{\delta W}{W} = 2\frac{\delta g}{g}. \qquad (1.37)$$

The spread of the relative velocity δg is determined by the velocity spreads δv_1 and δv_2 of the two beams, which are substantially reduced due to the use of fast beams (see (1.31)). Accordingly we have

$$\delta g = \frac{\delta v_1 + \delta v_2}{2} = \frac{1}{2}\left(\frac{\delta E_1}{m_1 v_1} + \frac{\delta E_2}{m_2 v_2}\right). \tag{1.38}$$

δE_1 and δE_2 are the energy spreads in the two beams. For the following discussion we assume that both beams have the same energy inhomogeneity, putting

$$\delta E_1 = \delta E_2 = \delta E. \tag{1.39}$$

Thus, we obtain

$$\frac{\delta W}{W} = \frac{\delta g}{g}\left(\frac{1}{m_1 v_1} + \frac{1}{m_2 v_2}\right). \tag{1.40}$$

Expressing the velocities v_i ($i = 1,2$) by the energies $E_i = m_i v_i^2/2$, (1.40) yields

$$\frac{\delta W}{W} = \frac{\delta E}{2}\left(\left(\sqrt{E_1 E_2 m_1/m_2} - E_1\right)^{-1} + \left(E_2 - \sqrt{E_1 E_2 m_2/m_1}\right)^{-1}\right). \tag{1.41}$$

For simplicity, we assume $m_1 = m_2$, since the general conclusion derived for this case applies equally well when $m_1 \neq m_2$. If the difference in beam energies $E_2 - E_1 = \Delta E$ is small compared to the average energy $E = (E_1 + E_2)/2$ of the two beams ($E_2 - E_1 \ll E$) (1.41) may be rewritten in the form

$$\frac{\delta W}{W} = 2\frac{\delta E}{\Delta E} \quad \text{and} \quad W = \frac{(\Delta E)^2}{8E}. \tag{1.42}$$

Accordingly, we have

$$\frac{\delta W}{W} = \frac{\delta E}{\sqrt{2EW}}. \tag{1.43}$$

Since the average laboratory beam energy E is large compared to the energy spread δE in the beams (of the order of 1 to 2 eV), the merged beams technique yields a high energy resolution W/δW even in cases where the relative energy W is small. A further advantage of this method is given by the fact that the reaction products, which in the center-of-mass system are deflected over the whole solid angle 4π, can all be measured since in the laboratory frame they are scattered into a small angular range in the forward direction.

In the following we discuss the experimental technique using measurements of total electron capture cross sections in collisions between multiply charged ions and hydrogen or deuterium atoms [Havener et al. (1989), Huq et al. (1989),

Folkerts et al. (1995), Havener et al. (1995), McCullough (1995), Pieksma et al. (1996), Pieksma and Havener (1998), Havener (1997)] according to the reaction

$$X^{n+} + H \rightarrow X^{(n-1)+} + H^+, \qquad (1.44)$$

where X stands for N, O, C, and Si; n lies between 3 and 5. An exact knowledge of these cross sections is required for the determination of the concentration of foreign atoms in plasmas using fast H atom beams (charge exchange recombination spectroscopy) (see Sect. 1.11.6) [Zinoviev (1992)].

Figure 1.31 shows a schematic diagram of the experimental apparatus. The neutral hydrogen beam is obtained by photodetachment of an H^- ion beam (energy 6 keV) (see Sect. 1.2.1). The energy of the mass selected beam of multiply charged ions can be varied between 40 and 93 keV (the energy of the H atom beam is fixed) in order to vary the relative energy between 1 and 800 eV. Both beams are electrostatically merged. After traveling together through an 80 cm interaction region, the primary and product beams are magnetically demerged. The H^+ ions formed in the reaction are sent through an energy analyzer and then detected by a channeltron (1), while the intensity of the X^{n+} beam, as well as the product $X^{(n+1)+}$ ions, is measured by a Faraday cup (2). The neutral beam is monitored by measuring secondary emission from a stainless steel plate (3). Ultrahigh vacuum is maintained in the interaction region (some 10^{-10} torr), to avoid a falsification of the H^+ signal due to ionization of H atoms in collisions with residual gas atoms. It is of advantage in this connection, that the cross sections for collisional ionization are about a factor of 100 smaller than the cross sections for electron capture ($\sim 10^{-14}$ cm^2).

The merged beams technique is not restricted to fast beams. To study state resolved ion–molecule reactions in the meV and sub-meV range, Gerlich and co-workers use a molecular beam machine in which a supersonic molecular beam is merged with a cold, low-energy guided ion beam [Gerlich (1993), Wick (1994)].

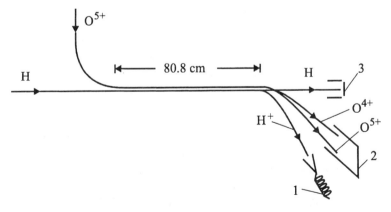

Fig. 1.31. Schematic view of a merged beams apparatus. (1) Channeltron, (2) Faraday cup, (3) stainless steel plate for secondary electron emission

1.11.5 Injection of Fast Beams into Fusion Plasmas

Many of the present problems concerning the world's energy supply could be solved by harnessing nuclear fusion. Therefore, for many years work has been done towards the practical usefulness of controlled nuclear fusion of hydrogen isotopes. The most promising fusion reaction occurs between deuterium and tritium

$$_{1}^{2}D + _{1}^{3}T \rightarrow _{2}^{4}He + n, \tag{1.45}$$

where 17.58 MeV is released per reaction in the form of kinetic energy of the reaction products, divided according to the mass ratio, with 3.52 MeV going to the α-particle and 14.06 MeV to the neutron. The neutrons are caught by a lithium wall where they react and produce tritium and ^{4}He according to

$$_{3}^{6}Li + n \rightarrow _{2}^{4}He + _{1}^{3}T + 4.8\,MeV, \tag{1.46}$$

while the α-particles dissipate their energy in the plasma.

The principle of thermonuclear fusion consists in the following: The fuel (deuterium and tritium) is heated to temperatures high enough to furnish a sufficiently high thermal velocity of the fusion partners in order to overcome the mutual Coulomb repulsion of the colliding nuclei. The necessary temperatures are of the order of 10^8–10^9 K. At these temperatures the fuel is a fully ionized plasma. This makes confinement by suitable magnetic fields a possibility. The goal is a "burning" plasma, where the energy losses of the plasma by radiation, convection, and heat conduction are compensated by the energy of the α-particles released in the fusion process, so that the temperature necessary for further nuclear reactions is maintained. Therefore, efficient heating methods for plasmas are needed. Today these are essentially realized by high frequency heating and heating by fast neutral particles.

Under radiofrequency heating, high frequency or microwaves at multiples of the cyclotron frequency (30–60 MHz for ion cyclotron heating, 140 GHz for electron cyclotron heating) are fed into the plasma. Under neutral particle heating, high energy neutral particles are injected into the plasma. They reach deep into the plasma without being hindered by the magnetic confinement fields and are then ionized by collisions with plasma particles. As ions they are then trapped by the confinement field and can release their excess energy to the plasma by further collisions and thereby heat it. The penetration depth of the neutral atoms into the plasma increases with their injection energy and thereby depends on the dimensions of the plasma. Besides the process of heating the plasma, neutral particle injection can also serve to control the "burning" of the plasma and to re-supply the necessary fuel.

The necessary heating power is very large (of the order of 100 MW) and large beam currents (of the order of several hundred Ampere) are required. While the power of the first injection systems was around 100 W, present-day systems reach several Megawatt [Gabovich et al. (1989)].

In realization of this principle, the production of fast molecular beams by charge transfer of highly energetic ion beams is applied on a large technological scale. H^+ and D^+ currents of the order of 100 A and energies up to several hundred keV (depending upon the type of reactor used) are generated in large-area ion sources and subsequently neutralized. In Tokamak configurations the beam energies used are typically around 100 keV. Units in the planning stage, like the International Thermonuclear Experimental Reactor "ITER" will require neutral beams of 50 to 100 MW power at an energy of 1 MeV [Ohara (1998)]. For beams up to 100 keV one uses a gas-filled chamber for neutralization by charge exchange. However, at higher energies this principle is no longer satisfactory because the efficiency decreases strongly with increasing energy (see Sect. 1.3.3). In this range one has to utilize negative ions and neutralization by collisional electron detachment. An energy-independent neutralization efficiency of about 60% can be reached. For this reason the development of high intensity beam sources for negative ions has been studied intensively for a number of years [Kwan et al. (1992), Wells et al. (1992)]. The production of intensive beams of negative hydrogen ions is much more difficult than of positive ions, because the negative hydrogen ion is not very stable with the low binding energy of its additional electron which, in turn, allows the advantage of energy-independent neutralization. Just recently promising progress has been made in the development of such sources [Okumura et al. (1996), (2000), Oka et al. (1998)]. In order to administer power levels of 50 to 100 MW to the plasma by neutral particle beams, the injection system of a reactor usually consists of a number of individual, simultaneously operating injectors.

Neutralization yields of 60% with collisional electron detachment in a gas-filled chamber still represents a large waste of energy. Of a planned injection system of 100 MW beam power this efficiency means that 40 MW of power is lost in suitable absorbers for the ions that failed to neutralize. Thus, efforts are underway to develop plasma targets of high density ($> 10^{13}$ cm^{-3}) [Tanaka et al. (1998)], which yield neutralization efficiencies up to 85% [Hershcovitch et al. (1984)].

1.11.6 Fast Beam Methods for Plasma Diagnostics

For the further development of fusion reactors, methods to determine the plasma parameters during operation, resolved in time and space, are vitally important. Towards this end, several methods have been developed employing atomic beams of thermal as well as hyperthermal energy. The latter methods are described in the following.

Impact Excitation Spectroscopy. In impact excitation spectroscopy [Boileau et al. (1989), Hintz and Schweer (1995), McCormick et al. (1996)), Fiedler (1995)], fast atomic beams of light particles (H, He, and most often Li) at energies between 10 and 100 keV and at beam intensities corresponding to some mA and several Ampere are utilized.

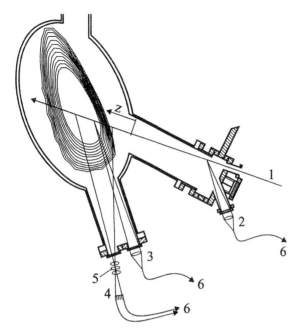

Fig. 1.32. Illustration of Li beam plasma diagnostics [Fiedler (1995)]. The plasma is indicated by the magnetic flux areas. (1) Li beam, (2) fluorescence detection for the determination of the neutral particle density, (3) output to charge exchange spectroscopy, (4) output to collision excitation spectroscopy, (5) interference filter, (6) optical fiber

If such a beam is shot into a plasma (see Fig. 1.32), collisions between the beam particles and those of the plasma (electrons, hydrogen ions, and ions of contaminants) may cause excitation of the atoms and subsequent emission of photons. For example, when using Li atoms, the 2P state is being excited and the resonance line (670.8 nm) is emitted subsequently. Measuring the relative dependency of the emission profile of the resonance line as a function of the beam coordinate yields absolute electron densities as functions of location and time [Schweinzer et al. (1992)]. The contribution of plasma ions to the excitation can be determined from measurements at various beam energies. Investigating the radiation in a region outside the plasma allows the neutral particle density during the discharge to be determined [Fiedler (1995)].

Charge Exchange Recombination Spectroscopy. Another diagnostic application of fast beams of light atoms follows from the spectroscopy of the highly ionized atoms of the contaminants, formed by charge transfer with the fast atoms [Schorn et al. (1992), Summers (1994)]. For example, when utilizing a fast H atom beam, collisions with the fully ionized atomic contaminants X^{z+} (e.g. He^{2+}, C^{6+}, O^{8+}) lead to recombination processes of the type

$$X^{z+} + H \rightarrow X^{(z-1)+}(n,l) + H^+. \tag{1.47}$$

The ensuing radiation is analyzed spectroscopically. From the Doppler-broadened line profiles one can determine local ion densities, ion temperatures, and plasma velocities. Furthermore, one obtains a quantitative detection method for atoms foreign to the plasma [Zinoviev (1992), Sarkissian et al. (1997), Marmar et al. (1997), Tabares et al. (1997)]. In order to separate the measured signal from the background, the fast beam is modulated.

Finally, fast beams also allow for measuring the magnetic field inside the plasma via the Stark effect of the fast moving atoms of the beam, which experience an electric field $\mathbf{E} = \mathbf{v} \times \mathbf{B}$ in their own coordinate system, causing splitting of the emission lines into several components. In addition, the lines are polarized, with the polarization depending upon the angle between $\mathbf{E}$ and the direction of observation [Marmar et al. (1997)].

1.12 Fast Beam Detection

The detection of fast neutral particles is much easier (depending on the beam energy) than the detection of thermal energy beams. At beam energies which are only slightly higher than those of thermal beams, the methods described in Chap. I.5 may be used for detection. With increasing energy, however, the detection probability of some detectors decreases, whereas it increases for others. An example for the first case is the universal beam detector based on electron impact ionization and mass selection. Its detection efficiency depends on the residence time of the neutral particles in the ion source, which decreases with increasing beam velocity v proportional to 1/v. This detector, therefore, is inappropriate for fast particle detection. On the other hand, the response of thermal detectors is proportional to the energy of the impinging particles and their signal increases with increasing beam energy. Consequently, thermal detectors have been used for fast particle detection for more than sixty years [Amdur and Pearlman (1939)].

In addition to the detection principles discussed in Chap. I.5, other physical principles may be used to detect fast particles, utilizing the increased energy of the beam particles. Especially, secondary electron emission from solid surfaces plays an important role in fast particle detection.

1.12.1 Surface Ionization (Langmuir–Taylor Detector)

As is discussed in detail in Sect. I.5.4, electron exchange between an atom or molecule and a metal surface results in the formation of positive or negative ions, depending on the direction of the exchange, which is determined by the work function of the metal and the ionization energy or the electron affinity, respectively, of the impinging molecule. If the temperature of the surface is sufficiently high, the ions evaporate from the surface and can be detected. Alkali metal atoms, which can be ionized with an efficiency of 100% at metal surfaces with high work function, are a typical example. Alkali metal atoms with kinetic energies far above thermal energy, can also be detected by surface ionization. At low energies (some eV), their sticking at the surface and their subsequent evaporation is not considerably changed as compared to thermal energy particles [Politiek and Los (1969)]. At higher energies, a fraction of the impinging atoms is reflected from a unheated surface as ions [van Amersfoort et al. (1985)]. The mechanism of ionization (tunnel effect in front of the surface) is the same as in the case of thermal atoms, but the energy of the fast particles is sufficient to reflect the ions from the surface. The fraction of impinging atoms which is reflected from the surface as ions increases rapidly with increasing energy between the threshold (about 7 eV) and about 30 eV [Helbing and Rothe (1969)] and reaches a maximum at energies between 30 and 100 eV with values close to 100%. With further increase of the energy the fraction decreases slowly (to about 20% at 1000 eV), since a part of the alkali atoms penetrate into the metal. At low penetration depths this fraction of alkali atoms can also be forced to leave the surface as ions by heating the surface to an appropriate temperature [Hollstein and Pauly (1966)]. In practical applications, however, this is usually not done, since the ionization at a cold surface is completely free of background and even beams of very low intensity can be detected by counting single ions [Gersing et al. (1973), (1976)]. The time constant for ionization at a cold surface is small (< 0.1 μs), so that this detector is well suited for time-of-flight measurements.

During the last decade it has been observed that numerous hyperthermal atoms and molecules (with energies of some eV) can be surface ionized in spite of the large energy gap between their ionization energy and the surface work function [Danon and Amirav (1987), (1990), Amirav (1990), Danon et al. (1988), Kuipers et al. (1991)]. Hyperthermal surface ionization, therefore, is driven by the kinetic energy of the particles and its yield increases substantially with increasing kinetic energy. Even mercury atoms, which possess the highest ionization potential among metal atoms (10.44 eV), have been ionized at a Pt(111) surface, maintained at a temperature of about 800 K to avoid mercury contamination [Danon et al. (1990)]. The ion yield increases between 3.6 and 8.2 eV by five orders of magnitude, attaining values between 10^{-5} and 10^{-6} at an energy of 8.2 eV. The ionization mechanism is explained by the extensive energy transfer to the surface creating a local hot spot. This increases the denominator in the exponent of the Saha–Langmuir equation (5.5).

Simple organic molecules with high ionization energies, such as the isomers of octane, benzene, toluene or aliphatic alcohols, can also be ionized very efficiently on continuously oxidized rhenium surfaces (also on Ir and Pt surfaces, but with less efficiency) [Amirav (1991), Kishi and Fujii (1995), (1997)]. The dissociative ionization results in a high yield of $(M-1)^+$ species (M is the mass of the parent molecule) and a few ionized fragments of low intensity. The mass spectrum depends both on the kinetic energy of the impinging molecules and the temperature of the surface. Polyatomic molecules such as anthracene and purine yield positive molecular ions [Danon and Amirav (1990)]. I_2 and IBr molecules result in the formation of negative molecular ions [Danon and Amirav (1988), (1989)], while alkyl halides, such as 1-iodopropane, are dissociatively ionized to I^-

and $C_3H_7^+$ [Amirav (1991)]. For large molecules, the ionization mechanism shows a strong dependence on the incident energy. As an example, at an energy of 5.3 eV the hyperthermal surface ionization mass spectrum of cholesterol is dominated by the parent molecular ion. Upon the increase of the molecular kinetic energy, a gradual increase in the degree of ion dissociation takes place, and at an energy of 22 eV the parent ion is completely dissociated and the mass spectrum is dominated by extensive fragmentation [Dagan et al. (1992)]. Nevertheless, the observed mass spectra are basically simpler than those obtained by electron impact [Dagan et al. (1995)], which opens new applications for surface ionization in analytical chemistry [Danon and Amirav (1994)].

Finally, the emission of positive and negative ions has also been observed for large, neutral clusters impinging with hyperthermal energies on a surface [Even et al. (1986), Vostrikov and Dubov (1991), Christen et al. (1992), Andersson and Pettersson (1997)]. This effect has been used to detect gas dynamically accelerated neutral carbon clusters (with an average number of 950 atoms/cluster) with a channeltron and to measure their time-of-flight distribution [Piseri et al. (1998)]. The estimated average detection efficiency for these clusters is 5×10^{-8}. For buckminsterfullerenes, C_{60}, seeded in a helium beam a detection efficiency of 5×10^{-6} at an energy of 9 eV has been found; this value decreases by an order of magnitude at an energy of 3 eV.

Thus, a channeltron is an inexpensive, versatile, and simple device to detect clusters of sufficient energy. It has the additional advantage of being selectively sensitive to the clusters of a seeded supersonic beam.

1.12.2 Bolometer and Pyroelectric Detectors

Thermal detectors of the bolometer or thermocouple type (see Sect. I.5.7) have already been used in the first scattering experiments utilizing fast neutral beams with energies between several hundred and several thousand eV [Amdur and Pearlman (1939), Amdur (1968)]. In contrast to the detection of thermal energy beams, which require cryobolometers with extreme sensitivity, fast beams have sufficient energy so that the thermal detector can be operated at room temperature. Under these conditions the detection limit of bolometers is of the order of 10^{-8}–10^{-10} $WHz^{-1/2}$, corresponding to beam fluxes of 10^8–10^6 particles/s at a particle energy of about 1000 eV.

The energy accommodation at the surface may be incomplete, and the fraction of particles reflected without having delivered their complete energy to the surface may depend on the particle energy. This is unimportant in the case of relative measurements, but absolute measurements require careful calibration of the detector. Calibration can be performed by using an ion beam of the same species and energy, permitting a comparison of the bolometer signal with the measured ion current. To obtain complete energy accommodation, a small gold cage is connected to the bolometer, in which the incoming particles are forced to make several wall collisions before they can leave the cage [van de Runstaat et al. (1970),

Kick (1974), Pauly and Schulz-Hennig (1975)]. This cage, however, increases the heat capacity of the bolometer and thus its response time considerably.

While the sensitivity of pyroelectric detectors (see Sect. I.5.7.6) is comparable to that of bolometers operated at room temperature, they are more robust and therefore less delicate to handle. Furthermore, they also have a much better time resolution. Due to these advantages, pyroelectric detectors should be preferred to bolometers for fast beam detection. Calibration is also performed by using an appropriate ion beam [Viehl et al. (1993), Tarnovsky and Becker (1993), Lagus et al. (1996)].

1.12.3 Laser-Induced Fluorescence

Laser-induced fluorescence with collinear or transverse excitation is another possibility to detect fast neutral beams. It is usually employed in experiments which already require a laser for beam excitation. A typical example is described in Sect. 1.11.3. In that case, a fast beam is excited by a laser (aligned perpendicular to the fast beam) and the beam intensity is monitored by a stationary fluorescence detector at some distance downstream of the excitation region. The light-collection system of the detector is rather similar to those described in Sect. I.5.9.2.

Another example is radiofrequency or microwave spectroscopy with fast beams, where laser-induced fluorescence (in a collinear geometry) is used for state-specific detection of the fast particles to observe the transitions induced by the microwave or radiofrequency field [Young et al. (1993)].

1.12.4 Secondary Electron Emission

The simplest and most widely used detector for fast neutral beams, especially at energies in the keV region, is based on the measurement of secondary electrons ejected by fast particle impact on metal surfaces. Similarly, as in the case of ions, the particle–electron conversion probability η (number of secondary electrons generated by one impinging neutral particle) strongly depends on the particle energy. Furthermore, it depends on the particle–surface interaction and on surface contamination by foreign atoms or molecules.

Qualitatively, the dependence of the secondary electron yield is similar to that observed for ions, particularly for energies above 1 keV [Stephen and Peko (2000)]. With increasing energy, η first increases, then it passes through a broad maximum at energies of about 10^5 keV, and finally decreases again. Typical values of η are of the order of 10^{-3} at an energy of 50 eV, while at energies of some keV values of $\eta \geq 1$ are achieved. Consequently, a multiplier (or a channeltron) provides a very simple and sensitive detector for fast neutral particles, even at low energies [Johnson et al. (1988)]. Using a microchannel plate, spatially resolved measurements are also possible. The determination of absolute intensities, however, suffers from the same drawbacks also encountered with ions: The dependence of the secondary electron yield η on the conversion electrode and of the multiplication factor on the surface conditions. Accordingly, absolute measurements of beam intensities require a careful calibration.

1.12.5 Collision-Induced Fluorescence

Another detection principle of high sensitivity, fast response, and temporal stability makes use of the electronic excitation and subsequent emission of fluorescence light in collisions of fast particles with a target gas. After calibration, this method can also be used to measure absolute intensities. As is shown schematically in Fig. 1.33, such a fluorescence detector consists, in essence, of a gas-filled target chamber which is traversed by the fast beam. The fluorescence light produced along the beam path (5) by collisions between beam and target particles is focused by a cylindrical mirror (1) through a glass window (3) onto a photomultiplier.

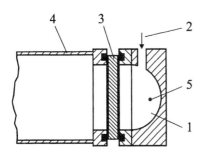

Fig. 1.33. Cross-sectional view of the target chamber of a fluorescence detector. The beam (5) propagates perpendicular to the plane of drawing. (1) cylindrical mirror, (2) gas supply, (3) glass window, (4) multiplier housing

Under single collision conditions, the fluorescence intensity I_f is proportional to the collision rate in the target chamber, which, in turn, is proportional to the number N of particles entering the target chamber per unit time and the product nL of target number density n and target chamber length L:

$$I_f = NnL\alpha(E). \tag{1.48}$$

$\alpha(E)$ is determined by the excitation cross sections involved, the solid angle of acceptance of the imaging optics, and the amplification of the signal. Both the solid angle and the amplification factor are independent of the particle energy.

For fast particle detection, the energy dependence of α must be determined by calibration. This can be done, for example, by using a calibrated bolometer detector. For a given geometry of the fluorescence detector, the calibration curve only depends on the interaction between beam and target molecules. Hence, the calibration curve is not subject to temporal variations and one calibration measurement is sufficient for a given beam–target combination.

The fluorescence detector may be operated in two modes:

(1) The detector is operated at low target gas pressures, so that only single collisions occur. According to (1.48), the signal is proportional to the number density n of the target gas, which must be kept constant during a series of measurements. Pressure fluctuations, which cause fluctuations of the beam signal, can

be corrected for by carefully monitoring the pressure of the target gas [Andresen (1978), Moritz (1978), Düren et al. (1980)].

(2) The detector is operated at pressures sufficiently high to completely decelerate the beam particles within the target chamber by multiple collisions. Under these conditions, the beam signal becomes independent of the target gas pressure and neither pressure fluctuations nor incorrect reproduction of the pressure have any influence on the signal. The calibration curve, which has to be taken in the same pressure range, increases linearly with energy, since the number of collisions required to decelerate the beam particles and thus the fluorescence intensity is proportional to the beam energy [Pauly and Schulz-Hennig (1975)].

The time constant τ of the fluorescence detector is determined by the residence time of the beam particles in the target chamber. Assuming a target chamber length L of a few centimeters, values of τ (depending on the beam energy) are of the order of 0.1 to 1 µs.

The threshold for collisional electronic excitation and thus the threshold of the fluorescence detector depends on the interaction potentials between projectile and target particles and the coupling elements between these potentials. Experimental values (in the laboratory frame) range from a few eV up to 100 eV [Andresen (1978)].

The detection limit is determined by the emission cross sections and their energy dependence. Empirical values have been determined for the detection of fast alkali metal atoms with N_2 and CO_2 as target gases. In these cases, predominantly alkali metal atoms are excited, which have rather large cross sections (about 10 $Å^2$). This results in a minimum detectable particle flux (at an energy of 1000 eV) of about 1000 particles/s.

The above-described technique to determine intensities of fast beams using the fluorescence induced by collisional excitation in a gas of given number density has been reversed in plasma physics: Here collision-induced fluorescence of fast lithium beams of known intensity is used to determine the number densities of neutral gases such as H_2 and D_2 [Fiedler (1995)].

2. Production and Diagnostics of Cluster Beams

Aggregates consisting of a few to a few million atoms or molecules, which may be held together by various types of binding forces with binding energies between a few tenths of an eV and several eV (see Table 2.1), are called clusters. The investigation of these aggregates represents an extremely interesting field for basic research in both physics and chemistry, which has developed rapidly during the last twenty years [Kappes and Leutwyler (1988), Kappes (1988), Bernstein (1990), Bjørnholm (1990), Rademann (1991), de Heer (1993), Haberland (1994), Schmid (1994), Klabunde (1994), Buck (1995), Baumert and Gerber (1995), Kondow et al. (1996), Martin (1996), Brack (1997), Sugano and Koizumi (1998)]. Clusters are often called a "new phase of matter", since they exhibit — due to their discrete energy-level structure and their large surface-to-volume ratio — new characteristic properties, which are neither those of the corresponding bulk nor those of their isolated constituents [Castleman and Keesee (1988), Jortner (1995)]. For example, clusters of nonmagnetic elements can be magnetic [Cox et al. (1993), Billas et al. (1994)]. Small clusters can change their properties continuously or abruptly by addition or removal of only a few constituents. This has been demonstrated by investigations of the chemical reactivity or the adsorption behavior of clusters, which may change by several orders of magnitude if the number N of cluster constituents is only changed insignificantly [Holmgren et al. (1996)]. Other properties, such as the ionization potential of metal or metal oxide clusters oscillate with increasing number N of constituents between even and odd values of N due to their electronic shell structure, while the average ionization potential correlates with the surface-to-volume ratio according to $N^{-1/3}$ [Lievens et al. (1997)]. Clusters, therefore, are an important subject to investigate the evolution from atomic and molecular to bulk properties with increasing cluster size [Jortner et al. (1988), Echt (1996)]. Their theoretical description requires new methods for treating many-body systems, since in the transition region neither the methods of molecular physics nor those of solid-state physics can be applied [Brack (1993), Yannouleas and Landman (1996)]. Some theoretical methods can be adapted from nuclear physics.

Clusters also represent a very attractive field for applied research in many areas. These include, for example, electronics, catalysis, lithography, carbon-chemical engineering, superconductivity, photography, aerosols and smoke, sintering of small particles, and semiconductor production. Due to their unique properties one might expect that clusters, if they could be synthesized in bulk quantities in such a way that they retain their structure, provide a new source of novel materials [Jena et al. (1996)]. Attempts in this direction are experiments in which neutral or ionized clusters are deposited on solid surfaces ("cluster beam deposition") [Takagi (1984), Ando et al. (1994), Mélinon et al. (1995), Perez et al. (1997)].

Chemical reactions may change clusters in numerous ways. Both adsorption reactions, in which the reactants remain at the cluster surface, and reactions, in which the reactants are bonded inside the cluster, are possible. The large surface-to-volume ratio of clusters makes cluster research particularly interesting for catalysis [Riley (1994)]. Another application is cluster impact lithography, which can be used to produce microstructures on solid surfaces [Gspann (1996)]. Cluster science also finds applications in astrophysics: The elucidation of the nature, the mechanisms of formation, and the properties of cosmic dust depends on our knowledge of clusters.

Work in cluster science and in the structural chemistry of astrophysical particles led to the discovery of three-dimensional, closed, cage-like structures of high stability in carbon clusters (fullerenes) [Kroto et al. (1985), Chibante and Smalley (1993), Curl (1997), Kroto (1997), Smalley (1997)] (see Fig. 2.1). The discovery of these clusters, which were produced in nozzle beams from laser ablation sources (see Sect. 2.4.2), opened a new, fast-expanding field in chemistry. Its importance was emphasized by bestowing the Nobel Prize for Chemistry on the three discoverers R.F. Curl, H.W. Kroto, and R.E. Smalley.

After having succeeded in producing the most prominent representatives of these clusters, C_{60} and C_{70}, in macroscopic quantities [Krätschmer et al. (1990)], their chemical and physical properties have been investigated in great detail, as well as their behavior in collisions with electrons, ions, neutral particles (fragmentation, ionization, atom capture, charge exchange), and with solid surfaces [Lorents (1997)]. New clusters with cage structure have been found [Tenne et al. (1992), Thilgen et al. (1993), Feldman et al. (1995), Piscoti et al. (1998)], demonstrating that fullerene-like clusters could not only be formed by carbon, but also by inorganic molecules such as WS_2, MoS_2, and $NiCl_2$ [Rapoport et al.

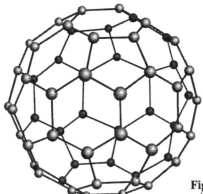

Fig. 2.1. Atomic structure of the C_{60} cluster (buckminsterfullerene)

(1997)]. Furthermore, foreign atoms, molecules, and metal ions have been attached both to the surface and to the interior of the hollow cage [Weiske et al. (1991), Chai et al. (1991), Campbell et al. (1992), Saunders et al. (1993), Schwarz et al. (1993), Tellgmann et al. (1996), Campbell et al. (1997), Gromov et al. (1997)]. Finally, many other structures (tubular and onion-shaped structures, nanotubes, nanowires) have been discovered [Smalley (1993), Ramirez et al. (1994)]. The experiences and results of this new branch of research are of great importance for many areas of science, ranging from structural chemistry and astrophysical dust particles to material science and nanotechnology [Sattler (1993), Ehrenreich and Spaepen (1994), Dresselhaus et al. (1996), Yoshimura and Chang (1998), Hirsch (1998)].

Investigations aimed at a first technical application of fullerene-like clusters have been reported by Tenne and co-workers [Rapoport et al. (1997)]. They have demonstrated that nanoparticles of tungsten disulfide, typically 120 nm in diameter, are an especially effective lubricant (2.5 times better than the best existing lubricants). The elasticity of these nanoparticles, their chemical inertness, their small adhesive power, and their ability to roll rather than to glide between surfaces are assumed to be responsible for the reduction of friction between moving parts.

Table 2.1 gives a survey of the nature of the chemical bonds observed in clusters. As can be seen, all types of bonds known in chemistry may be found.

Table 2.1. Classification of clusters according to their binding [Jortner (1984)]

Cluster type	Type of binding	Binding energy [eV]	Examples
van der Waals	Dispersive forces	≤ 0.3	$(Ar)_N$, $(N_2)_N$, $(CO_2)_N$
Molecular	Dispersive forces, weak electrostatic forces	$0.3 - 1$	$(org.molecules)_N$, $(I_2)_N$
Hydrogen-bonded	Intermolecular hydrogen bridges	$0.3 - 0.5$	$(HF)_N$, $(H_2O)_N$
Ionic	Heteropolar bonding	$2 - 4$	$(NaCl)_N$, $(CaF_2)_N$
Valence	Covalent bonding	$1 - 4$	C_N, S_8, As_4
Metal	Metallic bonding	$0.5 - 3$	Na_N, Al_N, Cu_N, W_N

Apart from characterizing clusters according to their chemical bond a classification according to their size or to the number N of their constituents, respectively, is often useful.

Agglomerates with $N \leq 10$ are often called *microclusters*. They usually show a well defined structure, although for larger microclusters several nearly isoenergetic isomers may exist. Furthermore, they show discrete vibrational spectra and also discrete electronic states. With respect to the classification given below in cluster constituents being located at the surface and those being situated in the volume, all constituents of microclusters are located at the surface. For a theoretical treatment of microclusters well known methods of molecular physics may be used. These methods, however, fail to work for *small clusters* characterized by $10 < N < 100$. Their molecular structures show a large number of isomers with closely spaced energy levels. Amorphous structures are also possible. The vibrational motion of small clusters is extremely complex and not reducible to normal modes.

Large clusters ($100 < N < 1000$) show a slow, gradual transition to bulk material with increasing N. The density of the electronic states and of the phonon states increases. Important differences to the bulk phase, however, are due to some unique characteristic properties: The large surface-to-volume ratio (see Table 2.2) and finite size effects, which can be either quantum size effects (resulting from the finite electron energy spacing in the valence band and the perturbed collective behavior of the confined electron gas) or thermodynamic size effects (which, for example, cause the dependence of thermodynamic properties such as surface tension on the sample size) [Jortner (1984)].

This holds also for *nanoclusters* with $N > 1000$, where the gradual evolution towards a solid-state structure is continued, which finally will be called *nanoparticles* (see Table 2.2), if the number of cluster constituents N becomes larger than about 10^6.

The number N_0 of particles within the surface of a given cluster can be easily estimated by assuming a spherical shape of the cluster. Under this assumption, the cluster radius R_N is given by

$$R_N = R_1 N^{1/3}, \quad \text{where} \quad R_1 = \left(\frac{3m_0}{4\pi\rho}\right)^{1/3} \tag{2.1}$$

is the radius of an atom (mass m_0) within the cluster. Usually, the cluster density ρ is approximated by its value in the solid or liquid phase. This yields for the number N_0 of surface particles

$$N_0 = \frac{4\pi R_N^2}{\pi R_1^2} = 4N^{2/3}. \tag{2.2}$$

Hence, the fraction N_0/N of surface particles becomes

$$\frac{N_0}{N} = 4N^{-1/3}. \tag{2.3}$$

Compared with the solid state, this is a very large number, as is shown by Table 2.2, which presents some values of N and N_0/N according to (2.1) and (2.3), calculated for an argon cluster ($R_1 = 0.208$ nm).

The dependence of the cluster vibrations on their size is obvious in the case of a one-dimensional cluster (linear chain of N particles). For each possible vibrational state the length of the chain must be an integral multiple of the corresponding wavelength to yield a stationary state. Accordingly, the eigen frequencies and

Table 2.2. Cluster sizes and surface-to-volume ratio

R_N (nm)	N	N_0/N
1	110	0.8
2	900	0.4
10	10^5	0.09
100	10^8	0.009
10^7	10^{23}	10^{-7}

energy distances are inversely proportional to the cluster size. Quantum size effects will be large, if the energy distances are of the order of the thermal energy kT.

Thermodynamic properties which depend on the cluster size, are the already mentioned surface tension, the vapor pressure, and the melting temperature.

2.1 Survey of Methods for Cluster Formation

Under thermodynamic equilibrium conditions, a gas or vapor contains a certain fraction of clusters, as has been discussed for dimers, for example, in Sect. I.2.1.7. Their intensity decreases roughly exponentially with cluster size, if the binding energy is a monotonous function of the cluster size. Well known deviations from this exponential behavior are bismuth, sulfur, selenium, and antimony, which yield substantial fractions of clusters in effusive beams of these elements (up to Bi_4, Sb_4, Se_8, S_8) [Mühlbach et al. (1981)]. Since an effusive beam source (regarding the Knudsen condition, see Chap. I.2) does not change the equilibrium state of the gas, such a source is the simplest way to produce clusters. Indeed, the first scattering experiments with dimers were performed using effusive beams [Rosenberg (1939), (1940)]. In spite of the low cluster intensities, effusive beams have also been used later, for example, to determine binding energies of small metal clusters (predominantly dimers) [Wu (1980), Gingerich et al. (1985)]. Nonmetallic clusters have also been investigated utilizing effusive beams [Scheuring and Weil (1985)].

In most cases, however, the cluster intensities are too low for experimental investigations, especially if larger clusters are required. Consequently, other cluster sources have been developed. A survey can be found in several review articles [Kappes and Leutwyler (1988), Mandich et al. (1990), de Heer (1993), Haberland (1994)]. Although a classification of the methods used today which allow the production of clusters of any element (and numerous molecules) is somewhat arbitrary, since often combinations of several methods are used, we divide the cluster sources into the following classes:

Supersonic jets: As described in Chap. I.3, the adiabatic expansion through a small nozzle into vacuum leads to a drastic temperature decrease in the beam. As a consequence, supersaturation and finally cluster formation occurs [Kappes and

Leutwyler (1988), Hagena (1992)]. The content of clusters in a free jet and the average cluster size, which may range from a few up to many thousand particles per cluster, is determined by the expansion conditions (stagnation pressure, stagnation temperature, and nozzle diameter), but the nozzle geometry and the use of carrier gases also have an important influence on cluster formation and average cluster size. The adiabatic expansion is particularly suited to produce clusters from gases and from species with high vapor pressure, yielding high intensity beams with narrow speed distributions. In general, little is known about the temperature of the clusters produced in a nozzle expansion, it is assumed that large clusters have a temperature close to the evaporation temperature discussed below.

Gas aggregation: Solid or liquid material is evaporated into a stationary or flowing cold gas, so that the atoms or molecules are collisionally cooled until condensation and cluster formation takes place [Sattler et al. (1980)]. The formation process, therefore, is similar to smoke, cloud, or fog formation in nature, and gas aggregation sources are often called "smoke sources". Evaporation may take place in an oven of the types described in Chap. I.4 or in an arc discharge burning in a rare gas. The latter method was successfully used to produce macroscopic quantities of C_{60} for the first time [Krätschmer et al. (1990)]. Pulsed discharges are also possible.

Gas aggregation produces mainly large clusters (N < 10 000), the cluster size can be influenced by the aggregation conditions (time, vapor pressure, temperature and flow velocity of the cooling gas) within rather large limits. The cluster beam intensities obtained by gas aggregation are considerably lower than those obtained from supersonic jets, but low cluster temperatures (T < 100 K) can be achieved. Due to many collisions with the aggregation gas, the clusters obtain a rather uniform translational velocity, so that clusters of different masses have different velocity distributions.

Surface erosion: These sources utilize the removal of atoms and clusters from a solid or liquid surface by heavy ion impact (sputtering, see Sects. I.4.5 and 1.7), intense laser radiation (laser ablation, see Sects. I.4.4 and 1.6), pulsed arc discharges or a high electric field.

Sputtering: Heavy ions, impinging on a solid target (which may be a solid, or a frozen gas or liquid) with sufficient energy, cause the ejection of atoms, molecules, and clusters from the surface [Begemann et al. (1986)]. The clusters produced are usually small, and their temperature is close to the evaporation limit. The energy distribution of the ejected clusters is rather broad (of the order of 10 eV) see Sect. 1.7).

Laser ablation: An intense, pulsed laser is focused onto a target, evaporating various atomic layers of the material. The hot plasma produced in the laser focus expands, and clusters are formed with sizes up to several hundred particles [Dietz et al. (1981)]. The temperature of the clusters, depending on the expansion conditions, is of the order of the temperature of the source. The time-averaged cluster beam intensities are rather low compared with those of nozzle beams, but the intensities per laser pulse are much higher. Laser ablation can be easily combined with gas aggregation. This improves the reproducibility of cluster formation, increases the average cluster size, and decreases the cluster temperature [Milani and de Heer (1990)].

Pulsed arc discharges: Instead of a laser a pulsed arc discharge can also be used to evaporate particles of the electrode material. These sources are called pulsed arc cluster ion sources (PACIS), since they produce predominantly ionized clusters. Combining this source with gas aggregation yields intense beams of clusters.

Liquid metal ion sources (LMIS): Electric field emission and ionization occurs from a cone of liquid metal which wets a tip of a refractory metal [Bhaskar et al. (1987)]. The temperature of the cluster ions is high, thus evaporation and fragmentation processes are observed after cluster formation.

Laser-induced pyrolysis: Another technique to produce large clusters of refractory elements is based on the CO_2-laser-induced decomposition of appropriate molecules containing a refractory element such as carbon or silicon (similar to chemical vapor deposition) in a gas flow reactor. The decomposition of the precursor molecules generates a supersaturated vapor of reaction products in the reaction volume, which is cooled in a rare gas flow, so that condensation and cluster formation occurs [Ehbrecht et al. (1993a)].

Diffusion through foils: Atoms, which diffuse through suitable foils kept at high temperature (Cs atoms and carbon foils, for example), may leave these foils as atoms, highly excited clusters, or ionized clusters [Åman and Holmlid (1991), Holmlid et al. (1992)].

Cluster aggregation: A recently discovered possibility to convert rare gas clusters into metal clusters [Rutzen et al. (1996)], which may be of interest for the production of clusters from refractory metals, is based on the capture of metal atoms by large rare gas clusters.

2.2 Supersonic Jets

The formation of clusters in gas dynamic expansions into vacuum was already observed at the beginning of nozzle beam experiments [Becker et al. (1956), (1962)]. Numerous experimental and theoretical research activities followed this early work, both in fundamental and in applied research. The latter efforts have been concerned, for example, with the production of "condensed beams" with high material transport in order to fuel fusion plasmas, or to use them as targets in high energy physics. The former activities were aimed at the understanding of the process of condensation and cluster formation under the nonequilibrium conditions of an expansion.

Gas dynamic expansions and the attempts to describe condensation and cluster formation as well as semiempirical scaling laws governing cluster formation are discussed in Chap. I.3. The following sections, therefore, are restricted to the discussion of experimental possibilities to influence cluster formation. Apart from the expansion conditions (stagnation pressure, temperature, and effective nozzle

78   2. Production and Diagnostics of Cluster Beams

diameter), these depend essentially on the geometry of the nozzle and the use of carrier gases.

2.2.1 Influence of Nozzle Shape

Figure 2.2 shows schematic cross sections of various nozzles which have been investigated partly by experiments [Hagena and Obert (1972)] and partly by numerical calculations based on the method of characteristics [Miller et al. (1985), Murphy and Miller (1984), Miller (1988)]. (1) is a short convergent nozzle, usually used to produce free jets (see Chap. I.3), (2) is an ideal (sharp-edged) aperture, and (3) a capillary aperture. In practice, apart from long capillaries, the differences in shape between these nozzles are not very pronounced, since it is rather difficult to control the shape exactly when machining small nozzles. This has, however, no considerable influence on the expansion, as has been shown by numerical calculations of the flow field for these three nozzle types. Free jets are, to first order, independent of source geometry [Murphy and Miller (1984)].

The essential difference between the nozzles 4–6 shown in Fig. 2.2 and the nozzles 1–3 is the additional divergent part of the nozzle, which is very important for cluster formation. (4) is a Laval nozzle, (5) a convergent-divergent nozzle and (6) a divergent nozzle.

Let us consider first the expansion from a simple, convergent nozzle. Three factors determine the content of clusters in free jets: stagnation pressure P_0, smallest aperture diameter d, and initial gas temperature T_0, as can be seen from the scaling parameter Γ^* introduced in Sect. I.3.9.2. In general, cluster content as well as average cluster size increase with increasing stagnation pressure and aperture diameter. Both quantities decrease with increasing source temperature T_0.

Figure 2.3 summarizes schematically the process of clustering in an expansion. Here, the logarithm of the pressure P is plotted as a function of the reduced distance x/d from the nozzle (dashed curves) for various stagnation pressures P_0 (in bar). The full curve represents the logarithm of the vapor pressure $P_v(x/d)$ along the centerline of the expansion. With decreasing stagnation pressure the points of intersection of the pressure curves with the vapor pressure curve $P_v(x/d)$, which mark the onset of supersaturation and cluster formation, are shifted to larger

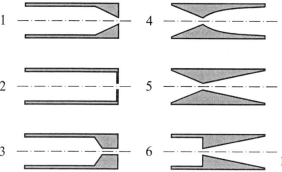

Fig. 2.2. Various nozzle shapes

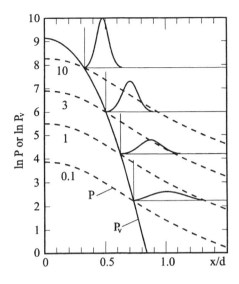

Fig. 2.3. Pressure P(x/d) (*dashed curves*) along the centerline of the expansion for different stagnation pressures P_0 (0.1, 1, 3, 10 bar) and vapor pressure $P_v(x/d)$ for Ar (*full curve*). The intersection points of these curves mark the onset of nucleation (schematically indicated by the nucleation rate J(x/d))

distances x/d. This is schematically indicated in Fig. 2.3 by the nucleation rate $J(\xi)$. Accordingly, for a given nozzle diameter d and stagnation temperature T_0, the region of supersaturation is shifted towards the nozzle with increasing stagnation pressure P_0. Additionally, for a fixed saturation ratio, the difference between pressure P and vapor pressure P_v increases. Both effects lead to an increased collision rate in the region of supersaturation, and thus the number of clusters and their average size increases with increasing stagnation pressure [Ryali and Fenn (1984)]. A reduction of the stagnation temperature T_0 also shifts the onset of clustering to smaller distances ξ from the nozzle and thus into the region of higher pressures, also favoring the formation and growth of clusters.

Cluster formation in an expansion from an ideal aperture (2) is rather similar to that from a short convergent nozzle (1), while a capillary aperture (3) enhances the formation and growth of clusters due to the increased time spent by the particles in the region of high pressure close to the nozzle [Murphy and Miller (1984)].

Convergent-divergent nozzles (5), Laval nozzles (4), and divergent nozzles (6) have a similar influence on cluster formation, as has already been shown in the early days of cluster research [Hagena and Obert (1972)] and confirmed by later investigations [Kappes et al (1986), Kappes et al. (1987)]. For a given stagnation pressure and nozzle diameter, high cluster intensities and largest average cluster sizes are obtained for divergent nozzles (6). As already discussed in the case of capillary apertures, this is due to the constrained expansion zone, which makes more collisions possible before the expansion terminates [Hagena (1981)]. The lowest cluster intensities, on the other hand, are obtained with free jet expansions from sonic apertures (1). In fact, these apertures are usually used in those cases where cluster formation must be avoided [Kappes and Leutwyler (1988)].

To compare cluster formation of standard convergent nozzles (sonic nozzles (1)), or of simple thin-walled apertures (2) with those of convergent-divergent (5)

Table 2.3. $c(\kappa)$ for axisymmetric expansions

$\kappa = 5/3$	$\kappa = 7/5$	$\kappa = 9/7$
0.736	0.866	0.986

or divergent (6) nozzles, the nozzle geometry must be accounted for in the dimensionless scaling parameter Γ^* (see (I.3.166)) [Hagena (1981)]. This can be done by introducing an "equivalent" nozzle diameter d_{eq} for divergent nozzles, which — under equal stagnation conditions — yields the same flow field as that obtained with a sonic nozzle. These equivalent nozzle diameter is given by

$$d_{eq} = c(\kappa) d / \mathrm{tg}\alpha, \tag{2.4}$$

where α is the half-opening angle of the divergent nozzle part. The constant c depends on κ and is different for axisymmetric and planar expansions, respectively. Table 2.3 shows some numerical values for the axisymmetric expansion. An angle $\alpha = 5°$, for example, yields an equivalent diameter d_{eq} which is 8.4 times larger than that of a sonic nozzle (assuming $\kappa = 5/3$). Hence, a sonic nozzle produces the same centerline flow field with a total flux reduced by a factor of 70 (as long as boundary layer effects are negligible.

2.2.2 Influence of Carrier Gases

The formation of clusters A_N in a supersonic jet can be enhanced by admixing a carrier gas (C) (rare gas) to the expansion of an atomic gas or vapor (A). The carrier gas mainly serves to cool the growing clusters and to carry off the heat of condensation. Thus it stabilizes the clusters against evaporation. In the case of insufficient cooling, the clusters are stabilized by the evaporation of one or more constituents. The same influence of carrier gases on cluster formation is also observed in expansions of molecules. This is often used in the production of molecular clusters [Bahat et al. (1987), Schlag and Selzle (1990)].

An optimum value exists for the partial pressure of the carrier gas P_C. At low partial pressures, the rate of collisions A_N–C is insufficient for cluster cooling. At high values of P_C, the rate of collisions A–A decreases and this compensates the effect of cluster cooling.

According to spectroscopic investigations of alkali metal dimers and trimers, small clusters achieve low temperatures in the expansion [Delacrétaz and Wöste (1985)]. The temperatures of large clusters, however, are rather high, of the order of the evaporation temperature [Bjørnholm et al. (1991)].

If the interaction energy A–A is not very different from that of A–C, the formation of mixed clusters, which have been studied in many experiments [Gough et al. (1983), Leutwyler (1984), Boesiger and Leutwyler (1987)], is also possible (see Sect. 2.6.1). Finally, cluster formation is also influenced by the mass of the carrier gas. A heavy gas is slow and provides, therefore, more time for the aggregation process, resulting in the formation of larger clusters in metal vapor expansions, for example [Kappes et al. (1982)].

2.2.3 Technical Realization of Cluster Sources

The technical design of cluster sources is identical to that of the sources described in Chap. I.4. To obtain microclusters or small clusters, often sonic nozzles ((1) in Fig. 2.2) are used, whereas divergent nozzles ((5) or (6) in Fig. 2.2) are usually used to produce larger clusters. Important for cluster formation is the dimensionless scaling parameter Γ^* defined in Sect. I.3.9.2, which can be calculated from the operation conditions of the source and the characteristic data of the beam material. Figure 2.4 illustrates this for the case of small clusters ($\overline{N} \leq 100$), presenting measured cluster sizes $\overline{N}$ as a function of Γ^* for expansions of Ar, Kr, and Xe [Hagena (1987), Karnbach et al. (1993), Buck and Krohne (1996)]. The average cluster sizes, determined with different methods (mass spectrometer, molecular beam diffraction) show good agreement and demonstrate the close correlation between average cluster size and scaling parameter Γ^*. In the range of cluster sizes shown in Fig. 2.4, this correlation may be expressed by a simple power law.

Figure 2.5 shows, as an example, a high-temperature source for the production of metal clusters (silver clusters) together with the corresponding differential pumping stages [Hagena (1991), (1992), Gatz and Hagena (1995)]. The oven, the nozzle, and the gas supply pipe are made of high-density graphite. The operation temperature of 2200 K is achieved by radiation heating using a double-wound coil of graphite and a radiation shielding, consisting of a number of tantalum sheets surrounded by a water-cooled cylinder. Alignment, mechanical support, and electrical insulation of the components of the heating system are achieved by concentric ceramic elements (boron nitride). Typical heating powers are of the order of 3 kW. At the operation temperature of 2200 K, the vapor pressure of silver is 200 torr. Argon is used as a carrier gas at a stagnation pressure of 3000 torr. The nozzles used are usually convergent-divergent nozzles ((5) in Fig. 2.2) with diameters

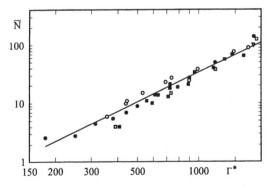

Fig. 2.4. Measured average cluster sizes as a function of the scaling parameter Γ^*. *Full circles* (Xe_N), *full squares* (Ar_N), *open squares* (Kr_N) [Karnbach et al. (1993)], *open circles* (Ar_N) [Buck and Krohne (1996)], *triangles* (Ar_N) [Hagena (1987)].

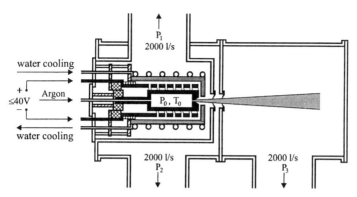

Fig. 2.5. Source for production of silver clusters [Hagena (1992)]

between 0.35 and 1 mm and opening angles α between 5° and 8.5°. The skimmer (⌀ 1.4–3 mm) is heated by a tungsten spiral to about 1300 K, well above the melting temperature of silver (1234 K). The oven chamber is evacuated by a refrigerator cryopump (its capacity is sufficient for 23 h operation at full load), while the differential pumping stage and the beam utilization chamber are evacuated by turbopumps.

Sources utilizing gas dynamic expansions with carrier gases belong to the most intense cluster sources available. As can be seen by the above example, these sources are operated at typical carrier gas pressures of the order of a few atmospheres, while the vapor pressures of the metal ranges from about 10 to 100 torr. Under these conditions the material consumption is of the order of 0.1 mol/h; about 10% of this material is used for cluster formation. Cluster sizes are of the order of a few hundred to a few thousand particles/cluster [Pederson et al. (1991), Lewerenz et al. (1993), Schütte (1997), Buck and Krohne (1996)], the cluster speed depends only slightly on their mass [Kresin et al. (1998)].

2.3 Gas Aggregation

Gas aggregation sources are also often called "smoke sources", since the formation of smoke, fog, or clouds in nature occurs according to the same principle. They represent a simple method to produce clusters with sizes that can be varied within wide limits by choosing appropriate operational conditions. The starting point is the evaporation of liquid or solid material into an aggregation chamber, which is filled with a cold gas at a pressure of several torr. The evaporated atoms or molecules are cooled in collisions with the aggregation gas and when their kinetic energy is low enough they form dimers in three-body collisions, which grow due to further collisions or may also be destroyed. These statistical processes cause a slow cluster growth, which is drastically accelerated when supersaturation is attained. The clusters formed in the early cooling off phase act as condensation nuclei. The average size of the clusters is mainly determined by their residence time in the aggregation chamber. However, it also depends both on the number of particles available for the condensation process (essentially determined by the vapor pressure at the oven aperture) and on the temperature of the aggregation gas, which is given by the temperature of the walls of the aggregation chamber. The number of condensation nuclei initially formed in the slow aggregation process is also of some importance.

The method of gas aggregation is rather old. It was first used in 1930 to generate metallic gold smokes for depositing adsorbing layers of fine gold particles ("gold black") on thermal detectors [Pfund (1930)]. The deposited films were found to consist of finely divided aggregates of gold (~ 100 Å $\varnothing$). Later, this method was used to produce nanoparticles [Hogg and Silbernagel (1974)] and finally to produce small clusters [Kimoto and Nishida (1977)].

To evaporate the cluster material, all types of ovens that have been described in Chap. I.4 may be used. Both effusive and nozzle beams are frequently employed. Using a carrier gas and different conditions for the expansion into the aggregation chamber yields further parameters to control cluster formation and average cluster size. In a second expansion, the beam is transferred from the aggregation chamber through a differential pumping stage into the beam utilization chamber. As an example, Fig. 2.6 depicts one of these sources for the production of metal clusters [Mühlbach et al. (1981), Recknagel (1984)]. An effusive oven (1) (with temperatures up to 1500 K) is used to evaporate the elemental material. The oven is surrounded by a cooled housing (2) (at liquid nitrogen temperature) and is located within the aggregation chamber (3) filled with helium (at pressures up to 40 mbar). The walls of the aggregation chamber are also cooled by liquid nitrogen. The growth process is interrupted when the clusters pass through a differential pumping stage (5) into the beam utilization chamber, where they are ionized by electron impact ionization (6). The ions can be analyzed with a time-of-flight spectrometer. A quartz microbalance (7) is used to monitor the neutral beam.

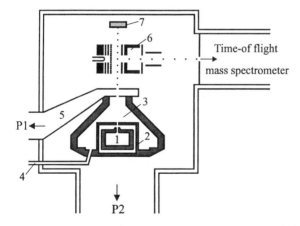

Fig. 2.6. Gas aggregation source for metal clusters [Recknagel (1984)]. (1) Oven with heat shielding (2), (3) aggregation chamber with gas supply (4) and cooled walls, (5) differential pumping stage, (6) ion source, (7) quartz microbalance, P1 and P2 turbopumps

Many variants of this basic principle are possible and have been applied [Abe et al. (1980), Frank et al. (1985), Bréchignac et al. (1991), Urban et al. (1996), Goldby et al. (1997), Baker et al (1997)]. Ovens capable of achieving temperatures up to 2000 °C have been used [Riley et al. (1982)]. Figure 2.7 shows a more recent design for the generation of large alkali metal clusters [Zimmermann et al. (1994)]. The material to be evaporated (1) (see insert of Fig. 2.7) is contained in an open metal crucible (4) (at a vapor pressure of about 0.1 mbar) placed inside a ceramic tube (6) (Al_2O_3) which is wrapped in tantalum wire (2) for ohmic heating. This unit is placed inside a second ceramic tube (6) for thermal insulation, which is surrounded by several radiation shields of tantalum sheet to reduce heat losses. The temperature of the crucible is measured by a thermocouple (5) at its base. Because the oven assembly is mounted on a linear motion feedthrough (9), the oven–nozzle distance can be continuously changed to optimize cluster formation. The oven is situated in the aggregation chamber (14). The inert gas flow into this chamber through the gas supply (13) can be precisely adjusted with a calibrated needle valve. Typical gas pressures range from 1 to 3 mbar. The gas achieves low temperatures (approximately 90 K) in the aggregation chamber, since the walls (including the nozzle (10)) are cooled with liquid nitrogen. The nozzle (10) is a tube of 30 mm length and 4.5 mm inside diameter. At the upstream end the nozzle is conical with a sharp-rimmed aperture of 3 mm diameter. This design reduces condensation of the cluster material at the aperture. The cluster beam generated in the aggregation chamber passes through two differential pumping stages P1 and P2, separated by skimmer-shaped apertures (11), into the beam utilization chamber, where they are photoionized and mass analyzed by a reflectron mass spectrometer. The cluster source shown in Fig. 2.7 provides four parameters to optimize the cluster yield and to vary the cluster size distribution: inert gas pressure, oven temperature (vapor pressure of the cluster material), inert gas type, and oven–nozzle distance.

To produce mixed clusters, for example C_{60} covered with alkali metal atoms, two separate ovens, each containing one of the elements to be mixed, can be placed inside the aggregation chamber. Slightly tilting the openings to each other, the evaporating vapor clouds will mix, producing mixed clusters whose stoichiometry can be adjusted by the relative temperatures of the ovens.

Instead of evaporating the cluster material from an oven, evaporation may also be achieved by using an arc discharge. As already mentioned, this method was the first to produce macroscopic quantities of C_{60} and C_{70} [Krätschmer et al. (1990)]. Other possibilities to produce the primary particles are laser ablation or sputtering.

The details of the condensation process in these sources are more complex than indicated above, since cluster formation is influenced by a large number of interdependent parameters. Aside from the vapor pressure of the metal and the pressure, temperature and flow velocity of the aggregation gas, the following parameters have also been found to be important: The mass of the cluster constituents and the aggregation gas, the geometry of the aggregation chamber, the diameters of the apertures used, and the available pumping speed.

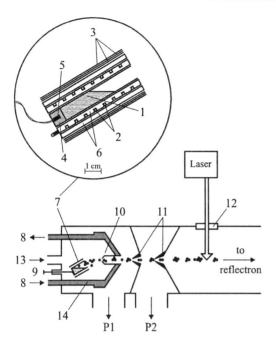

Fig. 2.7. Schematic of a gas aggregation source. (1) Cluster material, (2) heating coil between ceramic tubes (6), (3) radiation shields, (4) metal crucible, (5) thermocouple. (8) liquid nitrogen supply, (9) linear motion feedthrough, (10) source aperture, (11) apertures, (12) window for ionization laser, (13) gas inlet, P1 and P2 pumps [Zimmermann et al. (1994)]

The reproducibility, especially with respect to relative cluster intensities, is not always satisfactory [Schaber and Martin (1985)]. These authors discuss also the correlation of some experimental parameters with the cluster intensity. Since the transfer of the cluster beam from the aggregation chamber into the differential pumping stage is usually a more or less effusive process, or at most a very weak expansion, the clusters have a broad speed distribution. Depending on the source conditions, low cluster temperatures (T < 100 K) can be achieved. For further details, the reader is referred to the literature [Mühlbach et al. (1981), Frank et al. (1985), Broyer et al. (1987), Kappes and Leutwyler (1988), de Heer (1993), Haberland (1994)].

Gas aggregation sources have been extensively used to generate metal clusters (Pb, Bi, Sb, Zn, Cu, Ni, Ag, Li, Na, Cs) [Sattler et al. (1980), Riley et al. (1982), Frank et al. (1985), Zimmermann et al. (1995), Goldby et al. (1997)]. They have also been used to produce various composite clusters ($(NaCl)_N$, $(CuBr)_N$, $(AsS)_N$, and mixed clusters As_NS_M, P_NS_M, As_NO_M, Sb_NO_M (see Sect. 2.6.2).

2.4 Surface Erosion

Other methods to generate clusters make use of surface erosion: Material from a solid or liquid target is removed either by heavy particle impact (sputtering) intense laser radiation (laser ablation) or high electric fields. Cluster formation by sputtering was first observed in a radiofrequency spark source [Dornenburg and Hintenberger (1959)], while the first discovery of clusters in the vapor ejected by a focused, Q-switched ruby laser from a carbon target dates back to Berkowitz and Chupka (1964). Liquid metal ion sources, which may also be considered to be based on surface erosion, date back to 1961 when charged droplets and atomic ions were observed to emanate from liquid metals in high electric fields [Krohn (1961)]. These three types of sources are discussed in the following sections.

2.4.1 Sputtering

Sputtering sources were developed at the end of the 1960s to produce fast beams in the 0.5 to 20 eV energy range. They have been applied in many scattering experiments. Some fundamentals of sputtering as well as details of the experimental techniques are described in Sect. 1.7. It was soon discovered that not only fast atoms or molecules are produced by sputtering sources, but also neutral and ionized clusters [Honig (1963), Baede et al. (1971), Richter and Trapp (1981)]. Further investigations led to the development of cluster sources, which utilize high-energy ion beams (Ar^+ or Xe^+ ions with energies between 3 and 30 keV), which are focused onto the surface of a solid target (solid body or frozen liquid) [Fayet and Wöste (1985), Begemann et al. (1986), Katakuse (1987), Maguera et al. (1990)]. Similar devices are described in Sect. 1.7. Another method utilizes the energetic ions of a gas discharge, using the cathode as a target. Hollow cathode discharges [Xenoulis et al. (1996)] or magnetron devices [Haberland et al. (1991), (1992), Goto et al. (1997)] have found widespread application and are often combined with gas aggregation. The main application of the last-mentioned sources is the production of thin films with specific properties.

The primary clusters produced in the sputtering process are rather small. Sputtering of polycrystalline copper surfaces, for example, yields neutral clusters up to Cu_{15} [Coon et al. (1991)], while sputtering of indium produces clusters up to In_{32} [Ma et al. (1994), Wucher et al. (1994)]. The relative yield Y(N) of neutral clusters of a given size N, referred to the number of emitted atoms, can be represented by a power law. For example, in the case of indium, the relative sputtering yield Y(N) can be empirically described by

$$Y(N) = 1.7 N^{-5.6}. \tag{2.5}$$

The exponent of (2.5) depends on the sputtered element. For Cu_N an exponent of -7.8 is found, while the exponents for Al_N and Ag_N are -9.3 and -6.5, respectively [Coon et al. (1993), Wucher et al. (1993)]. The energy distributions of the sputtered clusters, which have been measured, for example, for copper dimers and trimers, range from 0.2 to 10 eV and have a broad maximum between 0.5 and 1.3 eV [Coon et al. (1991)]. Evidence such as this suggests that the clusters formed by ion sputtering do not arise from gas phase recombination of emitted atoms.

Instead, the primary clusters arise from the momentum transfer and electronic excitations which occur in the collision cascades produced by the energetic ions.

In general, the yield of ionized clusters is smaller than that of neutral clusters. In the case of indium, for instance, the ratio of ionized to neutral clusters is 10^{-4}.

Clusters have also been generated using energetic neutral atom bombardment of targets [Devienne and Roustan (1982)]. This has the advantage that electric surface charges are avoided when nonconducting surfaces are used, which may lead to instabilities in the case of sputtering with ions.

A disadvantage of sputtering cluster sources is the fact that the ejected cluster usually have high rotational and vibrational temperatures [Fayet et al. (1986)]. Furthermore, as was already mentioned above, they have broad distributions of kinetic energies [Richter and Trapp (1981)]. In addition, extensive unimolecular decomposition of sputtered clusters has been observed by measuring the cluster distribution as a function of the time following their initial ejection from the surface [Begemann et al. (1986)]. Hence, the cluster size distribution is shifted to smaller cluster sizes with increasing time delays. The temperature of the internal degrees of freedom can be reduced by passing the clusters through a cell filled with a cold gas, as has been shown, for example, in the case of ionized clusters [Hanley and Anderson (1985), Hanley et al. (1987)].

2.4.2 Laser Ablation

As has already been mentioned in Sects. I.4.4 and 1.6, laser ablation, which is caused by focusing the radiation of a pulsed laser (10 to 20 mJ within 10 ns) to a surface area ≤ 1 mm^2, results not only in the vaporization of atoms and molecules but also of small clusters. Particularly, high fractions of clusters are found for the elements C, Si, and Ge, especially if targets of suitable polymers are used. High intensities of small carbon clusters, for example, have been obtained by using captan foils as a target for laser ablation [Campbell et al. (1990)]. Besides neutral clusters, positive and negative cluster ions are also emitted [Ulmer et al. (1990)]. The clusters are very "hot" and cool down by the evaporation of constituents. Some examples of laser ablation sources are described in Sects. I.4.4 and 1.6.

A further decrease of the cluster temperature can be obtained by combining laser ablation with a gas dynamic expansion or with gas aggregation (or with both methods). In addition, this combination offers the possibility of varying the cluster size within large limits. This makes laser ablation a versatile and universal source for cold clusters.

In this combination the laser ablation source (usually a rotating and translating rod of the material to be evaporated) is placed between a pulsed gas valve and a divergent nozzle. The time delay between the gas pulse and the laser pulse is adjusted in such a way that the laser ablation occurs when the gas pulse has reached its maximum [Powers et al. (1982), Hopkins et al. (1983), Rohlfing et al. (1984), Riley et al. (1984), Heath et al. (1985)]. Continuous sources for the carrier gas have also been used [Pedersen et al. (1998)].

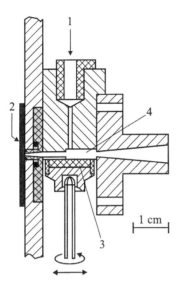

Fig. 2.8. Laser ablation source [Mayurama et al. (1990)]. (1) Laser beam, (2) leaf spring of a pulsed valve, (3) linear moving and rotating target disk, (4) "waiting room"

Figure 2.8 shows schematically a very compact version of these sources [Maruyama et al. (1990)] with low pumping requirements (170 l/s). Instead of a rotating rod, a rotating and longitudinally moved disk (3) is used as the target, so that the focus of the fixed laser (1) follows a spiral-shaped curve on the target disk. The material removed by a laser pulse is not immediately evaporated into the maximum of the gas expansion, but is expanded into a small "waiting room" (4) directly above the disk, postponing the gas pulse as late as possible after the vaporization laser has been fired. This allows the vapor plume to expand unimpeded for a while before it is entrained in the rising density of an extremely fast-rising gas pulse and expanded through the divergent nozzle [Milani and de Heer (1991)]. The fast rise time of the gas pulse is achieved by using a valve according to the current-loop principle (see Chap. I.4). The temperature of the clusters can be further reduced by cooling the nozzle or the complete source [Andersson et al. (1996), Mélinon et al. (1997)].

The cluster size depends on the operational conditions of the source and covers a region from a few to several thousand atoms per cluster. The material consumption depends on the material used, but is generally very low, of the order of 10^{15} atoms per laser pulse. This corresponds to about 10^{-3} mol/h. The intensity per pulse is much higher than that of a conventional nozzle beam.

Figure 2.9 shows in its left half another example of a laser ablation source utilizing both gas aggregation and gas dynamic expansion [Sugai and Shinhara (1997)]. The central part of this source is the evaporation and aggregation chamber (1) with a rectangular sample plate (2) containing the target material (3). This plate can be moved in z and y directions by two stepping motors. During operation, the sample plate is periodically displaced in a zigzag motion, so that each laser pulse finds an unused part of the surface. The preload valve (5) controls the aggregation gas supply, while the main valve (6) produces the pulsed carrier gas beam. The

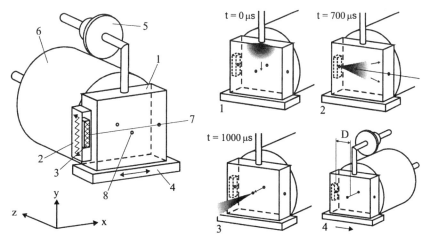

Fig. 2.9. Laser ablation source for large clusters [Sugai and Shinhara (1997)]. Left half: (1) evaporation chamber, (2) movable plate with a cavity (3) for the target material, (4) translation stage for the evaporation chamber, (5) preload valve, (6) main valve, (7) laser beam, (8) source aperture. Right half: Different steps of operation (see text)

laser beam enters the evaporation chamber through a lateral hole (7), the seeded cluster beam leaves the source through the aperture (8) in the front plate of the evaporation chamber. Three different steps of source operation are illustrated in the right half of Fig. 2.9:

(1) At first (t = 0), the aggregation gas, typically helium, is admitted to the evaporation chamber through the preload valve up to pressures between 1 and 10 torr.

(2) Then (t = 700 µs), the YAG laser (532 nm) hits the target. The evaporated particles are cooled in collisions with the aggregation gas atoms and cluster aggregation takes place, until this process is interrupted by the next step:

(3) (t = 1000 µs). Now the main valve is opened and a gas dynamic helium beam (stagnation pressure 10 atm) expands (in the z direction) together with the clusters through the chamber aperture.

By shifting the evaporation chamber in the x direction, the distance D between target and carrier gas beam can be varied (see (4) in Fig. 2.9). In addition to the time delay between step (2) and step (3), the distance D is a further parameter to influence cluster formation.

2.4.3 Pulsed Arc Discharges

Replacing the laser by a pulsed arc discharge, as is schematically shown in Fig. 2.10, similar cluster sources can be realized [Ganteför et al. (1990), Siekmann et al. (1991), Cha et al. (1992), Lu et al. (1996), Seeger et al. (1998)]. The pulsed arc is initiated by discharging a capacitor which is charged and discharged by delayed triggering of two thyristors. The electrode material evaporated by the discharge is clustered in the following carrier gas expansion from a pulsed valve. These sources, which can produce both neutral and ionized clusters, are named "pulsed arc cluster ion source" (PACIS).

As is illustrated in Fig. 2.10, the electrodes (1) and (2) are mounted in an insulator block (3) made of boron nitride or Macor. This block is supported by two aluminum plates (4). One of these plates supports the pulsed carrier gas source (5), the other one supports a "mixing chamber" (6) and the divergent nozzle (7) (30 mm long, 15° total cone angle) for final expansion.

The discharge (duration about 10 μs) is fired during the 0.5–2 ms long carrier gas pulse, the resulting plasma is flushed through a channel (10 mm long, 2 mm ∅) into the mixing chamber, which serves for thermalization and aggregation while effectively mingling the hot helium/plasma mixture with the colder portions of the carrier gas.

Finally, the resulting mixture is again compressed to a channel (1.5 mm ∅, length 5 mm) and expanded (and cooled) through a divergent nozzle.

The advantages of pulsed arc discharges compared to laser ablation are a higher cluster intensity per pulse and a higher repetition rate than laser ablation sources operated with commercial excimer or Nd:YAG lasers.

These advantages are achieved at a cost which is only a fraction of that of a laser ablation source. A substantial difference from laser ablation, however, is caused by the large amount of energy delivered from the discharge to the carrier gas. Accordingly, more care must be taken with the cooling of the metal–carrier gas mixture. Cha et al. (1992), therefore, use long divergent nozzles or long channels with water-cooled or even liquid nitrogen-cooled walls. The intensity of the neutral clusters increases with increasing length of the cooling zone, since ionized clusters are neutralized by charge exchange collisions. Thus, high yields of cluster ions require a compromise between cooling and neutralization.

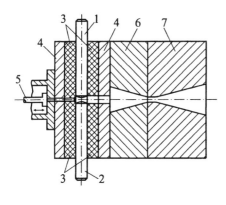

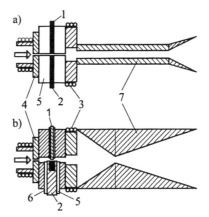

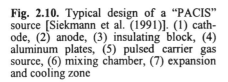

Fig. 2.10. Typical design of a "PACIS" source [Siekmann et al. (1991)]. (1) cathode, (2) anode, (3) insulating block, (4) aluminum plates, (5) pulsed carrier gas source, (6) mixing chamber, (7) expansion and cooling zone

Fig. 2.11. Two "PACIS" sources [Cha et al. (1992)]. a) for conducting materials, b) for alkali metals and liquids. (1) anode, (2) cathode, (3) cooling, (4) pulsed carrier gas source, (5) insulation, (6) metal block, (7) mixing chamber and cooling zone

Only the cathode is eroded by the arc discharge, indicating that the essential mechanism of erosion is ion sputtering. Semiconducting materials, liquid metals, or mixtures of two species can be used, if the cathode is hollowed out and filled with the material to be used for cluster formation [Cha et al. (1992), Lu et al. (1996)].

Figure 2.11 shows two variations of PACIS sources [Cha et al. (1992)]: a) has a very long channel, where cooling is obtained using water or liquid nitrogen; b) shows a device with a hollow cathode containing the material to be evaporated. It may be used for semiconductors, liquids, and mixtures of several components.

2.4.4 Liquid Metal Ion Sources (LMIS)

Another method for production of high intensities of atomic and cluster ion beams is based on field emission and ionization occurring from a liquid metal which wets a needle-shaped electrode in a high electric field. Figure 2.12 illustrates the principle [Clampitt and Jefferies (1978), Saunders and Fredrigo (1989)]. The needle consists of a filament tip of a refractory metal, which is electrochemically edged to achieve quite a small radius of curvature at the tip. Often, tungsten tips are used, but the choice of the needle material depends on the liquid metal to be used. After removing any metal oxide skin from the tip, it is dipped into a molten reservoir of the liquid metal of interest. Applying a voltage (3 to 10 kV) between the needle and a counter electrode with a central hole, field ion emission occurs, and an intense ion beam is obtained which consists both of atomic ions and singly or multiply charged cluster ions. These sources are designed to operate in vacuum and do not use a carrier gas.

Important for a successful operation is the ability of the liquid metal to wet the needle, since the wetting process maintains the material transport to the tip. Moreover, the liquid metal must not corrode the tip or alloy with it. Elements or alloys which easily sublime are also unsuitable since their vapor pressure would create a discharge. Elements such as Cs, Ga, In, Ge, Sn, Pb, Bi, Au, and Al fulfill the above discussed requirements. Other elements, which cannot be used directly because their melting point or their vapor pressure is too high, may be used by alloying them with another carrier metal. For example, to produce cluster ions from phosphorus, Cu–P alloys [Clark et al. (1987), Ishitani et al. (1984)] have been used. Another example is cobalt clusters which have been generated by using Co–Nd alloys [Bischoff et al. (1996)].

In some cases, mixed clusters are also emitted when using alloys. PdAsB alloys, for instance, which have been used to generate arsenic and boron clusters will also produce compound clusters such as $AsPd^+$, PdB^+, As_2Pd^+, and Pd_2As^+ [Clark et al. (1987)]. Instead of liquid metals, conducting solutions such as water–methanol mixtures have also been used [Mahoney et al. (1994)]. The energy spread of the emitted cluster ions is rather broad (100 to 160 eV) [Barr (1987), van de Walle and Joyes (1987)].

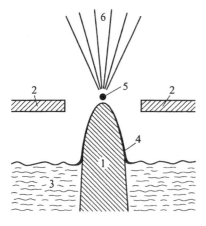

Fig. 2.12. Schematic of a liquid metal ion source. (1) needle-shaped electrode, (2) counter electrode, (3) liquid metal, (4) metal film, (5) plasma ball, (6) emitted cluster ions [Clampitt and Jefferies (1978)]

Many measured energy distributions show a bimodal structure, which may be explained by assuming two different processes for cluster production. One set of clusters arises from highly charged large droplets which "explode" almost immediately after leaving the tip. The other clusters stem from fragmentation of larger metastable clusters during their flight time to the analyzer. In either case, the internal energy of the clusters is expected to be extremely high since they are created in explosions of highly excited species.

Since the effective area of the emitter tip is very small (about 50 nm), these ion sources yield beams of high brightness (of the order of 10^{24} to 10^{25} ions/s sr cm^2) in comparison to conventional ion sources, which can be focused to spot sizes as low as 10 nm and are used in focused ion beam systems with energies up to 100 keV [Adamczewski et al. (1997)].

These devices are similar to scanning electron microscopes, except that a beam of ions is scanned across the sample. They have found many applications; these include compositional imaging via secondary electrons, direct etching of material, micromachining, transmission electron microscopy specimen preparation, localized deposition and implantation of metal and insulator structures, and ion beam lithography. For these applications, liquid metal ion sources are commercially available for various ions such as Ga, In, Au, Si, As, Ge, and others.

2.5 Laser-Induced Pyrolysis

An alternative method to produce large clusters of refractory materials is based on laser-induced pyrolysis of appropriate gaseous precursors containing the desired element in a flow reactor. This technique, commonly referred to as chemical vapor deposition, is frequently used to produce ultrafine particles (with diameters smaller than 100 nm) and thin films [Cannon et al. (1982), Batson and Heath (1993), Tamir and Berger (1996)]. When combined with a supersonic expansion of the nascent clusters into a high vacuum molecular beam apparatus, a cluster source is obtained, which yields intense beams of clusters over a large range of sizes (with

N varying from a few tens to several thousands) [Ehbrecht et al. (1995), (1997a,b), Ehbrecht and Huisken (1999)]. For a large number of species, suitable gaseous precursor molecules containing the atoms to be clustered are available. In the case of carbon clusters, for example, hydrocarbon molecules can be used, while for the generation of silicon clusters any of the various silicon hydrides (silanes) such as SiH_4 are well suited precursors. Many metal clusters can be formed by laser decomposition of their penta- or hexacarbonyls. This technique is also capable of producing beams of mixed clusters, such as silicon carbide (SiC) clusters by pyrolysis of silicon- and carbon-containing molecules such as silane and hydrocarbons, respectively [Huisken et al. (1999a)]. The precursor molecules are heated by the absorption of pulsed or CW CO_2 laser radiation and excited beyond their dissociation limit. If the precursor molecules are transparent in the frequency range of the CO_2 lasers (such as C_2H_2) an absorbing gas may be used as a sensitizer (SF_6, for example). After decomposition of the precursor molecules, a zone of supersaturated vapor of the dissociation products is obtained within the reaction volume. These are effectively cooled by collisions with rare gas atoms so that condensation and cluster growth becomes possible. The clusters are extracted from the reactor through a conical nozzle, and pass through a skimmer and a differential pumping stage into the beam utilization chamber.

This method offers a variety of possibilities not available in the case of the cluster sources discussed before. Both pulsed and CW CO_2 lasers can be used, yielding pulsed or continuous cluster beams, respectively, as has been demonstrated in the case of carbon clusters [Ehbrecht et al. (1993a)]. Heterogeneous clusters can be easily achieved by mixing the precursor gases. The parameters determining the composition and the growth of the clusters, such as, for instance, the concentration of the different gases, the reaction pressure, and the laser power can be easily adapted to the requirements of a given experiment. Furthermore, there is a strong correlation between cluster mass and cluster velocity, so that a narrowing of the size distribution of the clusters by means of a velocity selection becomes possible [Ehbrecht et al (1997a)].

2.5.2 Multiphoton Infrared Dissociation and Photosensitization

The first step of the above-described cluster formation process is the dissociation of the precursor molecules. This is achieved by absorption of many identical photons from the radiation of an infrared laser (multiphoton dissociation). For a qualitative understanding, Fig. 2.13 shows schematically the potential energy as a function of the normal coordinate q of a given molecular vibration. At low energies, a polyatomic molecule has discrete vibrational and rotational states, and due to the anharmonicity of the vibration it usually requires fairly intense laser radiation to excite a molecule over these discrete states via a coherent, resonant absorption process. The anharmonicity causes an unequal spacing between the vibrational levels in a polyatomic molecule; the spacings become closer together as one goes to higher vibrational levels. This poses a problem in the multiphoton excita-

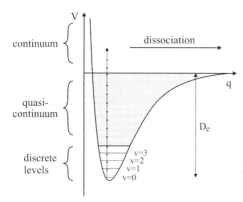

Fig. 2.13. Schematic illustration of multi-photon dissociation. For a description see text

tion since a highly monochromatic laser tuned to the lowest vibrational transition ($v_0 \to v_1$) gets out of resonance for higher transitions. Various mechanisms to overcome this problem, which is extremely sensitive to the laser power, frequency, and bandwidth, have been proposed. Using a laser wavelength that is red-shifted with respect to the lowest vibrational transition, for example, a P-branch ($\Delta J = -1$) transition may be used for the first photon, whereas a Q-branch ($\Delta J = 0$) and a R-branch ($\Delta J = +1$) transition may be used for the second and the third photon, respectively.

The density of states increases very rapidly with increasing energy, and once the required number of photons (typically 3 to 4) are absorbed, the molecule reaches a region called quasicontinuum (shaded area in Fig. 2.13). In this regime, there is such a high density of states, that the anharmonicity problem vanishes since there is always a rotational–vibrational level in resonance with the monochromatic laser light. Once in the quasicontinuum — provided that sufficient laser power is available — the molecule can continue to absorb energy up to and beyond the dissociation threshold.

The absorption of the first photons is only possible for those molecules which are in appropriate initial states. Only a fraction of the molecules, therefore, reaches the quasicontinuum and follows the excitation ladder up to the dissociation continuum. To increase the dissociation yield, an additional mechanism is required to bring the molecules into suitable initial states or into the quasicontinuum. In the gas phase of the reactor, this is achieved by inelastic intermolecular collisions. Collisional broadening of the energy levels compensates for the detuning between laser- and molecule frequencies, an energy-transfer collisions may directly excite molecules to the quasicontinuum.

The importance and effectiveness of energy-transfer processes for the excitation into the quasicontinuum becomes particularly evident when using molecules which cannot be brought to the quasicontinuum by direct absorption of radiation. In these cases, the admixture of an absorbing gas or vapor is sufficient to collisionally transfer the excitation energy of the absorbing molecules to the nonabsorbing molecules and to transfer them into the quasicontinuum. Once in the quasicontinuum, they can absorb energy from the laser field and finally dissociate. This process is called "photosensitization".

2.5.3 Source Design and Applications

Figure 2.14 shows a schematic of the flow reactor (1) [Ehbrecht et al. (1995), (1997a)]. It is built from a standard stainless steel cross vacuum fitting (with six flanges of 40 mm ∅) and incorporated into the source chamber of a molecular beam machine. The reaction gas is admitted to the center of the flow reactor through a 3 mm diameter stainless steel tube. In order to confine the reaction gas to the flow axis, rare gas is flushed as a buffer gas through an outer tube of 12 mm diameter, concentric about the reaction gas inlet tube. All gases and reaction products are pumped by a 4.5 m^3/h mechanical pump through a funnel-shaped tube opposite to the gas inlet tubes. The pumping speed can be adjusted by a needle valve. A filter in the pump line collects solid particles during operation.

The reaction leading to cluster formation is initiated by the infrared radiation of a CO_2 laser. Both pulsed and CW lasers may be used. The radiation is coupled into the reaction zone via a molybdenum mirror and a NaCl window and focused to the flow axis by means of a ZnSn lens. This window as well as the exit window for the laser beam are continuously flushed with rare gas to prevent solid particles from coating them.

The flow velocities and concentrations are regulated and controlled independently with a flow meter system. The reaction gases are mixed in a separate vessel before entering the flow reactor. Typical operating conditions are: reaction gas flow rate: 20 sccm (standard cm^3 per minute) ; buffer gas flow rate: 1100 sccm; total pressure: 350 mbar; pulse energy of the CO_2 laser: 50 mJ.

The reaction products are extracted perpendicularly to both the gas flow and the CO_2 laser beam through a conical nozzle (25 mm long, 0.3 mm ∅, opening angle 60°). After a flight path of 26 mm, the beam is collimated by a skimmer (aperture 0.5 mm ∅) and passes through a differential pumping stage to the beam

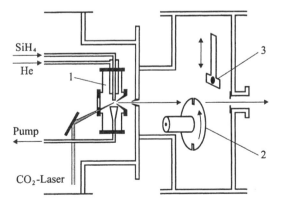

Fig. 2.14. Laser reaction source [Ehbrecht et al. (1997a)]. (1) Flow reactor, (2) chopper wheel for time-of-flight measurements, (3) surface for cluster deposition

utilization chamber (not shown in Fig. 2.14). The reactor is mounted on an x,y translation stage and can be moved in two dimensions perpendicular to the molecular beam axis for adjustment during operation. The differential pumping stage contains a beam chopper (2) for time-of-flight measurements. Clusters may be deposited on a substrate (3) by moving the sample holder into the beam.

When the reactor is operated under conditions where large particles are produced, the nozzle may become clogged after a short time (~ 10 min). In such cases, the 532 nm radiation of a frequency-doubled Nd:YAG laser is focused through a window in the rear of the flow reactor onto the nozzle to keep it open or to reopen it when clogged during operation.

Laser-induced pyrolysis has been used so far to produce carbon, silicon, and iron clusters and nanoparticles, as well as clusters from silicon carbide and silicon nitride [Ehbrecht et al. (1995), (1996), Ehbrecht and Huisken (1999)]. Carbon clusters and fullerenes are produced by using a gaseous mixture of C_2H_2 and SF_6, where SF_6 is used as a sensitizer. The precursor for silicon clusters is silane (SiH_4), which absorbs the laser radiation and thus does not require a sensitizer. Iron clusters are obtained from laser pyrolysis of iron pentacarbonyl ($Fe(CO)_5$) using SF_6 as a sensitizer. In this case, direct laser photolysis of $Fe(CO)_5$ by UV radiation is also possible, using either a Nd:YAG laser (3rd and 4th harmonics, 255 or 266 nm, respectively) or an ArF excimer laser (193 nm) [Huisken et al. (1999c)].

When using a CO_2 laser, besides pure iron clusters, fluorinated iron clusters have also been observed, indicating that the SF_6 sensitizer molecules were partially decomposed by the intense laser field. As a contrast, high purity iron clusters are found under conditions where no sensitizer was required, namely with the fourth harmonic of the Nd:YAG laser. Higher photon energies, however, delivered from an ArF laser, result in a partial dissociation of the carbonyl group and the subsequent oxidation of the iron clusters to Fe_NO or Fe_NO_2. Silicon carbide clusters and silicon nitride clusters, finally, are obtained by admixing C_2H_2 or NH_3 into the reaction gas silane (SiH_4) [Huisken et al. (1999c)].

Laser-induced pyrolysis in a gas flow reactor is one way to synthesize clusters in the region of 2 to 5 nm, which find an increased interest in material science, motivated by the possibility of designing nanostructured materials with novel electronic, optical, magnetic, chemical, and mechanical properties.

2.6 Doping of Clusters and Production of Mixed Clusters

Clusters consisting of two (or more) different constituents are called mixed clusters. A special case are large clusters (host clusters) with only one or a few foreign atoms or molecules, which may stick to their surface or penetrate into their interior. These are called doped clusters. The foreign particles may also coagulate to form small clusters (guest clusters) on or within the host clusters. The latter have been used to study solvation processes. Another application is high-resolution molecular spectroscopy of small van der Waals clusters embedded in large liquid helium clusters at very low temperatures [Hartmann et al. (1995), (1996), Bartelt et al. (1996), (1997)].

2.6.1 Formation of Mixed Clusters by Coexpansion

A simple possibility to generate mixed clusters is the coexpansion of two premixed gases or vapors through a nozzle with or without a carrier gas [Penner and Amirav (1993), Desai et al. (1994)]. When the foreign atoms or molecules are coexpanded with the clustering gas, they may be used as condensation nuclei for the latter and will be found inside the host clusters [Alexander et al. (1988), Papanikolas et al. (1991)].

2.6.2 Gas Aggregation

This technique has been widely used to produce mixed clusters [Sattler (1985)]. In many cases the desired chemical compound can be evaporated from an oven into the cold aggregation gas (CuBr, for example, to generate $(CuBr)_N$ clusters or arsenic sulfide As_2S_2 or As_2S_3 to produce As_NS_M clusters) [Martin (1984)]. Another possibility is the evaporation of the two components from separate ovens and the common condensation in a rare gas flow [Martin (1986)]. This method has the advantage, that the composition of the mixed clusters can be influenced by the partial pressures of the two vapors. If a gaseous species is used, it may be simply added to the cooling gas to obtain mixed clusters. Antimony, for example, has been evaporated into a He/O_2 or He/NO_2 mixture and subsequently cooled in a reaction zone to generate Sb_NO_M clusters [Kinne et al. (1997)]. In cases where the initial species already evaporates in the form of clusters, such as copper halide CuX, which evaporates as Cu_3X_3 and Cu_4X_4, two-stage ovens have been used for evaporation. In the first stage the species is heated to the temperature required for evaporation, while in the high-temperature, second stage the aggregates are thermally dissociated [Martin and Kakizaki (1984)].

2.6.3 Particle Capture ("Pick-up" Sources)

If a cluster beam is crossed by an effusive atom or molecule beam, or if the cluster beam passes through a gas-filled chamber, the clusters may pick up one or more dopant atom or molecule and form mixed or doped clusters. If the clusters are produced in a supersonic expansion, for example, and if the interaction occurs close to the nozzle within the expansion, the foreign particles can become nuclei of condensation so that they are finally situated inside the cluster. However, if the interaction occurs far away from the nozzle, where the formation of the host clusters (or substrate clusters) has already terminated, there is a high probability that the guest particles (or chromophores) become attached to or adsorbed on the surface of the host clusters. Because of the large mass of the host clusters, they are deflected only negligibly from their original trajectory by the inelastic collisions. The

98   2. Production and Diagnostics of Cluster Beams

excess energy is dissipated within the cluster and eventually released by evaporating one or a few cluster constituents.

This procedure to dope clusters with foreign atoms is called the "pick-up" method and was introduced by Gough et al. (1985). If the density of the foreign atoms is sufficiently high, two or more guest particles may be picked up by the host clusters and combine to form dimers [Levandier et al. (1990)] or small guest clusters on the surface of the substrate clusters [Huisken and Stemmler (1993)].

Figure 2.15 shows schematically an example of a pick-up source [Stienkemeier et al. (1996)]. A beam of large helium clusters is prepared by supersonic expansion from a cold nozzle (1) (10 μm ⌀, stagnation pressure $P_0 = 50$ bar, nozzle temperature 12 K, and N ≈ 5000 atoms/cluster). Since the He–He binding energy is very small (see Sect. I.3.6.4), high stagnation pressures are required, resulting in a high gas throughput despite the small nozzle diameter and the low gas temperature. Thus, the source chamber is evacuated by a pump (P1) of high pumping speed (32 000 l/s). The He cluster beam is chopped (3), skimmed, and passed through a pick-up cell (4) where a low (of the order of 10^{-4} mbar) vapor pressure of alkali metal atoms can be established. In this cell the heavy helium clusters adsorb alkali atoms without considerable deflection. Upon capturing a hot alkali atom, the cluster rapidly evaporates He atoms from its surface, and its temperature returns to the initial cluster temperature (about 0.4 K) [Hartmann et al. (1995)].

Optical excitation spectra of the doped clusters are obtained by scanning the frequency of a CW dye laser, which is fed into the apparatus through a single-mode fiber. The laser beam (7) crosses the cluster beam at the center of a laser-induced fluorescence detector (6) (see Sect. I.5.9.2) equipped with a photomultiplier (10) and a monochromator (11). A Langmuir–Taylor detector (9) is located further downstream of the cluster beam.

For helium clusters, which are the only clusters which are definitely liquid, the question whether the foreign atoms are attached to the surface of the host clusters or embedded inside has been studied theoretically. In a simple model the foreign

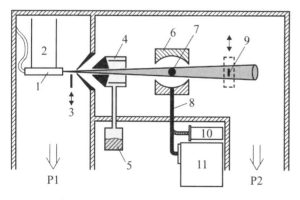

Fig. 2.15. "Pick-up" source [Stienkemeier et al. (1996)]. (1) He nozzle source with cold trap (2), (3) beam chopper (tuning fork), (4) pick-up cell, (5) alkali metal oven, (6) elliptical mirror, (7) laser beam (perpendicular to the drawing plane), (8) glass fiber bundle, (9) Langmuir–Taylor detector, (10) photomultiplier, (11) monochromator, P1 and P2 pumps

particle is assumed to interact with the helium environment via a Lennard–Jones pair potential characterized by a well depth ε and an equilibrium distance r_m [Ancilotto et al. (1995)]. The tendency of a foreign particle to be solvated in helium is expressed in terms of a dimensionless parameter λ which is essentially the ratio of the attractive potential energy to the surface energy of the bubble surrounding the guest particle

$$\lambda = n \frac{\varepsilon r_m}{2^{1/6} \sigma}, \qquad (2.6)$$

where σ is the surface tension of ^{4}He, and n is the number density. For λ > 1.9, foreign atoms and molecules localize at the surface, while for λ < 1.9 they are embedded inside the clusters. This has been found to be in agreement with the available experimental results [Bartelt et al. (1996)], based on spectroscopic investigations of the chromophore particles [Gu et al. (1990)]. In the case of molecules, the rotational temperature of the embedded or attached molecules can be measured, yielding a method to determine the cluster temperature [Hartmann et al. (1995)]. The spectroscopy of small guest clusters adsorbed on large host clusters, finally, yields information on their structure, which may be different from that of the free clusters [Huisken and Stemmler (1993)].

Large rare gas clusters (He_N, Ar_N) have been used so far as substrate clusters [Goyal et al. (1991), Scheidemann et al. (1993), Hartmann et al. (1996)]. Apart from various molecules and small molecular clusters (CH_3F, SiF_4, CH_3N, CF_3Cl, SF_6, $(SF_6)_N$, $(CH_3OH)_N$, $(CH_3F)_N$) atoms such as Ar, Ne, alkali metal, silver, indium, and europium atoms have been used as chromophores for absorption and emission spectroscopy [Visticot et al. (1992), Bartelt et al. (1996), (1997)]. Chemical reactions (Ba–N_2O) have also been investigated on the surface of substrate clusters [Lallement et al. (1992)].

Comparing the electron impact ionization of guest molecules and clusters on large substrate molecules with that of the corresponding free molecules and clusters, a different fragmentation behavior is observed. The fragmentation of SF_6^+ ions, for instance, is substantially reduced by the helium host clusters [Scheidemann et al. (1993)]. In the case of $(CH_3OH)_N$ and $(CH_3F)_N$ clusters on large argon substrate clusters the protonation reactions observed in free clusters are completely suppressed [Ehbrecht et al. (1993b)], since the excess energy is used to evaporate the host cluster. Instead, the parent ions of the clusters $(CH_3OH)_N^+$ and $(CH_3F)_N^+$ (N = 2 and 3) are observed, which do not appear in the mass spectra of the corresponding free clusters.

2.6.4 Cluster Aggregation

The capture of foreign atoms and the formation of guest clusters on or in large rare gas clusters described in the foregoing section may also be used as a source for new clusters, for instance for clusters from refractory metals [Rutzen et al. (1996)].

Large rare gas clusters are produced in a gas dynamic expansion, and the cluster beam is passed through a pick-up chamber (or an effusive beam) containing a given density of atoms. The rare gas clusters capture successively a rather large number of foreign vapor atoms, which in turn coagulate to form guest clusters on or within the rare gas host clusters. Their heat of condensation is delivered to the host clusters, which, if the guest clusters are sufficiently large, evaporates completely. Hence, a low-density vapor is effectively transferred into clusters.

2.6.5 Laser Ablation

To produce mixed clusters by laser ablation, either alloys or chemical compounds are used as targets, or the foreign particles are added before or during the carrier gas expansion. These methods have been employed in numerous cluster chemical investigations [Rohlfing et al. (1984), Richtsmeier et al. (1985), Geusic et al. (1985), Whetten et al. (1985), Liu et al. (1998)]. Laser ablation sources utilizing two targets and two laser beams, in which the vaporized materials are expanded through a common channel, have also been used [Nakajima et al. (1993), (1996)]. By adjusting the intensities of the two laser beams, bimetallic clusters with different composition have been produced [Wagner et al. (1997), Bouwen et al. (2000)]. Metal carbide clusters have been generated by ablation of carbon targets coated (by vacuum deposition or electrochemical techniques) with the desired metal [Reddic and Duncan (1997)].

Another method to generate mixed clusters, which has some similarities to the "pick-up" technique, utilizes a combination of a laser ablation source and a flow reactor [Kurikawa et al. (1995), (1997), Kaya et al. (1996)]. Figure 2.16 presents a schematic drawing of such a device. From a conventional rotating and translating rod (3), material is vaporized by laser irradiation (4) and cooled down in a carrier gas (helium) flow (from a pulsed valve (1)) through a channel. Then the clusters enter a flow reactor, where they collide with molecules (benzene, for example) admitted together with a carrier gas by a second pulsed valve (2), which is synchronized with the laser pulse and the first carrier gas pulse. Accordingly, mixed clusters $M_N(C_6H_6)_M$ (M = V, Ti, Cr, and Co), are formed in the flow reactor, which finally pass through a skimmer (5) and a differential pumping stage into the beam utilization chamber. Similar devices with two flow reactors in series have been used to deposit further molecules onto the mixed clusters. In this manner,

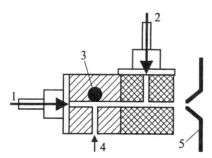

Fig. 2.16. Laser ablation source with a flow reactor. (1) and (2) pulsed sources, (3) target rod, (4) laser beam, (5) skimmer

$V_N(C_6H_6)_M(CO)_K$ clusters (with N, M, and K < 10) have been produced. An analogous method utilizes a laser ablation source with a long cooling channel similar to those shown in Fig. 2.11. The reaction gas is admitted to the channel (which serves as the flow reactor) at a location where the thermalization of the clusters has been terminated [Hintz and Ervin (1994), Schulze Icking-Konert et al. (1996), Parks et al. (1996)]. By an appropriate choice of the partial pressure of the reaction gas (CO, for example), doped clusters can be generated, which have one or more CO molecules (up to a full layer) chemisorbed on their surface.

The reactive gas may also simply be added to the carrier gas to generate mixed clusters during the expansion. This method has been used, for example, to produce Li_NO and $(Li_2O)Li_N$ clusters with a laser ablation source [Lievens et al. (1997)], adding 0.6 ppm H_2O or 0.1 ppm O_2 to the carrier gas helium.

2.6.6 Pulsed Arc Discharges

As already mentioned in Sect. 2.4.3, hollow cathode rods filled with mixtures of materials can be used to generate mixed clusters in pulsed arc discharges. Composite cathodes, consisting of mixtures of graphite with metals or metal oxides, for example, have been used to produce fullerenes and carbon clusters with built-in metal atoms [Kikushi et al. (1993), Shinohara et al. (1996), Kojima et al. (1996)].

2.7 Generation of Excited Clusters

When atoms diffuse through a carbon foil at high temperature, they may leave the foil surface in an excited state. Additionally, a large fraction of the diffused species leaves the foil as highly excited clusters [Åman and Holmlid (1991)]. This has been observed, for example, in the diffusion of Cs atoms through foils of graphite [Holmlid et al. (1992)]. In this case it is assumed that thermal ion emission from the foil and subsequent recombination with a thermal electron outside the surface is responsible for the excitation. This assumption is based on the fact that the energy barrier connected with the diffusion through the outer layer of the graphite surface agrees well with the energy of the Rydberg states at large distances from the surface [Möller and Holmlid (1988)].

The cesium is heated to temperatures between 400 and 700 K in a conventional, resistively heated source, which is closed by a graphite foil (0.125 mm thick, 4 mm ∅) instead of having an aperture. The foil is kept at 1100 to 1500 K by radiation heating. Under these conditions, excited clusters $(Cs)_N^*$ with broad distributions of N (the average value of N is several thousand), and excited Rydberg atoms Cs^* are emitted. The excited clusters can be easily detected by field ionization at field strengths of less than 400 V/cm. Ion flux densities measured in this way are of the order of 10^{15} ions/cm^2.

2.8 Determination of Size Distributions in Cluster Beams

All cluster sources described in the foregoing sections yield, essentially due to the statistical processes during cluster formation, a distribution of various cluster sizes. These size distributions may also exhibit relative maxima or minima, which can be interpreted in terms of thermodynamic stability or instability of special cluster sizes and the corresponding configurations of their constituents. To obtain size-specific information about the properties of neutral clusters, it is highly desirable to separate the clusters according to their mass before or during a given experiment. At least, the average cluster size in a given beam and the size distribution should be known. The selection of clusters of a definite size, however, is still an unsolved problem. Solutions are only available in a few special cases. Just like this, no standard method exists to analyze the size distribution of a given cluster beam. A large variety of methods has been applied, and are discussed in the following sections.

2.8.1 Electron Diffraction

To investigate the diffraction of electrons from cluster beams, well collimated electron beams (100 µm ∅ in the plane of detection) with energies of the order of 50 keV (and electron currents of a few µA) are used. The electrons, which traverse the cluster beam at right angles, yield a diffraction pattern which is recorded on a photographic film [Raoult and Farges (1973)]. Since the scattered intensity decreases drastically with increasing angle, a densiometric analysis of superimposed interference structures is impossible. To compensate the rapid intensity drop, therefore, a rotating sector of special shape is placed between the diffraction zone and the detecting film. This is a standard technique in gas electron diffraction [Raether (1957)]. The rotating sector is shaped in such a manner that the time of film exposure increases with increasing scattering angle according to the decrease in the diffracted intensity. Due to this compensation, the interferences occur on a rather constant background and can be accurately evaluated densiometrically.

We consider first two limiting cases, which can be treated by well known methods: The electron diffraction by atoms and the diffraction by large clusters.

The scattering amplitude A_s of electrons scattered by a single atom is given by [see Messiah (1969), for example]

$$A_s = A_0 \frac{2me^2}{\hbar^2 q^2}[Z - F(q)]. \qquad (2.7)$$

m and e are the rest mass and the charge of the electron, respectively, A_0 is the amplitude of the incident wave, $\hbar = h/2\pi$ Planck's action quantum, Z the nuclear charge, and F(q) the form factor of the electron density, which describes the shielding of the nucleus by the electron cloud of the atom. q can be expressed by the scattering angle θ

$$q = 2k\sin(\theta/2), \qquad (2.8)$$

where $k = 2\pi/\lambda$ is the wave number and λ the de Broglie wavelength of the electron. Due to the factor $1/q^4$, the differential cross section

$$\frac{d\sigma}{d\omega} = \frac{4m^2 e^4}{\hbar^4 q^4}[Z - F(q)]^2 \qquad (2.9)$$

is a monotonous and rapidly decreasing function of the scattering angle (see lowest curve on the left side of Fig. 2.17).

If the electrons are scattered by many atoms instead by one atom (target number density n and scattering volume V), the differential cross section according to (2.9) must be simply multiplied by nV, provided that the individual scattering processes superimpose incoherently. This condition is fulfilled as long as the atomic density is not too high, or, more accurately, as long as the average distance between the atoms is large compared to their diameter.

On the other hand, the diffraction pattern recorded from large clusters ($\overline{N} \geq 1000$) consists of two components. One component is due to the scattering of the electrons by the atoms of the clusters, as discussed above, and shows no interference structure. A second component is superimposed due to the characteristic distances of the periodic crystalline structure, which causes a distinct interference pattern correlated with the Bragg crystal planes. The situation is rather similar to the case of electron diffraction from a powder of small, randomly oriented crystals (Debye–Scherrer rings). To evaluate the interference structure, well known crystallographic methods can be applied [Kittel (1976)]. From the position of the interference rings the crystal structure and the lattice parameter can be determined, while the width of the interference maxima is related to the cluster size. Finally, the damping of crystalline lines with respect to the diffraction angle is related to the mean atomic displacement and thus yields information on the cluster temperature [Farges et al. (1973), (1981), DeBoer and Stein (1981), Kim and Stein (1982), Torchet et al. (1990)].

Between these two limiting cases, which means for medium-sized and small clusters, the diffraction patterns observed are difficult to evaluate. The scattering amplitude is a superposition of contributions from the various atoms of the cluster and consequently given by

$$A = \sum_k A_k \exp(i(\mathbf{k}_e - \mathbf{k}_s)\mathbf{r}_k), \qquad (2.10)$$

where $\mathbf{r}_k$ are the position vectors of the cluster atoms which are numbered by the index k (referred to an arbitrary field point), $\mathbf{k}_e$ and $\mathbf{k}_s$ are the wave vectors of the incident and the scattered wave, respectively, and A_k is the scattering amplitude of the atom k. This expression yields for the scattered intensity

$$A^2 = \sum_i \sum_k A_i A_k \exp(i(\mathbf{k}_e - \mathbf{k}_s)\mathbf{r}_{ik}), \tag{2.11}$$

with $\mathbf{r}_{ik} = \mathbf{r}_i - \mathbf{r}_k$. Since the spatial position of the clusters is not fixed in space, a spatial averaging is required, which leads to

$$A^2 = \sum_i \sum_k A_i A_k \frac{\sin qr_{ik}}{qr_{ik}} \exp(-q^2 \langle L_{ik}^2 \rangle / 2). \tag{2.12}$$

For each distance r_{ik} an average fluctuation $\langle L_{ik}^2 \rangle$ has been assumed.

As an example, Fig. 2.17 shows measured angular distributions for electron diffraction from argon nozzle beams at different stagnation pressures P_0 in the source. The rotating sector yields the measured quantity

$$I(q) \propto q^3 \frac{d\sigma}{d\omega}. \tag{2.13}$$

The two limiting cases discussed above can be clearly distinguished: At a stagnation pressure of $P_0 = 0.8$ bar, the beam contains only argon atoms and the structureless pattern I(q) observed represents Ar atomic scattering according to (2.9). At high pressure ($P_0 = 15$ bar), however, large clusters are formed in the beam producing crystallographic patterns with sharp lines which can be uniquely associated with the crystalline structure of the clusters.

Intermediate patterns show broader structures in the angular distribution I(q) with smaller amplitudes and cannot be accounted for by crystalline models. The line broadening is due to the small, finite size of the clusters, while the low amplitudes are a consequence of the thermal motion of the cluster atoms and a larger fluctuation of their distances in noncrystalline aggregates. A quantitative interpretation of the observed patterns in terms of cluster size and structure can only be obtained by elaborating noncrystalline models. Such models may follow from molecular dynamics simulations, or they are constructed on the basis of a homogeneous atom distribution of given geometry. The latter procedure is usually used, since molecular dynamics simulations are very time-consuming. The cluster size N, the distances r_{ik} and their average fluctuation L_{ik} (see (2.12)) are used as parameters. For a given set of parameters, the diffraction pattern is calculated according to (2.12). The experimental and computed patterns are then compared and optimized in a trial and error procedure until an acceptable fit is obtained. Cluster size, structure, and temperature of the so-determined model cluster is then assumed to be identical with the experimentally investigated clusters.

Correctly, this fitting procedure should take into account the size distribution in a cluster beam. Since the numerical effort of the above procedure is rather high, this has not been done so far. The calculations, therefore, are performed for a given cluster size which is assumed to be the average cluster size in the beam

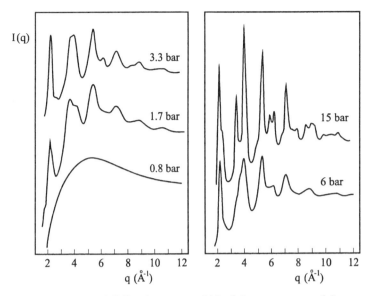

Fig. 2.17. Measured diffraction patterns I(q) of electrons scattered from an Ar nozzle beam at various stagnation pressures P_0. For details see text

[Farges et al. (1981), (1983), (1986), Lee and Stein (1985), (1987), Torchet et al. (1990), Kovalenko et al. (1998)].

2.8.2 Light Scattering

Already in 1967, Rayleigh scattering was used to investigate the condensation of water in a planar expansion of air and to determine the average radius of the clusters (20–60 Å) as a function of the specific moisture [Stein and Wegener (1967)]. The theory of Mie for the scattering of electromagnetic waves (of wavelength λ) from homogeneous, spherical particles (radius R) with an index of refraction n and the magnetic permeability $\mu = 1$ yields in the limit

$$\frac{2\pi R}{\lambda} \ll 1 \quad \text{and} \quad n\frac{2\pi R}{\lambda} \ll 1, \tag{2.14}$$

an angular independent differential scattering cross section [Newton (1966), Kerker (1969)]:

$$\frac{d\sigma}{d\Omega} = \frac{8}{3}\left(\frac{2\pi R}{\lambda}\right)^4 \pi R^2 \left[\left(\frac{n^2-1}{n^2+2}\right)\right]^2, \tag{2.15}$$

which has to be averaged over the cluster size distribution function f(N).

Since the differential scattering cross section for Rayleigh scattering is small under the above assumptions (2.14), Stein and Wegener (1967) used the intracavity radiation of a He–Ne laser (632.8 nm). Klingelhöfer and Moser (1972) studied very large hydrogen clusters (up to 10^9 atoms/cluster) by light scattering. Under these conditions the cluster size is comparable to the wavelength of the light (532.8 nm); accordingly, the differential cross section is angular dependent, as follows from Mie's theory [Denman et al. (1966)]. To evaluate the measurements of the angular distribution of the scattered light, a size distribution f(N) is assumed and the agreement between the experimental and computed differential cross section is optimized by trial and error.

Finally, Abraham et al. (1981) have used Rayleigh scattering to investigate the onset of clustering in nozzle expansions and to compare the results with those of nucleation theory.

As an example, Fig. 2.18 shows the intensity of the light scattered from the expansion of a pulsed nozzle as a function of the stagnation pressure P_0 for xenon and krypton beams [Ditmire et al. (1998a), (1998b), Smith et al. (1998)]. Vertically polarized 532-nm light from a frequency-doubled Nd:YAG lasers (100 µJ in a 10-ns pulse) is focused onto the beam (6 nozzle diameters downstream of the nozzle). The 90° side-scattered light is detected by a photomultiplier fitted with a narrow-band, 532-nm interference filter. A convergent-divergent nozzle (d = 500 µm, $\alpha = 45°$) at a stagnation temperature of $T_0 = 295$ K is used.

The experimental results indicate that clusters of a detectable size begin to form at pressures of about 1000 mbar in xenon and 2000 mbar in krypton. This corresponds to a value of the scaling parameter $\Gamma^* \approx 2400$ (see (I.3.166) and (2.4) and Tables I.3.7 and 2.3).

Moreover, the scattered signal S_{RS} displays a power law dependence with increasing stagnation pressure (full lines in Fig. 2.18):

$$S_{RS} \propto P_0^k. \tag{2.16}$$

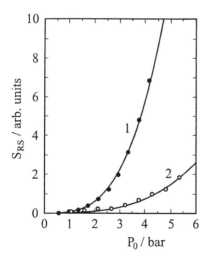

Fig. 2.18. Measured Rayleigh scattered light signal as a function of the stagnation pressure P_0 for Xe (1) and Kr beams (2) [Ditmire et al. (1998a,b)]. The full lines follow a power law (see text)

The exponent k is close to 3 (k = 3.3 for xenon and k = 3.2 for krypton). This is consistent with a simple scaling argument. The scattered signal S_{RS} is proportional to the product of cluster density n_C and Rayleigh scattering cross section. According to (2.15), the latter is proportional to the sixth power of the cluster radius R. Hence it follows

$$S_{RS} \propto n_{Cl} R^6. \qquad (2.17)$$

Assuming that all the atoms have condensed into clusters, the cluster density can be expressed by the monomer density n_M before clustering:

$$n_{Cl} \propto n_M / N. \qquad (2.18)$$

On the other hand, the monomer density before clustering n_M is proportional to the stagnation density n_0 in the source. Because $R \propto N^{1/3}$, we obtain finally

$$S_{RS} \propto n_0 N \propto P_0 N. \qquad (2.19)$$

Together with the empirical result (2.16) the cluster size N can be represented by

$$N \propto P_0^{k-1} \propto \Gamma^{*k-1}. \qquad (2.20)$$

With the experimental values for k given above (k ≈ 3) relation (2.20) shows excellent agreement with other experimental data [Wörmer et al. (1989), Hagena (1992)]. Accordingly, the Rayleigh scattering is a relative method to determine cluster sizes as a function of the stagnation pressure P_0 in the source or, more generally, as a function of the scaling parameter Γ^*. To determine absolute values of cluster sizes N, Rayleigh scattering results must be compared with other measurements of cluster sizes, for example with the measurements presented in Fig. 2.4.

In the above considerations we have tacitly replaced the size distribution of the clusters in a nozzle beam by a uniform cluster size, which is identified with the average cluster size.

2.8.3 Gas Scattering

The scattering of a cluster beam traversing a gas-filled target chamber can be used to determine the average cluster size in a beam. If a target beam which crosses the cluster beam under a given angle, is used instead of a target chamber, both the average cluster size and the size distribution can be determined if the clusters are large. In the case of microclusters, it is even possible to select clusters of a given size by scattering.

Various scattering methods have been proposed and experimentally investigated, ranging from simple total cross section measurements to investigations of the beam deflection due to scattering. They are discussed in detail in the following section.

Measurements of Total Cross Sections. The effective total cross section σ_{eff} (see Sect. I.2.1.3) is measured using the cluster beam attenuation in a target chamber according to Beer's law

$$I = I_0 \exp(-nL\sigma_{eff}), \qquad (2.21)$$

where I_0 is the intensity of the unscattered beam and I is the beam intensity attenuated by scattering; n is the target number density and L is the length of the target chamber. Assuming a total cross section σ_t which is independent of the relative velocity of the colliding particles, the effective cross section is given by

$$\sigma_{eff} = \sigma_t Fa_0(\infty, x). \qquad (2.22)$$

$x = v_{Cl}/v_w$ is the ratio of the cluster velocity v_{Cl} and the most probable velocity v_w of the target gas. $Fa_0(\infty,x)$ is a function which accounts for the averaging over the Maxwellian velocity distribution in the target gas (see Chap. I.2). Thus, the total cross section is given by

$$\sigma_t = \frac{1}{nLFa_0(\infty, x)} \ln(I_0/I). \qquad (2.23)$$

If the cluster is assumed to be a hard sphere, the geometrical cross section σ_g can be written as

$$\sigma_g = \pi R_N^2 = \pi R_1^2 N^{2/3}, \quad \text{where} \quad R_1 = \left(\frac{3m_0}{4\pi\rho}\right)^{1/3} \qquad (2.24)$$

is the effective radius of an atom within the cluster (see (2.1)). m_0 is the mass of a cluster constituent, ρ the density of the macroscopic liquid or solid phase. Assuming $\sigma_g = \sigma_t$, the cluster size N becomes

$$N = \left(\frac{\ln I_0/I}{nLFa_0(\infty, x)\pi R_1^2}\right)^{3/2}. \qquad (2.25)$$

With $N = \overline{N}$ a simple estimate of the average cluster size in a given beam is obtained. More accurately, the total cross section according to (2.22) contains a weight function due to the size dependent detection efficiency, which can be taken into account if the size distribution in the beam and the dependence of the detection probability on the cluster size is known [Schilling (1993)].

The total cross section can also be measured in a crossed beam experiment. If the target beam is monochromatic, the averaging function $Fa_0(\infty,x)$ in (2.22), (2.23), and (2.25) has to be replaced by g/v_{Cl}, where

$$g = \sqrt{v_{Cl}^2 + v^2 - 2vv_{Cl}\cos\alpha} \tag{2.26}$$

is the relative velocity of the two beams and α is the angle of intersection [Schilling (1993)]. In the case of a target beam, however, the determination of the product nL of target number density n and target length L requires an additional scattering experiment for calibration. This can be done by measuring the attenuation of a beam (according to Beer's law (2.21)) of atoms of known cross section (or known interaction potential) with respect to the target particles. This is the most accurate and simplest method to determine the product nL also in the case of a scattering chamber [van den Biesen (1988)].

Velocity Loss Measurements. Let us consider a cluster beam (cluster mass M, velocity vector $\mathbf{v}_{Cl}$), which traverses a gas-filled scattering chamber (mass of the gas particles m, velocity vector $\mathbf{v}$).

The momentum transferred to a cluster in a completely inelastic collision with a gas molecule (in which the gas molecule is absorbed by the cluster) is μg, where μ is the reduced mass and $\mathbf{g} = \mathbf{v} - \mathbf{v}_{Cl}$ is the vector of the relative velocity. This follows immediately by separating the center-of-mass motion from the relative motion of the collision partners. Accordingly, the velocity change $\delta \mathbf{v}_{Cl}$ of the cluster due to a collision (in magnitude and direction) is given by

$$\delta \mathbf{v}_{Cl} = \mu \mathbf{g}/M \approx m\mathbf{g}/M, \tag{2.27}$$

since $m \ll M$ is presupposed. The velocity loss δv_{Cl} of a cluster is determined by the component of the relative velocity vector $\mathbf{g}$ in the direction of the cluster velocity $\mathbf{v}_{Cl}$. This component (see Fig. 2.19) is $g\cos\vartheta$, where g is the magnitude of the relative velocity. Thus we have

$$\delta v_{Cl} = \frac{m}{M} g \cos\vartheta. \tag{2.28}$$

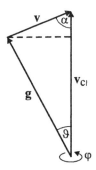

Fig. 2.19. Velocity space coordinates

2. Production and Diagnostics of Cluster Beams

If dW is the probability that a cluster of size N collides during the time interval dt with a gas molecule having the velocity **v**, the velocity loss of a cluster during the time dt

$$\Delta v_{Cl} = \delta v_{Cl} dW = \frac{m}{M} g \cos\vartheta \, dW. \tag{2.29}$$

The probability dW may be expressed by

$$dW = n\sigma_N g f(v) dv dt = n\sigma_N \frac{g}{v_{Cl}} f(v) dv dL. \tag{2.30}$$

n is the number density of the scattering gas, σ_N the total cross section for a cluster with N constituents, dL the path element traversed by a cluster during the time $dt = dL/v_{Cl}$, and f(v) is the distribution function of the translational energy according to (I.2.11).

Assuming a velocity-independent cross section σ_N, the total velocity loss on the cluster path L through the scattering chamber follows by integration of (2.30) over the distribution function f(v) and the path L:

$$\overline{\Delta v_{Cl}} = \frac{n\sigma_N}{v_{Cl}} \int_0^L \int \Delta v_{Cl} g f(v) dv dL = n\sigma_N L \frac{m}{M v_{Cl}} \int g^2 \cos\vartheta f(v) dv. \tag{2.31}$$

The integration over the angles can be easily performed using appropriate polar coordinates (see Fig. 2.19). Expressing the velocity v by the relative velocity and the cluster velocity according to

$$v^2 = v_{Cl}^2 + g^2 - 2v_{Cl} g \cos\vartheta, \tag{2.32}$$

and taking into account that the velocity space volume element dv is given by

$$dv = g^2 \sin\vartheta \, d\vartheta \, d\varphi \, dg, \tag{2.33}$$

we obtain, using the reduced quantities

$$x = \frac{v_{Cl}}{v_w}, \quad z = \frac{g}{v_w}, \tag{2.34}$$

$$\frac{\overline{\Delta v_{Cl}}}{v_{Cl}} = n\sigma_N L \frac{m}{M} F(x) \tag{2.35}$$

with

$$F(x) = \frac{2}{\sqrt{\pi}x^2} \int_0^\infty z^4 \exp(-(x^2+z^2)) \int_0^\pi \exp(2xz\cos\vartheta)\cos\vartheta\sin\vartheta\,d\vartheta\,dz. \qquad (2.36)$$

The result of the integrations is

$$F(x) = \left(\frac{1}{x} + \frac{1}{2x^3}\right)\frac{\exp(-x^2)}{\sqrt{\pi}} + \left(1 + \frac{1}{x^2} - \frac{1}{4x^4}\right)\operatorname{erf} x. \qquad (2.37)$$

In the limit of large x we have

$$\lim_{x\to\infty} F(x) \longrightarrow 1 + x^{-2} - x^{-4}/4. \qquad (2.38)$$

Introducing $M = Nm_0$, where m_0 is the mass of a cluster atom, into (2.35), and assuming a hard sphere cross section for σ_N

$$\sigma_N = \pi(R_N + R_0)^2, \qquad (2.39)$$

where R_N is the cluster radius according to (2.1) and R_0 is the radius of a gas atom, (2.35) can be written as

$$\frac{\overline{\Delta v_{Cl}}}{v_{Cl}} = nL\frac{m}{m_0 N}\pi(N^{1/3}R_1 + R_0)^2 F(x). \qquad (2.40)$$

Cuvellier et al. (1991) used computer simulations to estimate R_0; gas kinetic cross sections may also be used. If the relative change of the cluster velocity is measured (by a time-of-flight method, for example) as a function of the number density n of the scattering gas [Cuvellier et al. (1991), Schilling (1993)], a straight line is obtained (for fixed expansion conditions). Its slope depends on the cluster size. Since for large clusters the assumption $R_N \gg R_0$ is a good approximation, we obtain for the cluster size

$$N \approx \left(nL F(x)\frac{m\pi R_1^2}{m_0}\frac{v_{Cl}}{\overline{\Delta v_{Cl}}}\right)^3. \qquad (2.41)$$

This relation shows the main disadvantage of this method: The experimental errors are raised to the third power in the determination of the cluster size. The main source of error without doubt is the absolute determination of the number density n of the scattering gas. This problem occurs in all absolute measurements of total scattering cross sections and has been discussed extensively in the corresponding literature [Pauly and Toennies (1965), Toennies (1974), van den Biesen (1988)]. As has already mentioned, the determination of the product nL with the aid of a calibration measurement using a known cross section is the simplest and most accurate method to reduce this error. Another source of error in (2.41) is due to the determination of the effective radius R_1 of a cluster atom. Usually, the macroscopic density ρ of the liquid or solid phase is used to calculate R_1 (see (2.24)), which, however, may be different from the cluster density. Finally, it is important to use the exact averaging function $F(x)$. Cuvellier et al. (1991) approximate $F(x)$ by the function $Fa_0(\infty,x)$ (see Sect. I.2.1.3), which converges less rapidly to the same limiting value with increasing x. Another disadvantage is the neglect of the size distribution of the clusters. Changes in the cluster size due to evaporation of cluster atoms or due to capture of gas particles are also neglected, which is justified, however, if the clusters are sufficiently large.

The above-discussed problems occur only in absolute measurements of cluster sizes. Relative measurements with this method, such as measurements of the cluster size as a function of the stagnation conditions in the source, for example, can be performed with much higher accuracy. This has been demonstrated for beams of argon clusters, which confirm the dependence of the cluster size on the scaling parameter Γ^* and which agree very well with the results of numerous other measurements employing different methods [Cuvellier et al. (1991)].

An advantage of this method is the low experimental effort required, especially if the cluster apparatus used is already equipped with a time-of-flight spectrometer.

Broadening of the Beam Profile. Another method for the determination of average cluster sizes in nozzle beams is based on the broadening of the beam profile due to collisions in a scattering chamber [De Martino et al. (1993)]. Assuming that the collisionally broadened beam profile is a Gaussian distribution — an assumption which is justified for sufficiently large beam attenuations — the square of the average broadening of the beam profile $\overline{\Delta\theta^2}$, where $\Delta\theta$ is defined as the distance between the inflection points of the Gaussian distribution, can be readily calculated, provided that all collisions are completely inelastic.

First we calculate the deflection angle $\Delta\theta$ of a cluster, which collides with a gas particle crossing its direction at an angle α (assuming complete inelasticity). Using simple geometrical considerations (see Fig. 2.20), this angle can be expressed by the momentum of the gas atom p and the momentum of the cluster p_{Cl} [Lewerenz et al. (1993)]:

$$\text{tg}\Delta\theta = \frac{p\sin\alpha}{p_{Cl} + p\cos\alpha}. \tag{2.42}$$

Since the deflection angles are small and because $p \ll p_{Cl}$, (2.42) can be

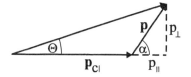

Fig. 2.20. Vector diagram to calculate the angular deflection in atom–cluster collisions

simplified to yield

$$\Delta\theta = \frac{p}{p_{Cl}}\sin\alpha = \frac{mv}{Nm_0 v_{Cl}}\sin\alpha, \qquad (2.43)$$

where m is the mass of the scattering gas particles, m_0 is the mass of a cluster constituent. If the square of the deflection angle $\Delta\theta$ (2.43) is averaged over the velocity distribution of the scattering gas (with respect to magnitude and direction), taking into account the collision probability dW according to (2.30), we obtain the following result:

$$\overline{\Delta\theta^2} = \frac{1}{N^2}\frac{1}{2\pi\sqrt{\pi}v_w^3}\left(\frac{m}{m_0}\right)^2 nL\sigma \int \left(\frac{v}{v_{Cl}}\right)^2 \frac{g}{v_{Cl}}\exp(-v^2/v_w^2)\sin^2\alpha\, dv. \qquad (2.44)$$

v_w is the most probable velocity in the scattering gas. To integrate over the angles we use (2.32) and the coordinates defined in Fig. 2.19. This yields

$$\overline{\Delta\theta^2} = \frac{A}{2\pi^{3/2}v_w^3}\int_0^\infty\int_0^\pi\int_0^{2\pi}\left(\frac{v}{v_{Cl}}\right)^2 \frac{g}{v_{Cl}} f(g,\vartheta)\sin^2\alpha\, g^2 dg \sin\vartheta\, d\vartheta\, d\varphi, \qquad (2.45)$$

using for abbreviation

$$A = \frac{m^2}{N^2 m_0^2}nL\sigma \quad \text{and} \quad f(g,\vartheta) = \exp\left(-(v_{Cl}^2 + g^2 + 2v_{Cl}g\cos\vartheta)/v_w^2\right). \qquad (2.46)$$

With (see Fig. 2.19)

$$v^2\sin^2\alpha = g^2\sin^2\vartheta, \qquad (2.47)$$

$$\overline{\Delta\theta^2} = \frac{A}{\sqrt{\pi}v_w^3 v_{Cl}^3}\int_0^\infty g^5 \exp\left(-\frac{v_{Cl}^2 + g^2}{v_w^2}\right)\int_0^\pi \exp\left(-\frac{2v_{Cl}g\cos\vartheta}{v_w^2}\right)\sin^3\vartheta\, d\vartheta\, dg \qquad (2.48)$$

is obtained, where the integration over the azimuthal angle has already been performed. The integration over ϑ yields, using the reduced quantities x and z according to (2.34),

$$\int_0^\pi \exp\left(-\frac{2v_{Cl}g\cos\vartheta}{v_w^2}\right)\sin^3\vartheta\, d\vartheta = \frac{1}{x^2 z^2}\left[\cosh(2xz) - \frac{\sinh(2xz)}{2xz}\right]. \qquad (2.49)$$

Thus we obtain

$$\overline{\Delta\theta^2} = \frac{A}{\sqrt{\pi}x^5} \int_0^\infty z^3 \exp(-(x^2+z^2))\left[\cosh 2xz - \frac{\sinh 2xz}{2xz}\right] dz, \qquad (2.50)$$

or, after integration,

$$\overline{\Delta\theta^2} = \frac{1}{N^2}\left(\frac{m}{m_0}\right)^2 nL\sigma \frac{F(x)}{2x^2}. \qquad (2.51)$$

The function $F(x)$ is defined by (2.37). Using (2.39) for the cross section σ, (2.51) may be written as

$$\overline{\Delta\theta^2} = nL\pi \frac{m^2}{m_0^2} \frac{F(x)}{2x^2}\left[R_1 N^{-2/3} + R_0 N^{-1}\right]^2. \qquad (2.52)$$

For large clusters we have $R_0 \ll R_N$, and the cluster size N is given by

$$N \approx \left[\frac{nL\pi m^2}{\overline{\Delta\theta^2} m_0^2} \frac{F(x)}{2x^2} R_1\right]^{3/4}. \qquad (2.53)$$

Experimental errors play a less important role than in velocity loss measurements. But similar to that method, the cluster size distribution in the beam is neglected. The beam profile broadening is calculated for a uniform cluster size, which is assumed to be the average cluster size. The averaging function $F(x)$ derived here deviates for $x < 1$ from the expression derived by De Martino et al. (1993), since they perform the averaging process approximately in two steps. In practical applications of this procedure the intrinsic beam profile measured without scattering gas must be accounted for in the analysis of the broadened beam profile data. The cluster size determination from measured angular distributions after scattering has also been used to determine the size of mixed van der Waals clusters, produced by coexpansion of a binary gas mixture [Fort et al. (1998)].

Determination of Capture Cross Sections. The cross section σ_c for foreign particle capture by clusters of a given size can be measured in a pick-up arrangement similar to the one shown in Fig. 2.15, using a mass spectrometer for detection. A cluster beam (Ar_N) passes through a gas- or vapor-filled target chamber (H_2O), and the intensity I of single-doped cluster ions ($Ar_N H_2O^+$) is measured as a function of the number density n of the vapor in the target chamber. Since capturing is a random process, the intensity I(n) follows a Poisson distribution [Lewerenz et al. (1995)]. This is illustrated in Fig. 2.21, which shows the measured intensity of $Ar_4 H_2O^+$ ions as a function of the H_2O pressure in the target chamber. The full

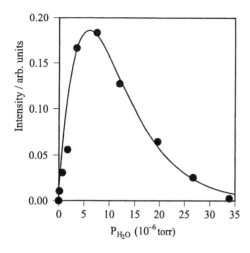

Fig. 2.21. Intensity of single-doped Ar_4 cluster ions as a function of the H_2O vapor pressure of target. *Full curve:* fitted Poisson distribution [Macler and Bae (1997)]

curve drawn through the measurements is a Poisson distribution fitted to the experimental data. Accordingly, the number density dependence of the intensity of the single-doped clusters is given by

$$I = K\alpha L \exp(-\alpha L), \tag{2.54}$$

where K is a constant, L is the length of the target chamber, and α is the capture coefficient. α is essentially the product of the capture cross section σ_c and the target number density n. It can be calculated by integrating the collision probability per unit time dW/dt according to (2.30) over the velocity distribution f(**v**) of the target molecules:

$$\alpha = n(\sigma_c/v_{Cl}) \int g f(\mathbf{v}) d\mathbf{v} = n\sigma_c F_{a0}(\infty, x). \tag{2.55}$$

σ_c is the capture cross section, $F_{a0}(\infty,x)$ is the averaging function defined in Sect. I.2.1.2, and $x = v_{Cl}/v_w$ is the ratio of the cluster velocity v_{Cl} to the most probable velocity of the target particles.

A velocity-independent capture cross section and a sticking probability of 1 for the dopant molecules is assumed. The fit of Poisson distributions to the experimentally determined intensity distributions I(n) yields the capture coefficient α for each cluster size and thus the capture cross section σ_c. Describing the clusters by the hard sphere model, the cluster size N can be readily calculated from σ_c.

The cluster sizes determined in this manner (between $N = 30$ and $N = 400$) are in good agreement with the data of other experiments (see Fig. 2.4, for example) [Macler and Bae (1997)].

Cluster Size Distributions from Collisional Beam Deflection. Let us consider a monochromatic molecular beam which crosses a cluster beam under an angle α. For a cluster mass $M = Nm_0$ with velocity v_{Cl} colliding with a secondary beam particle of mass m and velocity v, the angle of deflection is given by (2.43), assuming $M \gg m$, single collisions, small deflection angles, uniform velocities in both beams, and a complete momentum transfer in all collisions. Hence, the cluster size N as a function of the deflection angle θ may be written as

$$N(\theta) = \frac{mv}{m_0 v_{Cl}} \frac{1}{\theta} \sin \alpha, \tag{2.56}$$

leading to the following conversion of a measured angular distribution I(θ) to a size distribution f(N):

$$f(N)dN \propto I(\theta)\theta^2 d\theta. \tag{2.57}$$

Taking into account that the scattering probability is proportional to the cross section of the cluster ($\propto N^{2/3}$) and assuming further that the detection probability is also proportional to the cluster cross section, the conversion of I(θ) to the size distribution f(N) is finally given by

$$f(N)dN = \theta^{10/3} I(\theta) d\theta. \tag{2.58}$$

Figure 2.22 shows an example of a measured angular distribution I(θ) for the deflection of a helium cluster beam by a SF_6 target beam. The deflection angles are rather small and a detection system of high angular resolution is required.

Figure 2.23 presents typical cluster size distributions f(N) calculated from measured angular distributions I(θ) using (2.58) [Leverenz et al. (1993)]. The full lines are weighted least-squares fits to a log-normal distribution which was found to give the best agreement with the experimental data. This result has been confirmed by many other experiments [Gspann (1982), Schütte (1997), Ehbrecht et al. (1997a)].

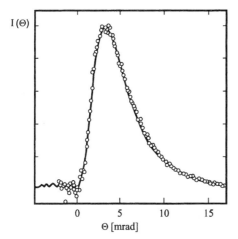

Fig. 2.22. Angular distribution of a He cluster beam deflected by a crossed beam of SF_6 [Lewerenz et al. (1993)]

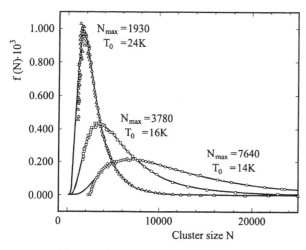

Fig. 2.23. Size distributions in a helium cluster beam for different stagnation temperatures T_0 in the source. The full lines are log-normal distributions [Lewerenz et al. (1993)]

Figure 2.23 confirms another well established empirical finding, according to which the halfwidth of a cluster size distribution is approximately equal to the average value of the distribution.

The normalized log-normal distribution function [Aitchison and Brown (1973)] has the following form

$$f(N) = \frac{1}{N\sigma\sqrt{2\pi}} \exp\left(-\frac{(\ln N - \mu)^2}{2\sigma^2}\right). \tag{2.59}$$

It is essentially a Gaussian distribution with a logarithmic-distorted abscissa. The mean value $\overline{N}$ and the standard deviation S are given by

$$\overline{N} = \exp(\mu + \sigma^2/2); \quad S = \overline{N}\sqrt{\exp(\sigma^2) - 1} \quad \text{with} \tag{2.60}$$

$$\mu = \ln\left(\overline{N}^2/\sqrt{S^2 + \overline{N}^2}\right); \quad \sigma^2 = \ln\left(S^2/\overline{N}^2 + 1\right).$$

The maximum of the distribution N_{max} and the functional value $f(N_{max})$ there are

$$N_{max} = \exp(\mu - \sigma^2); \quad f(N_{max}) = \frac{1}{\sigma\sqrt{2\pi}} \exp(\sigma^2/2 - \mu). \tag{2.61}$$

The halfwidth of the asymmetrical distribution is found to be

$$\Delta N_{1/2} = \exp\left(\mu - \sigma^2 + \sigma\sqrt{\ln 4}\right) - \exp\left(\mu - \sigma^2 - \sigma\sqrt{\ln 4}\right). \tag{2.62}$$

118   2. Production and Diagnostics of Cluster Beams

Size Selection by Scattering. In the upper part of Fig. 2.24 the method of cluster separation in a crossed beam experiment is illustrated. An atomic beam (velocity v, mass m) crosses the cluster beam (velocity v_{Cl}, mass $M = Nm_0$) at right angles. Under single collision conditions, the elastically scattered microclusters ($N \leq 10$) can be separated with respect to their mass due to the different angular and velocity distributions of each cluster size in the laboratory system after the collision [Buck and Meyer (1984), (1986), Buck et al. (1985)]. Assuming that both beams are monochromatic (a condition which is well satisfied if the clusters are generated in a gas dynamic expansion), the scattering kinematics (energy and momentum conservation) is determined by a single velocity vector (or Newton) diagram. Such a diagram is shown in the central portion of 2.24. In the laboratory system, the velocity vectors v and v_{Cl} before collision are perpendicular to each other. The connecting line of the two vectors is the relative velocity vector

$$\mathbf{g} = \mathbf{v} - \mathbf{v}_{Cl}. \tag{2.63}$$

The center-of-mass of a given collision pair (cluster of given N and atom) divides the connecting line according to the corresponding mass ratio. The magnitude of the final center-of-mass cluster velocity u_{Cl} is given by

$$u_{Cl} = \frac{m}{M+m} g = \frac{m}{Nm_0 + m}. \tag{2.64}$$

Thus, the final velocity vectors (in the center-of-mass system) end on a sphere with the radius u_{Cl}. With increasing cluster mass, the radii of these spheres become smaller and their centers come closer to each other. In the laboratory system, the final velocities v_{Cl}' are determined by the distance between the origin of the coordinate system and the intersection points with the various Newton spheres. In general, two intersection points exist for each Newton sphere (see Fig. 2.24), which correspond to two different scattering angles in the center-of-mass system and consequently to two different velocities in the laboratory system. This is due to the fact that the transformation from the center-of-mass system to the laboratory coordinate system is not unique, since the mass of the clusters is usually much larger than the mass of the atoms of the deflecting beam [Pauly and Toennies (1965), (1968)]. For each cluster mass, a limiting laboratory angle Θ_l exists, which defines the maximum deflection angle of a given cluster size in the laboratory system. Under the above assumptions, this limiting angle is given by

$$tg\Theta_l = \frac{2mv}{(M-m)v_{Cl}} = \frac{2mv}{(Nm_0 - m)v_{Cl}}. \tag{2.65}$$

Accordingly, at the laboratory angle shown in Fig. 2.24, only monomers, dimers, and trimers ($N = 1$, $N = 2$, and $N = 3$) can be detected. The limiting angles for larger clusters are smaller. From Fig. 2.24 it can be seen immediately that the

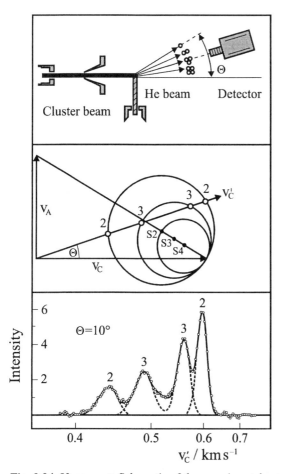

Fig. 2.24. Upper part: Schematic of the experimental setup. Central part: Newton diagram. Lower part: Cluster intensity, measured at a fixed laboratory angle on the dimer mass, as a function of cluster velocity. The maxima for dimers and trimers measured at low velocities correspond to large angle scattering in the center-of-mass system, whereas the maxima measured at higher velocities correspond to small angle scattering in the center-of-mass system [Buck et al. (1985)]

choice of the laboratory angle divides the mass spectrum of the clusters into two regions: clusters with a limiting angle smaller than the selected angle, are excluded from detection. To separate the smaller clusters from each other, which have limiting angles larger than the selected one, two possibilities exist:

One possibility is indicated in the lower part of Fig. 2.24. Since clusters of different size measured at a fixed laboratory angle have different velocities, they can be separated from each other by a velocity selection. This is illustrated in the lowest part of Fig. 2.24, which shows a time-of-flight spectrum of argon clusters (scattered by a helium atom beam), which have been detected under a laboratory angle of 10°. At this angle, the beam contains monomers, dimers, and trimers.

Since a universal beam detector with electron impact ionization and mass filter has been used with the mass filter adjusted to the dimer mass, the monomers in the beam are not detected. Hence, only dimers and trimers are observed in the time-of-flight spectrum, which can be clearly separated from each other. The mass-spectrometric detection of trimers is possible since trimers are partly decomposed to monomers and dimers during electron impact ionization. Thus, using a mechanical velocity selector, a cluster beam containing only clusters of a well defined size may be generated.

A simpler way to discriminate against the smaller cluster masses is to use a mass spectrometer. This has already been indicated above, where the contribution of the monomers in the beam has been suppressed by detecting the beam on the dimer mass. Correspondingly, the mass spectrometer detector can be used to discriminate against all smaller clusters, provided that the fragmentation of the selected cluster size is incomplete [Buck (1994)].

2.8.4 Atom Diffraction

The differential scattering cross section for elastic atom–atom and atom–molecule collisions exhibits characteristic interference effects, which are due to the fact that the interaction potential, which determines the scattering pattern, is composed by an attractive and a repulsive part [Pauly (1979)]. Two characteristic features may be distinguished: Rainbow scattering, which (at thermal collision energies), shows oscillations in the differential scattering cross sections having rather large angular distances, and diffraction scattering, which causes oscillations in the differential cross section with small angular spacing.

While the rainbow scattering is mainly determined by the attractive portion of the interaction potential, the diffraction scattering is decisively determined by the repulsive part of the potential. If, at a given collision energy, the repulsive part of the potential dominates the scattering process, the differential cross section exhibits only diffraction oscillations. In the limit of a vanishing attractive potential and an infinitely steep repulsive potential (hard sphere) these diffraction oscillations are identical with the interferences observed in the scattering of light from small spherical particles. In a simple approximation, the angular spacing $\Delta\vartheta$ of these oscillations is given by

$$\Delta\vartheta = \frac{\pi}{ka} = \frac{\lambda}{2a} \quad \text{with} \quad k = \frac{\mu g}{\hbar}, \tag{2.66}$$

where λ is the de Broglie wavelength (μ = reduced mass, g = relative velocity) and a is the radius of the sphere. Accordingly, the size of the particles can be determined from the experimental observation of these interference oscillations [Buck and Pauly (1968)].

The scattering of helium atoms from clusters is especially suited for this method. The light helium atoms have a comparatively large wavelength resulting in a rather large angular spacing $\Delta\vartheta$ of the oscillations. As a consequence of their low attractive interaction with other particles, the differential cross section exhibits only diffraction oscillations. Finally, intense and very monochromatic helium beams can be easily produced by gas dynamic expansion.

A more accurate evaluation of the measured differential cross sections requires quantum mechanical calculations assuming a potential model for the atom–cluster interaction. An analytical model potential has been derived by Gspann and Vollmer (1979). To derive this model, the cluster is represented by a sphere of homogeneously distributed centers of force. Each center of force interacts with the atom via a Lennard–Jones (12,6) potential. The integration over all contributions yields a potential that has the correct limiting values. For $N = 1$ it is the Lennard–Jones (12,6) potential for the interaction of a He atom with a single cluster atom. For $N \rightarrow \infty$, the potential for the interaction of a He atom with the plane surface of a solid body composed of the cluster atoms is obtained [Hill (1948)]. The analytical form of the potential is given by

$$V(R) = N \frac{4\varepsilon\sigma^6}{\left(R^2 - R_G^2\right)^3} \left[\frac{\sigma^6 \left(R^6 + (21/5)R^4 R_G^2 + 3R^2 R_G^4 + (1/3)R_G^6\right)}{\left(R^2 - R_G^2\right)^6} - 1 \right], \quad (2.67)$$

with $R_G = R_1(N^{1/3} - 1)$, where R_1 is the effective radius of a cluster atom according to (2.1). m_0 = mass of a cluster atom, ρ = density of the solid, ε, σ = well depth and zero of the atom–atom potential.

Figure 2.25 shows the model potential for the interaction of helium atoms with argon clusters of different size N. With increasing N, the repulsive core of the potential, characterized by the value R_0 for which $V(r) = 0$, increases approximately proportional to $N^{1/3}$, while the attractive part of the potential remains essentially unchanged.

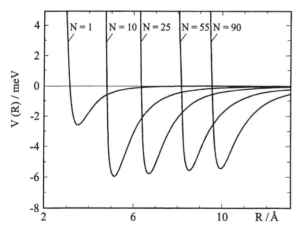

Fig. 2.25. He–Ar$_N$ potential for various cluster sizes N according to the model of Gspann and Vollmer (1979)

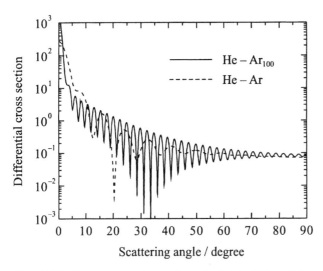

Fig. 2.26. Quantum mechanically calculated differential cross section for He–Ar$_N$ (N = 100) (*full curve*) and He–Ar scattering (*dashed curve*)

Figure 2.26 presents quantum mechanically calculated differential cross sections for He–Ar$_{100}$ (full curve) and He–Ar (dashed curve) scattering. In the first case the diffraction oscillations appear over the whole angular range and have a nearly constant angular spacing. They are practically independent of the detailed shape and the depth of the potential well, the main reason for the accuracy of this method in the determination of cluster sizes. Very important is the fact that the oscillations survive the averaging over a distribution of cluster sizes, since their angular distance changes only according to $N^{-1/3}$. Hence, the averaging over a given size distribution reduces only the amplitudes of the oscillations and may modulate them with beat oscillations depending on the size distribution.

By contrast, the angular spacings of the atom–atom differential cross section oscillations (dashed curve) are rather large and their amplitudes decay more rapidly with increasing angle. This is important for distinguishing experimentally between the two types of oscillations, since the measured differential cross section is always a superposition of the monomer and the cluster cross section.

Compared to the other scattering procedures for cluster size determination, the experimental effort connected with this method is rather high, requiring a universal scattering apparatus (see Chap. I.1) of high angular resolution. Experimental details and the evaluation of the measurements are described in the literature [Buck and Krohne (1996)].

2.8.5 Mass Spectrometric Methods

The first information on cluster size distributions in supersonic beams was obtained by mass spectrometric investigations [Bentley (1961), Henkes (1961), (1962), Greene and Milne (1963), Hagena and Henkes (1965)]. This method is straightforward. A cluster beam passes through an ion source, where the clusters are ionized by electron impact or by photoionization. A subsequent mass analysis yields the distribution of cluster ions produced in the ion source.

The ionization process, however, is rather complicated. Instead of only ionizing the parent clusters, extensive fragmentation may occur, depending on the energy deposited into the clusters and on the structural relaxation upon ionization. As a result, neutral and charged cluster fragments are obtained. Consequently, each cluster mass contributes to several peaks of the mass spectrum, and the measured size distribution may be severely distorted with respect to the size distribution of the original neutral beam. This problem is well known from mass spectrometric investigations of molecules. Even a simple three-atom molecule such as CO_2 yields a number of peaks in the mass spectrum. Apart from the parent ions CO_2^+ at mass 44, the fragment ions C^+, O^+, CO^+, and O_2^+ with masses 12, 16, 28, and 32, respectively, appear also in the mass spectrum. Polyatomic molecules produce even more lines in the mass spectrum and many clusters behave similarly.

For molecules, however, the fragmentation behavior can be studied using the pure species. For a given mass spectrometer and a well defined ionization (e.g. electron energy or photon wavelength) calibration spectra can be measured, so that the composition of gas mixtures can be obtained from the measured mass spectra, provided that the same ionization conditions are used. This procedure cannot be applied to clusters, since it is impossible to generate clusters of a given size. An exception are a few small clusters, which have been size selected using scattering. In these cases, the abundances of cluster fragments have been determined [Buck et al. (1985)].

As has already been mentioned above, two causes are responsible for fragmentation: structural relaxation and the excess energy deposited in the cluster. Structural relaxation occurs if the geometrical structure of a given cluster in its ground state is quite different from the geometric configuration of the corresponding cluster ion. In this case, the cluster ion will be produced with excess vibrational energy, which may exceed the cluster ion binding energy and thus lead to fragmentation. This mechanism plays an essential role in the fragmentation of microclusters and is independent of the kind of ionization.

With increasing cluster size, a growing localization of the structural changes is observed. In rare gas clusters, for example, the charge is localized at a dimer ion, and its energy of formation (of the order of 1 eV) is released to the rest of the cluster by vibrational relaxation, leading to the evaporation of one or more atoms [Jortner (1984)].

In the case of molecular clusters, the charge is initially localized in a single molecule of the cluster. Depending on the energy connected with the configurational changes of the cluster, homolytic fragmentation ($M_N^+ \rightarrow M_{N-1}^+ + M$) or ion–molecule reactions may occur.

For metal atom clusters, which have completely delocalized valence electrons and a little defined geometry, structural rearrangement upon ionization is usually unimportant. Fragmentation, therefore, is rather weak. This is also the case for rare gas microclusters, which are doped with large aromatic molecules.

The fragmentation of large clusters, for which the evaporation of some constituents is of no practical importance, is determined by the excess energy deposited by the ionization process. To make fragmentation negligible, the excess energy above the ionization threshold has to be as low as possible. This is difficult to realize for electron impact ionization due to experimental reasons: At low acceleration voltages, space-charge effects permit the production of electron beams of sufficient intensities. Accordingly, substantial fragmentation occurs. In the case of photoionization, however, it is possible to operate close to the ionization threshold by choosing an appropriate wavelength. The measured size distributions correspond (for large clusters) to the size distribution of the neutral clusters. A sufficiently low power density of the laser is important, since otherwise fragmentation may occur due to the successive absorption of several photons.

As an example, Fig. 2.27 shows two sets of size distributions measured with a NH_3 cluster beam under equal conditions for electron impact ionization and photoionization (resonance enhanced two-photon ionization at 193 nm) using a reflectron time-of-flight mass spectrometer [Schütte (1997)]. It can clearly be seen that

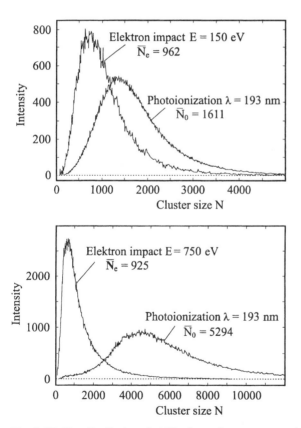

Fig. 2.27. Size distributions in NH_3 cluster beams measured under equal conditions with a mass spectrometer for electron impact ionization (with two different energies) and for photoionization. For a detailed description see text [Schütte (1997)]

for electron impact ionization the size distribution is substantially shifted to smaller masses due to fragmentation. The shift increases with increasing electron energy, as follows from the measurements shown in the lower half of Fig. 2.27.

Detailed investigations of electron impact ionization of ammonia and rare gas clusters show that the number of evaporating monomers increases linearly with the average cluster size. This can be interpreted as an indication that the fragmentation is caused by a quasicontinuous deceleration process due to multiple electron collisions within the cluster.

Early measurements of cluster size distributions by electron impact ionization utilized a retarding field technique and did not consider fragmentation [Falter et al. (1970), Hagena and Obert (1972)]. In more recent investigations, time-of-flight mass spectrometers (see Chap. I.5) were used to analyze cluster size distributions. Fragmentation has been taken into account by simple models for the mechanism of energy release from the electron to the cluster and the subsequent unimolecular decay [Engelking (1987), Karnbach et al. (1993)].

To investigate rare gas clusters with negligible fragmentation, they have been doped with sodium atoms by passing them through a pick-up stage. The sodium atoms have an ionization energy of 5.14 eV and can be ionized by a single photon from an ArF laser (193 nm, 6.42 eV). Hence, the excess energy is about 1.3 eV, and the number of evaporated cluster atoms can be estimated to be of the order of 10 to 20 atoms. This is negligible for large clusters (with $N > 1000$). Multiple photon absorption, which could increase the number of evaporated cluster constituents, does not occur; the average cluster size is independent of the laser power [Schütte (1997)].

In negative charging of large, liquid helium clusters (10^4–10^8 atoms/cluster) by electron attachment, much less energy is set free compared to positive charging by electron impact ionization, so that in this case a substantially smaller mass loss by evaporation is observed. Thus the size distribution of the ionized droplets is approximately proportional to the neutral size distribution. Furthermore, for cluster sizes of the order of $N \approx 10^5$–10^8, only single-charged negative ions are formed. and the attachment cross section is proportional to the geometrical cluster cross section. Accordingly, the distribution of negative ions has only to be divided by a factor $N^{2/3}$ to account for the increase in the attachment cross section [Fárník et al. (1997)].

2.8.6 Other Methods

Velocity Selection. Clusters generated in gas aggregation sources experience numerous collisions on their path through the aggregation area, so that they achieve a rather uniform translational temperature. Consequently, clusters of different mass exhibit different speed distributions. This has been confirmed experimentally, for example, by measurements of speed distributions of large sodium clusters [Zimmermann et al. (1995)].

Figure 2.28 presents speed distributions for different masses of sodium clusters formed in a gas aggregation source. A laser pulse with 0.3 µJ/mm² at 193 nm is used for ionization. The speed distribution of the ions of a given mass range — selected by a reflectron mass spectrometer — is measured with a time-of-flight spectrometer. The low-intensity "ridge" at the left foot of the dominant distributions (marked by ++) is due to doubly charged ions. As can be seen from Fig. 2.28, the velocity distribution depends on the cluster mass. Accordingly, a certain cluster size selection is possible by using a velocity selection (see Chap. 3) of the cluster beam [Arnold et al. (1985a), Roux et al. (1994)].

Velocity distributions of clusters generated by laser pyrolysis also show a considerable size dependence [Ehbrecht et al. (1993a), Ehbrecht and Huisken (1999)], since they are formed at a rather well defined temperature in the flow reactor. Figure 2.29 shows, as an example, the mean velocity of silicon clusters as a function of the cluster size N. The relative widths of the velocity distributions are of the order of 10 to 14%, so that a velocity selection does not yield a complete size selection, but the dispersion of the cluster size distribution can by substantially reduced.

This method has been used to perform size-selected low-energy silicon cluster deposition on various substrates (CaF$_2$ and LiF substrates, for example). The deposition yields single crystalline, spherical Si clusters almost completely free of planar lattice defects, which maintain their shape in the deposited film and do not agglomerate into larger clusters. These monodispersed silicon films have been further characterized by studying their luminescence and Raman scattering behavior. The peak of the luminescence curve shifts with decreasing cluster size to smaller wavelengths (higher energies), as is predicted by theory. Thus, lumines-

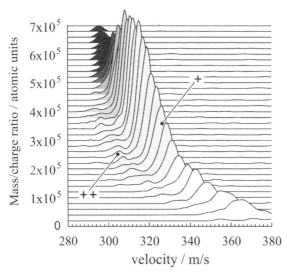

Fig. 2.28. Speed distributions for different masses of large Na clusters generated by gas aggregation. The low "ridge" on the left corresponds to doubly ionized clusters [Zimmermann et al. (1995)]

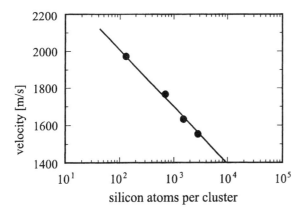

Fig. 2.29. Mean velocity of neutral clusters produced by laser pyrolysis as a function of the number of atoms/cluster [Ehbrecht and Huisken (1999)]

cence spectra show a strong dependence on the cluster size [Ehbrecht et al. (1997b), Werwa and Kolenbrander (1996), Ehbrecht and Huisken (1999), Huisken et al. (2000), Hofmeister et al. (1999)].

Neutralization of Mass Selected Cluster Ions by Charge Exchange. A neutral mass-selected cluster beam can also be generated by ionization, subsequent mass separation of the cluster ions, and reneutralization by charge exchange [Arnold et al. (1985b), Bréchignac et al. (1988), (1990)]. Charge exchange cross sections are rather large (typically 10^{-14} cm^2 for resonant and 10^{-15} cm^2 for nonresonant charge exchange), since the electron transfer occurs at large impact parameters and practically without transfer of momentum (see Chap. 1). Consequently, charge exchange is rather effective and does not increase the cluster beam divergence. Fragmentation of the clusters, however, cannot be ruled out completely. Since fragmentation depends strongly on the energy which is transferred to the neutral cluster in the neutralization process, it seems unlikely or can be avoided completely in the case of resonant or quasiresonant charge transfer. An experiment of Abshagen et al. (1991) may be used as an example. Pb_N^+ ions (with $N \leq 3$ to $N \leq 12$) are neutralized by passing through sodium vapor. The ionization energies of small lead clusters differ only by a few tenths of an eV from the ionization energy of sodium (5.14 eV). Thus the energy transferred to the clusters is very small and they are in their ground state after neutralization.

Deflection in Inhomogeneous Fields and by Resonance Photons. The deflection of cluster beams in inhomogeneous electric or magnetic fields utilizing the electric or magnetic moment of the clusters (see Chap. 4) has also been proposed for size selection. [Arnold et al. (1985a), Kappes and Schumacher (1985)]. Since the deflection is inversely proportional to the cluster mass, the resolution of this method decreases strongly with increasing cluster size. Another problem is due to the fact that the (state-dependent, permanent or induced) electric and magnetic moments of the clusters are usually unknown. The method is generally employed in the opposite way to determine electric or magnetic dipole moments of clusters [de Heer and Knight (1988), Apsel et al. (1996), Billas et al. (1996), Woenckhaus et al. (1996)]. Despite this, attempts in this direction have been carried out [Keesee et al. (1984), Hofmann et al. (1979)].

Deflection by photon recoil (see Chap. 5), which is also inversely proportional to the particle mass, can be used to select small clusters. An example is the separation of sodium dimers from a beam containing monomers and larger clusters [Herrmann et al. (1979), Schmiedmayer (1997)].

Diffraction of Cluster Beams by Transmission Gratings. Another method to select clusters of light atoms or molecules such as helium and hydrogen, is based on the diffraction of a highly monochromatic cluster beam by a transmission grating. This method, which enabled the first experimental proof of the existence of the extremely weakly bonded He_2 van der Waals dimer, is described in more detail in Chap. 5. Rather large helium clusters He_N up to $N \leq 29$ have been resolved with a setup of high angular resolution (70 μrad) using gratings (100 nm period) made of silicon nitride produced by achromatic interferometric lithography [Savas et al. (1995), Schöllkopf and Toennies (1996)]. Compared to other methods this technique has the advantage of being completely nondestructive as long as the slit width of the grating is large compared to the cluster size [Hegerfeldt and Köhler (1998)]. Otherwise the detection of the weakly bonded helium dimer would not have been possible.

Cluster–Surface Scattering. In addition to the cluster size distribution, a knowledge of the fraction of monomers in a given cluster beam is also very important for many experiments. A recently developed technique utilizes the scattering of cluster beams from solid surfaces to determine the monomer fraction in a beam [Fort et al. (1998)]. It is based on mass spectrometric measurements of the angular distribution of monomers reflected from a heated surface (440 K, highly oriented pyrolytic graphite) due to an impinging beam of clusters and monomers at a fixed angle of incidence (under ultrahigh vacuum conditions). This technique, applicable to weakly bound clusters, which essentially evaporate when impinging on the surface, makes use of the fact that the angular distribution of the monomers initially in the beam is quite different from the angular distribution of monomers from the evaporating clusters. This has been demonstrated by measurements of angular and velocity distributions of reflected monomers upon bombardment of graphite surfaces with large van der Waals clusters [Châtelet et al. (1992), Vach et al. (1994)]. The analysis of these data yields an analogy to the macroscopic Leidenfrost phenomenon in which the normal part of the kinetic energy of the incoming clusters is converted into thermal energy and used for the evaporation of the clusters, while the tangential part is nearly entirely conserved. Consequently, the clusters glide along the surface during the evaporation process. This simple model gives a good description of the angular distribution of the monomers resulting from the evaporation of clusters [Vach et al. (1994), (1995), Benslimane et al. (1995)]. Three different scattering components can be distinguished:

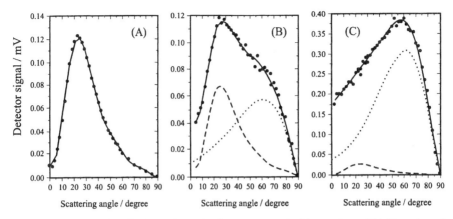

Fig. 2.30. Angular distributions of a N_2/Ar beam scattered off a surface of highly oriented pyrolytic graphite measured at Ar monomer mass for different stagnation pressures P_0 [Fort et al. (1998)]. The angle of incidence is 20°. (A) $P_0 = 1$ bar, the beam consists only of monomers, (B) $P_0 = 11$ bar, both monomers and clusters are in the beam, (C) $P_0 = 21$ bar, the beam is essentially composed of clusters. *Full circles*: experimental data, *full line*: total fit, *dashed line*: monomer component: *dotted line*: evaporated cluster component (see text)

- a diffuse component resulting from adsorption and desorption of impinging monomers. The resulting angular distribution can be described by Lambert's cosine law
- a thermal component of monomers due to the evaporation of monomers from clusters gliding along the surface. This gliding motion transfers an additional tangential velocity component to the evaporating monomers
- a grazing component of remnants of large clusters having survived the impact with the surface. This leaves the surface nearly parallel to the surface. This component increases with increasing angle of incidence, with increasing cluster size, and with increasing temperature of the surface [Pradère et al. (1997), De Martino et al. (1996)]. It can be neglected for small angles of incidence.

As an example, Fig. 2.30 presents typical angular distributions of Ar atoms reflected from graphite surfaces using beams (angle of incidence 20°) with different fractions of clusters [Pradère et al. (1994), Fort et al. (1998)]. The fraction of clusters is varied by using different stagnation pressures P_0 in the source (at a fixed nozzle diameter d). At low stagnation pressure (Fig. 2.30 (A)), no cluster formation occurs in the expansion. The measured angular distribution of monomers is characteristic of the reflection of monomers from the graphite surface. The shape of this angular distribution does not change with the stagnation pressure P_0, as long as the beam contains no clusters. Figure 2.30 (C) shows a measured angular distribution for the other limiting case. The beam consists almost completely of clusters, resulting in a characteristic angular distribution of the monomers produced by evaporation.

Figure 2.30 (B) shows the result of an intermediate case. The beam contains both monomers and clusters. The observed angular distribution is a superposition

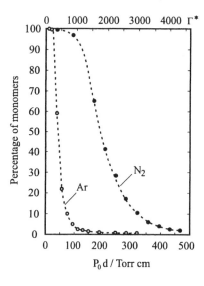

Fig. 2.31. Argon and nitrogen monomer percentage in Ar or N_2 beams, respectively, as a function of P_0d. The upper scale shows the scaling parameter Γ^* for the Ar beam [Fort et al. (1998)]

of the distributions for the two limiting cases. Consequently, the fraction of monomers in the beam (dashed curves in Fig. 2.30 (B) and 2.30 (C)) and the fraction of monomers due to evaporated clusters (dotted curves) can be determined from an analysis of the measured angular distribution. While the experimental curve for the monomers in the beam (Fig. 2.30 (A)) is used for this analysis, the angular distribution of the cluster component is modeled by a dynamic zone structure model [Vach et al. (1995)].

Figure 2.31 shows the percentage of monomers in a pure Ar or N_2 beam, respectively, determined as a function of P_0d (P_0 = stagnation pressure, d = nozzle diameter). For Ar, where the scaling law can be applied (see Sect. I.3.8.2), the upper scale shows the parameter Γ^*.

Electron-Induced Fluorescence. The fluorescence intensity in the spectral range between 200 and 600 nm induced by fast electrons (5.5 keV) exciting large CO_2 clusters, is negligible compared to the fluorescence intensity from CO_2 molecules, as has been shown by Khmel' and Sharafutdinov (1997).

These authors were able, therefore, to measure the velocity distribution of the monomers in a CO_2 cluster beam with a time-of-flight method by observing the electron-induced fluorescence of the beam [Khmel' and Sharafutdinov (1998)]. If the cluster beam impinges upon a quartz plate immediately after the excitation region, the clusters are decomposed and the fluorescence intensity increases. In this case, the time-of-flight spectrum exhibits a bimodal distribution, since the cluster velocity shows a velocity slip compared to the velocity of the monomers, which has been determined in this way.

Nanoscaled Sieves. The first application of nanoscale particle sieves has been aimed at the determination of the size of Rydberg atoms [Fabre et al. (1983)]. The same principle was used later to determine the average internuclear distance within the $^4He_2$ molecule [Luo et al. (1996)].

The basic idea is extremely simple. For spherical particles of diameter d, the transmission T of a sieve with n circular "illuminated" holes of diameter D (with D > d) within a total "illuminated" area A follows from simple geometric arguments to be

$$T = n\frac{\pi D^2}{4A}\left(1-\frac{d}{D}\right)^2 = T_0\left(1-\frac{d}{D}\right)^2 \cong T_0\left(1-2\frac{d}{D}\right), \quad (2.68)$$

where T_0 is the transmission for point-like particles. In the right part of (2.68) d << D has been assumed. T_0 can be determined by measuring the transmission of helium atoms. Accordingly, by measuring the relative transmission of helium dimers and helium atoms through arrays of many uniform holes in thin silicon nitride membranes with hole diameters in the range from 98 to 410 nm Luo et al. (1996) were able to determine the mean internuclear distance of neutral helium dimers to be 62±10 Å.

Schöllkopf et al. (1998) used nanofabricated free-standing transmission gratings with a 100 nm period and 50 nm wide slits (see Chap. 5) as sieves for helium clusters of 10^4–10^6 atoms. By rotating the grating around an axis parallel to the slits, the angle of incidence can be changed resulting in an effective slit width between 50 and 2 nm. In this case, the transmission T_0 for point-like particles is given by

$$T_0 = nD/B. \quad (2.69)$$

n is the number of illuminated slits, D the effective width of the slits, and B is the width of the illuminated part of the grating. The transmission of a single slit may be written as

$$T_1 = D - d, \quad (2.70)$$

where d = d(N) is the cluster diameter. The average transmission T follows by integration over the cluster size distribution P(N) and the detection probability Z(N):

$$T = n\frac{D}{B} \int_0^{N_{max}} (D - d(N))P(N)Z(N)dN. \quad (2.71)$$

N_{max} is the largest cluster size that fits through the slit so that clusters larger than N_{max} are filtered out entirely. For spherical clusters N_{max} follows from equating the cluster diameter $2R_N$ to the effective slit width D. Using (2.1), this yields

$$N_{max} = \frac{\pi D^3}{6m_0}\rho. \quad (2.72)$$

m_0 is the mass of a single cluster atom. With the assumption that the detection probability $Z(N)$ is proportional to the geometric cross section of the clusters,

$$Z(N) = c N^{2/3}, \qquad (2.73)$$

where c is an empirical constant, the measured transmission T can be calculated for an assumed size distribution (e.g. a log-normal distribution) from (2.71), yielding the mean cluster size [Schöllkopf et al. (1998)].

2.9 Some Cluster Applications

Although most of the cluster work done so far is basic research and aimed at a better understanding of the clustering process and of the cluster properties, there is a rapidly increasing interest in technological applications. These concentrate predominantly on the modification of surface properties, either by cluster beam deposition or cluster beam lithography. The extraordinary interest results from the importance of surfaces with well defined properties in various fields of modern technology, ranging from microelectronics and optoelectronic devices to electrochemistry and plasma display panel applications [Cho et al. (1999), Huisken et al. (1999b), Diederich et al. (1999), Cich et al. (1998)]. This is illustrated in the following by a few examples.

2.9.1 Cluster Beam Deposition

The deposition of layers on solid surfaces by cluster beams has met with great interest in the last few years, since one is able to create layers with properties which

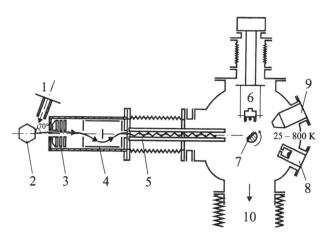

Fig. 2.32. Schematic of a UHV apparatus for ionized cluster beam epitaxy. (1) Laser beam, (2) sputtering target, (3) lens system, (4) energy selector, (5) mass filter, (6) scanning tunneling microscope, (7) rotatable surface, (8) thermal beam source, (9) ion source, (10) pump

were impossible to produce by conventional molecular beam epitaxy. A distinction is made between layer deposition by neutral clusters at low energy (LECBD "low energy cluster beam deposition") and the deposition of cluster ions (ICBD "ionized cluster beam deposition"), whose energy is variable within a wide range. Further, one can work with mass-selected clusters. Compared to layer deposition by neutral atoms, molecules or monomer ions one has the cluster size as an additional important parameter.

The processes involved when a cluster impinges upon a surface are extremely complex and are determined by impact energy, cluster size, and type of surface. In the limit of very high impact energies, to be dealt with in the next section, destruction of the solid surface dominates. Therefore one uses energetic cluster beams for surface erosion. Lowering the cluster energy reduces surface erosion. The energy released within the close neighborhood of impact results in significant fragmentation of the cluster, formation of local defects on the surface which may or may not heal again due to the high temperature, and implantation of cluster fragments into the surface. Simulation calculations show that pressures up to 100 GPa and temperatures of several thousand kelvin are reached. The extreme nonequilibrium nature of such collisions is the cause for growth of relatively smooth layers with good adhesion, already at low temperatures [Yamada (1989), Haberland et al. (1996)]. In the limit of low impact energy one can deposit the clusters on the surface without destroying them, so that they keep at least part of their individual properties. The "soft" deposition of clusters on a surface can be supported by adsorbed buffer layers of rare gas atoms which can be removed after cluster deposition by a rise in temperature of the substrate [Bromann et al. (1996)]. The multitude of processes between the two limiting cases, the numerous parameters characterizing the experimental conditions (type and size of the clusters, beam intensity, type and temperature of the substrate, additional buffer layers, vacuum conditions) are presently under intense investigation, both experimental as well as theoretical (molecular dynamics simulations) [Broyer et al. (1996), Perez et al. (1997), Hu et al. (1996), Rongwu et al. (1996), Kelchner and DePristo (1997), Insepov and Yamada (1997), Betz and Husinsky (1997)]. A typical setup is shown schematically in Fig. 2.32 allowing the investigation of layers deposited by size-selected cluster ions of varying energy by means of a scanning tunneling microscope. Under the same conditions layers from a thermal molecular beam can be grown and investigated in order to establish the differences between the two fabrication methods [Bromann et al. (1996)].

2.9.2 Cluster Impact Lithography

In connection with the detection of heavy cluster ions in Chap. I.5, the processes involved when a large cluster impinges upon a solid surface were indicated. The resulting erosion of the surface can be used for machining, smoothing, and polishing of surfaces of very hard materials, especially also for the fabrication of microstructures on solid surfaces (nanolithography) [Henkes and Klingelhöfer (1989), Gspann (1995), (1996)].

The impact velocities used for erosion by clusters are of the order of 20 km/s. Thus, they are higher than the velocity of sound in solids whose maximum is 18 km/s in diamonds. Macroscopic particles of this velocity cannot be produced in a laboratory. Consequently, there are no comparable surface impact studies for macroscopic particles. Only the impact by meteors on the Earth's surface is comparable, with velocities around 25 km/s.

When a large cluster impacts on a solid surface at supersonic velocity, a shock wave is generated both in the cluster as well as the solid [Insepov and Yamada (1996), (1997), Tombrello (1995)]. This results in a high compression and an extreme rise in temperature within a very short time (50 to 500 fs) [Raz and Levine (1996)]. A plasma cloud of high density is formed of cluster material and material of the solid. The shock wave comes to rest and then returns as an expansion wave and expels at least some of the material of the plasma from the crater that had been formed. The diameter and depth of the crater (proportional to $E^{1/3}$) depend upon the cluster energy E [Aoki et al. (1997)] and the hardness of the material [Gehring (1970), Gspann (1992), (1996), Insepov and Yamada (1997)].

A complete theoretical description [Dienes and Walsh (1970)] is difficult and not yet available, but there are numerous molecular dynamics simulation calulations [Landman and Charles (1992), Haberland et al. (1995), Schek and Jortner (1996), Betz and Husinsky (1997)].

The high temperatures reached upon impact have led to a discussion of the possibility of cluster impact fusion due to the impact of heavy clusters upon solid surfaces of deuterium or tritium [Fortov et al. (1995), Yamamura and Muramoto (1995)].

2.9.3 Examples of Experimental Results

Figure 2.33 shows the schematic for a setup for lithography by bombardment of solid surfaces by cluster ions ($(CO_2)_N$, $(SF_6)_N$) [Gspann (1996)]. The neutral

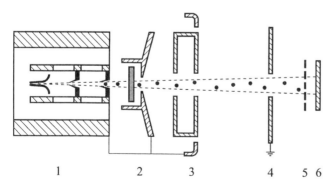

Fig. 2.33. Cluster impact lithography [Gspann (1996)]. (1) cluster source, (2) electron impact ionization, (3) ion lens, (4) acceleration path (100 kV), (5) mask, (6) target

clusters are produced by adiabatic expansion through a Laval nozzle (1) and are transferred into high vacuum through a skimmer and collimating apertures. They pass an electron impact ionization source (2) where they are ionized and subsequently focused (3) and are finally accelerated to 100 kV (4). Passing through a mask made of nickel, the accelerated clusters impinge upon the target to be eroded (6).

If the mask and the target are moved to the side, the average cluster size N (of the order of 1000) can be determined with a time-of-flight mass spectrometer. With the target at ground potential, the cluster source and the ion source are on a potential of −100 kV. Such cluster ion sources are now commercially available. Compared to atomic or molecular ions, the cluster ions deliver their energy on the immediate surface without "channeling" effects. Furthermore, at the same sputtering rate the electric buildup of charges on the substrate is smaller by orders of magnitude.

As Fig. 2.34 shows, erosion through bombardment by clusters results in very smooth surfaces and sharp edges. In this case, hexagonal blind holes have been eroded into a surface of Pyrex glass, whose bottom surfaces are just as smooth as the original surface itself [Gspann (1995)]. Closer studies suggest that the target material melted on its surface [Blanckenhagen et al. (1997)].

The erosion can be accelerated by a suitable reaction between cluster and surface material [Gspann (1996)]. For example, the erosion of diamond by CO_2 clusters proceeds just as fast as for silicon, despite its much greater hardness.

At the high temperature developed at the impact of the cluster, the CO_2 dissociates and the nascent oxygen reacts with carbon to form CO. The volatile CO ensures a fast and complete deportation of the eroded carbon. The eroded surfaces are transparent and show no presence of any carbon coverage. The erosion of silicon by SF_6 clusters also turns out to be reactively accelerated since the volatile SiF_4 is formed.

Even cluster ion beams of very much higher energies (in the MeV range) have been utilized for a few years for studies of cluster interactions with the solid state phase at several particle accelerators [Tomaschko et al. (1996a)]. Due to the enormous energy released by a high energy cluster in a very small volume one expects to find new energy loss mechanisms and changes in the material [Tomaschko et al. (1996b)]. The latter could be of major practical importance.

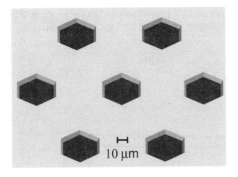

Fig. 2.34. Hexagonal blind holes, eroded into Pyrex glass by bombardment with CO_2 cluster ions employing a nickel mask [Gspann (1995)]. The side length is 25 μm

2.9.4 Cluster Beams in High Energy Physics

Hydrogen cluster beams have found an important application in high energy physics, e.g. as internal, windowless target for studies on meson production in proton–proton collisions at energies in the GeV range [Dombrowski et al. (1996), (1997)]. A target of high particle density can be realized with a cluster beam, simultaneously keeping the pressure in the scattering chamber low. In addition, the target is of the highest purity and its density can be varied over orders of magnitude, depending upon the requirements of the experiment.

The schematic of such a setup is shown in Fig. 2.35. The precooled hydrogen gas is expanded through a Laval nozzle, with the conditions chosen to produce large clusters (up to 10^6 molecules). A skimmer (700 µm ∅) transfers the central part of the beam into the first differential pumping stage, a second skimmer serves as collimator which limits the divergence of the beam to less than 1°.

After passing through a second differential stage pumped by a specially designed cryopump, the cluster beam arrives in the scattering chamber which is part of the proton storage ring. Afterwards the beam disappears again quantitatively in a beam absorber consisting of three differentially pumped stages, two pumped by cryopumps, one by a turbo-molecular pump. Two UHV valves on the scattering chamber allow sealing off the vacuum system of the storage ring from the vacuum system of the cluster beam device, in case of malfunction.

Target densities of $\rho \geq 10^{14}$ particles/cm^3 at pressures below 10^{-7} mbar in the scattering chamber are achieved.

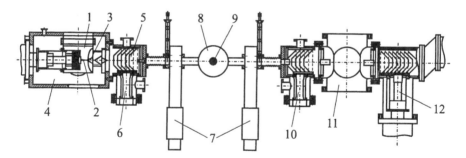

Fig. 2.35. Cluster beam as target for a storage ring. (1) source chamber, (2) Laval nozzle, (3) skimmer, (4) differential pumping stage, (5) differential pumping stage with cryopump (6), (7) UHV-Valves, (8) scattering chamber, (9) proton beam (perpendicular to plane of drawing), (10) and (12) differential pumping stages with cryopumps, (11) differential pumping stage with turbo-molecular pump

3. Velocity Measurement and Selection

To measure the velocity of neutral particles, different techniques in many variants are used in present-day molecular beam experiments. Two of these techniques are based on the definition of the velocity, requiring, therefore, a time and distance measurement. The slotted disk velocity selector, which has been frequently employed since the beginning of molecular beam research [Eldridge (1927)], makes use of the Fizeau principle. The distance to be measured is determined by the space between the first and the last disk, while the time is given by the rotational frequency. This method cannot only be used to measure velocities of molecular beams, but may also select particles of a desired velocity interval from a beam with a broad velocity spread (e.g. a Maxwellian velocity distribution). In time-of-flight measurements, the beam is chopped into narrow bunches of particles and the time required for the particles to travel the distance between chopper and detector is measured. The large dead time of this single-pulse method, which results from the condition that successive particle bunches are not allowed to overlap, causes a great loss in intensity, which can be overcome by using multiple-pulse methods (cross-correlation methods). Optical methods utilize a velocity-dependent physical effect (Doppler effect). Since the availability of intense, tunable, and narrow-band lasers, this method has found widespread application [Hammer et al. (1976)], in particular in experiments which already require a laser. The deflection in inhomogeneous magnetic fields has also often been used to measure velocities. This method requires a magnetic dipole moment for the particles under investigation [Cohen and Ellett (1937), Bassi et al. (1976), Este et al. (1978), (1983), Jaduszliwer and Chan (1994)]. This method can also be employed for velocity selection, and a substantial intensity increase can be achieved by using magnetic hexapole fields, which have focusing properties. The photon recoil, experienced by an atom during resonance absorption, also yields a velocity-dependent beam deflection, which can be used for velocity measurements [Düren et al. (1975)]. Finally, the free fall of atoms (deflection in the gravitational field) was already used many years ago to measure and select velocities of molecular beams [Estermann et al. (1947a), (1947b)]. However, in the two latter cases the deflections, which can be achieved with thermal energy molecular beams, are very small. Consequently, these methods have gained increasing practical importance since it is possible to produce slow atoms by the interaction with laser light (see Chap. 5).

A new method, which follows one of the classical experiments of Stern and co-workers [Estermann and Stern (1930), Estermann et al. (1931)], has been introduced rather recently: The diffraction of molecular beams using nanofabricated free-standing transmission gratings or appropriately prepared crystal surfaces. While the diffraction from crystals is a direct continuation of the old experiments, which is practically applicable only to helium atom beams, the recent progress in nanotechnology allows the fabrication of transmission gratings with periods down to 100 nm [Keith et al. (1991a), (1991b), Carter et al. (1992), Ekstrom et al. (1992)], which can be universally applied. Optical gratings, produced by standing laser waves, have also been realized (see Chap. 5) [Pritchard (1993), Abfalterer et al. (1997)]. By measuring the de Broglie wavelength, these methods enable velocity measurements for molecular beams with an unrivaled precision [Chapman et al. (1995), Schöllkopf and Toennies (1996), Schief et al. (1996), Buckland et al. (1997)].

Finally, it should be pointed out that velocity selection of molecular beams has become superfluous in many cases since molecular beams with narrow velocity distribution can be produced by gas dynamic expansions. By using carrier gases, the beam velocity may be varied within wide limits. Especially in the case of helium beams, reciprocal velocity resolutions $\Delta v/v$ of a few tenths of a percent are achievable (see Sect. I.3.6.4). On the other hand, new applications of velocity selection have been established. Clusters of different masses, which can be generated in gas aggregation or laser pyrolysis sources with a rather uniform translational temperature, have different velocities. Hence velocity selection may be used for a coarse cluster size selection [Roux et al. (1994)]. Similarly, a velocity selector may be used to separate the beam molecules from slower or faster contaminating species (e.g. in beams produced by a chemical reaction) [Porter and Grosser (1980)].

Concerning the experimental methods discussed in this chapter, a number of review articles is available in the literature [Fluendy and Lawley (1973), van den Meijdenberg (1988), Auerbach (1988), Bergmann (1988), Schmiedmayer et al. (1997)].

3.1 Mechanical Selectors

In 1920 Stern made a first rough measurement of the velocity distribution in a molecular beam. He used a device which essentially consisted of two rotating concentric cylinders. The inner cylinder carried the collimation slit, the outer one served as condensation detector. The beam source (a silver-sheathed platinum wire) was situated in the center of both cylinders [Stern (1920)]. A few years later, these measurements were refined by Eldridge (1927) and Lammert (1929). Their design was based on the principle that was used by Fizeau to measure the velocity of light. A molecular beam travels through a system of slotted disks mounted on a rotating shaft, with the disks adjusted in such a way that only particles of a desired velocity interval can pass through the slits of all disks at a given rotational frequency. This method, which has also often been used for velocity selection of thermal neutrons [Fink (1936), Zinn (1947)], was further refined during the 1950s in two directions: Rotating cylinders with helical slots cut into their surface [Dash and Sommers (1953), Miller and Kusch (1955), Greene et al. (1960), Bally et al. (1961), Höhne (1961)], and slotted disk devices, which have achieved a high degree of perfection and reliability and which have found widespread application [Moon (1953), Bennett (1954), Bennewitz (1956), Hostettler and Bernstein (1960), Pauly (1960), Politiek et al. (1968), Trujillo et al. (1962), (1975)].

3.1.1 Slotted Disk Velocity Selector (Fizeau Principle)

To derive the characteristic properties of a slotted disk velocity selector we consider first two disks of radius R and thickness d, located at a distance L on a common shaft which rotates with the frequency ν. Each disk has a slit of width a, the slit in the last disk being twisted by an angle $\phi = D/R$ (pitch angle) against the slit in the first disk, where D is the circumferential displacement between the two slits. Particles passing the slit in the first disk parallel to the selector axis can pass through the slit in the second disk, if their velocity v_0 satisfies the relation

$$v_0 = \frac{2\pi RL}{D} \nu. \tag{3.1}$$

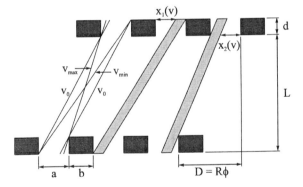

Fig. 3.1. Circumference of a two-disk velocity selector unrolled onto the drawing plane and illustration of the quantities necessary to derive the transmission T(v)

If the circumference of the disks is unrolled onto a plane, Fig. 3.1 is obtained, which can be interpreted as a time–distance diagram. In this diagram the trajectory of a particle having the nominal velocity v_0 corresponds to a straight line with the slope L/D. It can be seen further from Fig. 3.1, that not only particles of the nominal velocity v_0 are transmitted, but also particles with smaller and larger velocities between the limiting velocities v_{min} and v_{max}, which are given by

3. Velocity Measurement and Selection

$$v_{min} = v_0 \frac{L+d}{D+a} = v_0 \frac{1+d/L}{1+a/D} = v_0 \frac{1+\beta}{1+\gamma}, \text{ and} \tag{3.2}$$

$$v_{max} = v_0 \frac{L-d}{D-a} = v_0 \frac{1-d/L}{1-a/D} = v_0 \frac{1-\beta}{1-\gamma}. \tag{3.3}$$

In (3.2) and (3.3) we have used the dimensionless quantities

$$\beta = d/L \text{ and } \gamma = a/D, \tag{3.4}$$

for abbreviation. Furthermore, we introduce the fraction η of the circumference of the disks which is transparent to the beam. In the case discussed here of only one slit in each disk we have

$$\eta = a/2\pi R. \tag{3.5}$$

Simple geometric considerations using Fig. 3.1 yield the transmission curve

$$T = \eta\left(1 - \frac{x_1(v)}{a}\right) \quad \text{for } v_{min} \leq v \leq v_0,$$

$$T = \eta\left(1 - \frac{x_2(v)}{a}\right) \quad \text{for } v_0 \leq v \leq v_{min}, \tag{3.6}$$

in the first interval, v is given by

$$v = v_0 \frac{1+\beta}{1+x_1(v)/D}, \tag{3.7}$$

and in the second interval we have

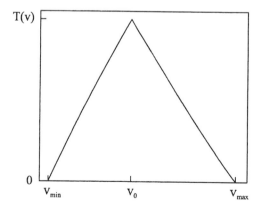

Fig. 3.2. Transmission curve of a slotted disk velocity selector according to (3.9). The sides of the triangle have a weak hyperbolic curvature

$$v = v_0 \frac{1-\beta}{1-x_2(v)/D}. \tag{3.8}$$

Calculating $x_1(v)$ and $x_2(v)$ from (3.7) and (3.8) and introducing the result into the equations (3.6) yields the transmission curve $T(v)$

$$T(v) = \frac{\eta}{\gamma}\left[(1+\gamma)-(1+\beta)\frac{v_0}{v}\right] \quad \text{for } v_{min} \leq v \leq v_0,$$

$$T(v) = \frac{\eta}{\gamma}\left[-(1-\gamma)+(1-\beta)\frac{v_0}{v}\right] \quad \text{for } v_0 \leq v \leq v_{max}, \tag{3.9}$$

$$T(v) = 0 \quad \text{for } v \leq v_{min} \text{ and } v \geq v_{max}.$$

The transmission curve $T(v)$ is a triangle with hyperbolically curved sides (see Fig. 3.2). We define the resolving power R_S of a selector in the following manner:

$$R_S = \frac{v_{max}+v_{min}}{v_{max}-v_{min}}. \tag{3.10}$$

Other definitions can also be found in the literature. Due to the weak asymmetry of the transmission curve, these yield expressions which differ slightly from each other. Hostettler and Bernstein (1960), for example, use the definition

$$R_S = \frac{2v_0}{(v_{max}-v_{min})}, \tag{3.10a}$$

while van den Meijdenberg (1988) defines the resolving power by

$$R_S = \frac{v_0}{\Delta v_{1/2}}, \tag{3.10b}$$

where $\Delta v_{1/2}$ is the halfwidth of the transmission curve. All expressions, however, converge with increasing resolution to the same value (for $R_S = 20$, for example, the deviations between the different definitions are of the order of a few tenths of a percent), so that the different definitions have no significance in practice. Using (3.2) to (3.4), the resolving power follows from (3.10) to be

$$R_S = \frac{(1-\gamma\beta)}{\gamma-\beta} \approx \frac{1}{\gamma-\beta}. \tag{3.11}$$

If f(v) is the velocity distribution of a given molecular beam, the apparent velocity distribution F(v) measured with a slotted disk velocity selector is given by the convolution product

$$F(v) = T(v) * f(v) = \int_{-\infty}^{\infty} T(v-u) f(u) du. \tag{3.12}$$

For $R_S \gg 1$ the convolution integral can be approximated by

$$F(v) = f(v) \int_{-\infty}^{\infty} T(u) du = f(v) T_i(v), \tag{3.13}$$

where $T_i(v)$ is the integral transmission, which, using (3.9), may be written as

$$T_i(v) = \frac{\eta}{\gamma} \cdot v \left[(1+\beta) \ln\left(\frac{1+\beta}{1+\gamma}\right) + (1-\beta) \ln\left(\frac{1-\beta}{1-\gamma}\right) \right], \tag{3.14}$$

or, for $\beta \ll 1$ and $\gamma \ll 1$

$$T_i(v) = v(\eta/\gamma)(\gamma - \beta)^2. \tag{3.15}$$

Accordingly, the integral transmission is

$$T_i(v) = v B(\eta, \beta, \gamma), \tag{3.16}$$

where $B(\eta,\beta,\gamma)$ is a factor which depends only on the geometry of the selector. Therefore, the apparent velocity distribution F(v) measured with the selector is given by

$$F(v) = v f(v) B(\eta, \beta, \gamma). \tag{3.17}$$

These results apply to a beam whose particle trajectories are parallel to each other and to the selector axis. A divergence of the beam reduces the resolution [Frankl (1974), van den Meijdenberg (1988)]. While the divergence in radial direction has a negligible influence on the resolution this is not true for the beam divergence in the tangential plane of the selector. If this divergence (total divergence angle α) is symmetric with respect to the selector axis, and if α satisfies the relation

$$\alpha \le a/2L \tag{3.18}$$

(a = slit width, L = length of the selector), the resolution may be approximated by

$$R_s = \frac{1}{\gamma - \beta + \alpha/\phi}, \qquad (3.19)$$

where α/ϕ is the ratio of the divergence angle α and the twisting angle ϕ. In many molecular beam experiments, the beam divergence α is essentially determined by the distance between source and detector (usually of the order of 1 m) and the size of the detector (of the order of a few mm) in the tangential plane of the selector. This means $\alpha/\phi \ll 1$, and the influence of the beam divergence on the resolution can be neglected in these cases.

Elimination of Sidebands. The selector composed of two disks with one slit in each disk discussed so far transmits not only particles of the nominal velocity $v_0 = 2\pi R \nu L/D$, but also particles having velocities which are given by

$$v_k = \frac{D}{k2\pi R + D} v_0, \qquad (3.20)$$

with $k = 1, 2, 3 \cdots$ (higher harmonics). These particles are so slow that the selector has rotated by an angle $k2\pi + \phi$ during their travel from the first to the last disk. For a finite thickness d of the disks, however, these particles can only pass through the selector slits if the thickness d fulfills the condition

$$d < \frac{La}{k2\pi R + D}. \qquad (3.21)$$

In the case of a selector with two disks with one slit in each disk, already the first sideband ($k = 1$) is generally closed, since the condition (3.20) would require very thin disks. Despite the fact that a selector without sidebands can be easily realized with two disks (and one slit in each disk), such a selector is not very useful in practical applications due to its very low transmission η (of the order of 10^{-3} (see (3.5)). The transmission may be increased by about an order of magnitude, if each disk carries several slits on its circumference with enough space between adjacent slits so that no sidebands exist for disks of appropriate thickness. Such a two-disk selector with 6 slits in each disk has been described by Marcus and McFee (1959), for example. In this case each disk is rotated by a separate synchronous motor in contrast to the selectors discussed so far. A phase shifter varies the phase of one motor relative to the other, so that a velocity distribution can be measured at a fixed rotational frequency by varying the phase (and thus the twisting angle ϕ) between the two disks.

In a modern version of this principle, which is often used for velocity selection of beams produced by pulsed laser ablation or by pulsed laser pyrolysis, only one disk is required since the first disk is replaced by the pulsed laser. Using a light barrier and a variable delay generator, the laser is synchronized to the rotating disk. By choosing the appropriate time between firing the laser and the moment when the particles pass through the slit of the disk, it is possible to select a given velocity interval from the velocity distribution of the beam [Weaver and Leone (1995), Ehbrecht et al. (1997), Laguna et al. (1999)]. Another version of this principle, well approved for many years, utilizes a single chopper disk and replaces the second disk by a pulsed tunable laser for laser-induced fluorescence detection [Pasternack and Dagdigian (1977)].

To achieve maximum transmission, the width b of the opaque part between successive slits (tooth width) has to be equal to the slit width a. This results in $\eta = a/(a + b)$ yielding the maximum transmission of 50%. As a safety precaution against adjustment errors and machining imperfections, the width of the teeth is usually made somewhat larger than the slit width in practice. This and the finite disk thickness, which also increases the effective width of the teeth, reduces the transmission (see (3.14)), so that achievable transmissions are always lower than 50%. In such a device, however, the sidebands are very close to each other and additional disks must be inserted between the first and the last disk, so that particles with other velocities strike a tooth on their way through the selector. This is illustrated in Fig. 3.3 [Hostettler and Bernstein (1960)].

For a given nominal velocity v_0, a corresponding rotational frequency v and a desired resolving power R_S, it is rather easy to find an arrangement of disks without sidebands, if arbitrary displacements of the disks are permitted. This can be simply achieved with a minimum number of disks if these subdivide the selector length L according to a geometrical progression. The adjustment of such a design, however, is rather complicated. Arrangements, in which the slits of the various

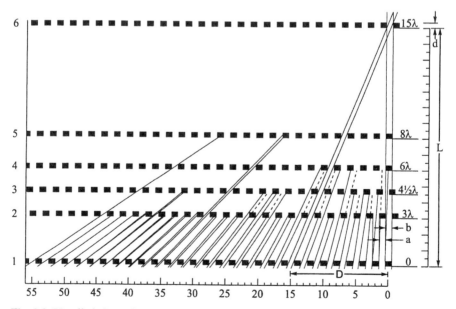

Fig. 3.3. Unrolled circumference (time–distance diagram) of a slotted disk velocity selector with additional disks to eliminate higher harmonics. The position of the disks is given at the right in units of $\lambda = L/n$ [Hostettler and Bernstein (1960)]

disks either coincide (peripheral displacement = a + b) or fall exactly between one another (displacement = a) are much easier to adjust. This symmetry of a selector requires the fulfillment of the condition

$$R\phi = \frac{n(a+b)}{2}, \qquad (3.22)$$

where n is an integer. The pitch angles ϕ_k of the intermediate disks have to satisfy the corresponding conditions

$$R\phi_k = \frac{m(a+b)}{2}, \quad \text{with} \quad m = 1, 2, 3 \ldots, n-1. \qquad (3.23)$$

For a symmetric selector, therefore, the number of locations for the intermediate disks is limited to n − 1. These locations are usually expressed in units of length, λ, along the selector axis

$$\lambda = 2L/n. \qquad (3.24)$$

A symmetric selector facilitates not only the adjustment, but also has the advantage that the direction of rotation can be reversed. In this way, it can be controlled as to whether the selector axis is correctly aligned parallel to the molecular beam, which is necessary for an accurate absolute measurement of velocities. This aspect is considered below in more detail.

If it is not possible to find a symmetric device for a given set of selector data, an arrangement, which requires only one disk with a displacement of a/2, can usually be found. This complicates the adjustment only insignificantly. A very simple elimination of sidebands can be achieved by augmenting the thickness of the first disk. This can be easily seen from a time–distance diagram such as shown in Fig. 3.3.

In many cases, selector configurations have been found by trial and error, either graphically, using a diagram similar to Fig. 3.3 [Hostettler and Bernstein (1960)] or by an appropriate computer program [Helbing (1966), van Steyn and Verster (1972)]. It has been found by van Steyn and Verster, who investigated a large number of possible selector configurations, that some configurations are very insensitive to small adjustment errors, while in other cases small errors can easily lead to an incomplete elimination of sidebands. They defined a criterion to characterize selector configurations in this respect. Whenever possible, the first mentioned arrangements should be used. A systematic procedure to eliminate sidebands for selectors having helical slits has been proposed by Kinsey (1966). Later, it was extended to apply also to disks with straight slits [Wykes (1969), Bolzinger (1975)].

To compare different selectors, a figure of merit G may be introduced, which summarizes the important properties of a selector in one formula [Pauly and

Toennies (1968)]. It is the product of transparency η, resolving power R_S, and the velocity v_{max} at the maximum rotational frequency v_{max} of which the rotor is capable

$$G = \eta R_S v_{max} = \frac{2\pi RL}{a+b} v_{max} = ZLv_{max}. \tag{3.25}$$

Z is the number of slits of the disks. Since the length L of the selector should be as small as possible to obtain maximum intensity, a large value of G requires a large number, Z, of slits and a high maximum frequency, v_{max}. Neglecting small, compact rotors with low velocity resolution, values of G are of the order of 10^6 to 10^7 cm/s.

Technical Details. Various techniques to produce the slits have been applied. Slits as narrow as 0.3 mm may be cut into disks of aluminum alloys with standard slitting saws. The disks (between 5 and 15, depending on the required slit width and the thickness of the disks) are stacked on a mandrel, clamped between two thick end disks (about 1 cm thick), and fixed on a dividing head. The slits are machined on a jig borer, using high-speed slotting cutters of appropriate thickness. When assembling the selector, it is important to mount the disks in the same sequence as has been used for sawing, the desired paths of the particles being defined by slits which have been machined on the same pass of the cutter. In this manner, small machining errors can be compensated and have no influence on the nominal rotor data.

Aside from sawing, a number of other tooling methods have also been used for slit production. Among these are punching [Moore and Opal (1975)], chemical etching [Cardillo et al. (1971)], and spark erosion [Kristensen et al. (1967), Kim (1991)]. The latter technique yields highly accurate slits (with an accuracy of a few thousandth of a millimeter) and has the advantage, compared with edging methods, that the thickness of the disks is not limited. The smallest slit and tooth widths, finally, are achieved by photoetching [Giles and Logan (1972), Trujillo (1975)], a method that has found widespread application. In this case, however, the thickness of the disks must be smaller than or equal to the slit width; thus, an elimination of sidebands by the finite thickness of the disks is impossible.

The choice of the disk material is determined by its tensile strength if high rotational speeds are required. A further criterion for selecting the material is its aptitude for the machining method employed. The maximum admissible stress within a disk of radius R (in cm) due to the centrifugal forces is given by

$$\sigma_R = 1.67 \times 10^{-7} \rho R^2 v^2 \text{ kp/mm}^2, \tag{3.26}$$

where ρ is the density of the disk material (in g/cm^3) and v is the frequency of rotation (in Hz). The tensile strengths of hardened steel alloys are of the order of 100 to 150 kp/mm^2. Special titanium alloys have comparable tensile strengths (110–130 kp/mm^2). Assuming a disk radius of R = 10 cm and a tensile strength of about 130 kp/mm², a maximum rotational frequency of the order of 900 Hz (for steel) and about 1180 Hz (for titanium), respectively, is feasible, if a safety factor of 10% is allowed for. A constant stress within the disks can be achieved by a special shape of the disks (see Sect. 3.2.3).

While the length L of the rotor axis is determined by the desired velocity range according to (3.1), its diameter d follows from the following considerations: To avoid the excitation of eigen frequencies of the rotor axis, the frequency of its lowest bending oscillation should be higher than the maximum rotor frequency used. For a simple estimate, we use the eigen frequencies of an unloaded axis of circular cross section, which can be written as

$$v_k = \frac{d}{3L^2} \frac{\beta_k^2}{2\pi} \sqrt{\frac{E}{\rho}}. \qquad (3.27)$$

E is the modulus of elasticity and ρ the density of the axis. The coefficients β_k (k = 1, 2, $\cdots$) depend on the bearing of the axis. For a rod which is free at the ends, or which is supported on flexible mounts, the coefficients β_i are the solutions of the equation $\cos\beta_i \cosh\beta_i = 1$. For the fundamental frequency v_1 this yields $\beta_1 = 4.730$. This is approximately 2.3 times higher than that of the same rod simply supported at both ends, such as by rigidly mounted bearings ($\beta_k = k\pi$). Using the latter value for β_1, the estimate (for an axis made of steel) is

$$d \geq \frac{vL^2}{2} 10^{-5} \, cm, \qquad (3.28)$$

where v must be inserted in Hz and L in cm. Hence a rather large diameter is required.

Different methods have been used to assemble and adjust the disks on the rotor axis, which, to some extent, are described in the literature [Hostettler and Bernstein (1960), Trujillo et al. (1962), Kinsey (1966), van Steyn and Verster (1972, Frankl (1974), Bolzinger (1975)].

A simple method, which is especially suited to adjust short selectors with relatively large slits (≥ 0.1 cm), is the following: The disks are stacked on a short mandrel in the same sequence as has been used for machining the slits. After having made the necessary displacements, the disks are clamped together between two screw-nuts, and adjustment holes (e.g. four on the circumference of the disks) are bored through the packet of disks. Then the disks are assembled on the axis with appropriate spacers in between. Four rods in the adjustment holes with appropriate spacers guarantee the adjustment. After having fixed the disks on the axis between two screw-nuts, the adjustment rods are removed.

A more accurate procedure can be used to align arbitrary displacements of the disks. The axis with the unadjusted disks and the spacers is chucked into a turning lathe. One slit in each disk, machined on the same pass of the cutter, serves as "reference slit". The disks are adjusted one after the other, using an optical thread

148   3. Velocity Measurement and Selection

tool gauge to fix the twist angles of the individual disks with the turning lathe. The selector axis has two longitudinal groves to fit corresponding tongues of the spacers, so that the disks cannot rotate when the whole set of disks is screwed together. A rather similar method has been described in the literature [Trujillo et al. (1962)].

After being assembled, the selector should be dynamically balanced to a high degree of accuracy. For rotational speeds not too high (below 300 Hz), a precisely machined selector may be operated without balancing. In this case, however, it is necessary to allow the rotor to find its own axis of rotation by supporting the ball bearings at the ends of the axis by compliant coil springs or rubber bands [Pauly and Toennies (1968)].

An approved design of a selector drive is shown in Fig. 3.4, which presents a cross-sectional view of a slotted disk velocity selector. A squirrel-cage rotor of an induction motor is mounted on the axis between two disks, the stator with its windings is in a casing of Araldite cast into an aluminum block, so that the windings cause no vacuum problems. This design has the advantage that the overall length of the selector is kept at a minimum. Furthermore, no problems occur due to the coupling between the motor and the selector axis. In many cases, however, velocity selectors have also been driven by commercially available motors designed for use in vacuum. The motor is connected to the velocity selector rotor by a short wire [Trujillo et al. (1962)]. To measure the rotational speed of the rotor, one of the disks contains a number of boreholes, and the resulting light pulses due to illumination by a small bulb or photodiode are detected by a photocell and counted on an electric counter.

Commercial precision ball bearings, suited for the desired rotational speed and for use in vacuum, are usually used. Well approved are self-lubricated bearings

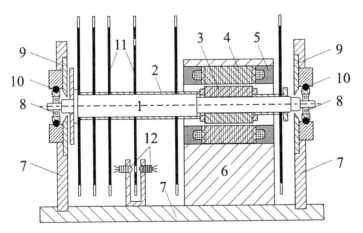

Fig. 3.4. Cross-sectional view of a slotted disk velocity selector. (1) rotor axis, (2) spacers, (3) laminated rotor, (4) stator winding, (5) Araldite vacuum casing, (6) aluminum block, (7) aluminum support plates, (8) ball bearings, (9) emergency bearings (brass), (10) rubber O-rings for soft bearing support, (11) slotted disks, (12) light source and photodiode for rotational frequency measurement

with cages of special material that offer reliable performance in vacuum up to temperatures of 300 °C. The cage material essentially consists of Teflon-coated, minute glass fibers impregnated with molybdenum disulfide. But other bearings which must be lubricated, have also often been used. If the rotor is precisely balanced, the bearings at the end of the axis may be rather rigidly mounted, using a shock-absorbing rubber O-ring between the bearings and the supporting plates, as shown in Fig. 3.4. As already mentioned above, unbalanced or not precisely balanced rotors are supported by coil springs or rubber bands.

The condition of the ball bearings can be easily checked. At a given angular frequency of the rotor its drive is switched off and the time required to come to rest is measured, since at pressures below 10^{-4} torr the friction torque due to the residual gas can be neglected compared to the friction torque caused by the bearings. The angular frequency ω of the rotor (after having switched off the motor at a time $t = 0$) obeys the simple differential equation

$$\Theta \frac{d\omega}{dt} + D\omega = 0, \tag{3.29}$$

where Θ is the moment of inertia of the rotor and D is the friction torque of the bearings. Thus, the decrease of the angular frequency with time is given by

$$\omega = \omega_0 \exp(-(D/\Theta)t). \tag{3.30}$$

$t_0 = \Theta/D$ is the time constant of the deceleration process. For large rotors, empirical values of t_0 are of the order of several hours starting at initial angular frequencies of the order of 300 Hz.

Having determined the time constant t_0 with new bearings, their condition may be checked from time to time by re-measuring t_0. This is especially important if bearings are used which must be lubricated.

Close to the ball bearings, at the tapered ends of the axis, there are also "emergency bearings", which are simple plates of brass or Teflon having a hole with a diameter about 1 mm larger than the diameter of the axis at this location. This is an approved safety precaution. In the case of a sudden unbalance of the rotor due to a damage of a disk or if a bearing freezes, the energy of the rotor is converted into deformation energy of the emergency bearings (and into heat) in a controlled way.

A slotted disk velocity selector offers no possibility for optical adjustment, which is necessary to align the rotor axis parallel to the beam. Therefore, a straight passage at a given position of the selector is usually made by removing a few teeth. The resulting sidebands are negligible due to the very low transmitted intensity. To keep the rotor in the position required for optical alignment, a small brake, which can be actuated electromagnetically, touches one of the disks. In other designs, the rotor frame is hinged. In the rest position, the rotor is just below the atomic beam to allow optical adjustment. A mechanical or electromagnetic device lifts the selector into the beam path.

The moment of inertia of a slotted disk velocity selector is rather high. Assuming a design according to Fig. 3.4, with an axis of 30 cm length, 10 aluminum disks of 0.15 cm thickness and a radius of 8 cm, results in a moment of inertia of the order of 50 kg cm^2. At high rotational speeds, therefore, a considerable amount of kinetic energy is stored in a velocity selector. At a rotational frequency of 500 Hz, the kinetic energy (in the above example) is about 2.5×10^{11} erg. This corresponds to the kinetic energy of a car (with a weight of one ton) moving at a velocity of 25 km/h. In the case of a defect of the rotor or its ball bearings, this energy is converted into deformation energy of the emergency bearings shown in Fig. 3.4.

3.1.2 Calibration

The relationship between the rotational frequency v of the velocity selector and the nominal transmitted velocity v_0 is given by (3.1). To check the adjustment (both of the arrangement of the disks on the axis and the parallelism between the rotor axis and the beam) the selector must be calibrated. Some possible calibration methods are discussed in the following.

Measurement of a Known Velocity Distribution. A well known velocity distribution is the Maxwellian distribution $f(v)$, which is obtained by using an effusive source with an ideal aperture and which is uniquely determined by the temperature of the source (see Sect. I.2.1.1). The measured "apparent" distribution function $F(v)$, taking into account the modification due to the velocity selector, is described by (3.17). Furthermore, the measured distribution function $f(v)$ depends on the quantity which is measured by the detector (particle flux (I.2.111), particle number density (I.2.14), or energy flux (I.2.119)). Table 3.1 gives a survey of the modifications of the Maxwellian distribution function due to the different types of detectors and due to the velocity selector.

Table 3.1. Normalized modifications of the Maxwellian velocity distribution due to the detector ($f(x)$) and due to the selector ($F(x)$) ($x = v/v_w$)

Detector	$f(x)$	$F(x)$
Number density detector (e.g. universal beam detector)	$\frac{4}{\sqrt{\pi}} x^2 \exp(-x^2)$	$2x^3 \exp(-x^2)$
Particle flux detector (e.g. Langmuir–Taylor detector)	$2x^3 \exp(-x^2)$	$\frac{8}{3\sqrt{\pi}} x^4 \exp(-x^2)$
Energy flux detector (e.g. bolometer detector)	$x^5 \exp(-x^2)$	$\frac{16}{15\sqrt{\pi}} x^6 \exp(-x^2)$

Instead of an effusive beam with Maxwellian speed distribution, a supersonic atomic beam of high Mach number may also be used for calibration. The speed distribution of such a beam is completely determined by the source temperature and the Mach number (see Sect. I.3.6.4). Although the speed distribution of nozzle beams may be subject to some uncertainties, these are smaller than 1% if beams of monoatomic gases (rare gases) with sufficiently high Mach numbers ($M_a > 10$) are used. By employing beams from different rare gases at different temperatures, a rather accurate calibration is possible [Bredewout (1973), Linse (1977)].

Symmetric Selectors. Symmetric selectors, which transmit a velocity-selected beam (at the same nominal speed) rotating in either a clockwise ("forward") or counterclockwise ("reverse") sense, have already been mentioned. Selectors which transmit different nominal velocities in the forward and reverse mode, respectively, have also been constructed [Grosser 1967), Cardillo et al. (1971)]. Both types of selectors are especially suited for an accurate velocity calibration. This is accomplished by measuring the velocity v_0 at which the maximum of a narrow velocity distribution is observed (e.g. the distribution of a nozzle beam) with the selector rotating in forward and reverse directions. For perfect alignment, the speeds of the maximum determined from the measured frequencies and the design parameters of the selector should be the same (v_0), independent of the direction of rotation. A discrepancy in the agreement can be ascribed to a misalignment of the selector axis with the beam direction (with a misalignment angle $\alpha = D'/R$). The angle between the selector axis and the beam direction has the effect of an additional twisting D' (see (3.1)), which is added to the twisting D_f in the forward direction of rotation and subtracted from the twisting D_r in the reverse direction of rotation. If v_f and v_r are the corresponding frequencies at which the maximum of the distribution is measured, and taking into account that the measured velocity v_0 must be independent of the sense of rotation, the following equation is obtained

$$v_0 = \frac{2\pi RL}{D_f - D'} v_f = \frac{2\pi RL}{D_r + D'} v_r. \tag{3.31}$$

The ratio of the measured rotational frequencies at the maximum of the distribution gives immediately the misalignment twisting D'

$$D' = (v_r D_f - v_f D_r)/(v_r + v_f), \tag{3.32}$$

and the correct velocity can be easily determined.

Comparison with Time-of-Flight Measurements. A slotted disk velocity selector may also be calibrated by comparison with time-of-flight measurements. This is especially useful to check the absence of velocity sidebands [Johnson et al. (1970), Bolzinger (1975)]. An accurate absolute calibration, however, requires a considerable effort and a precise time-of-flight spectrometer (see Sect. 3.2).

Diffraction from Transmission Gratings. As described in detail in Sect. 3.6, the diffraction of molecular beams from crystals or from free-standing transmission gratings is the most accurate method to determine velocities of molecular beams. Although this method has not been used to calibrate mechanical velocity selectors, it yields the highest achievable precision, since crystal data are accurately known and the data of transmission gratings can be precisely determined by diffraction of monochromatic light. This has been demonstrated by Buckland et al. (1997), who used the diffraction of helium beams to calibrate a time-of-flight spectrometer. An inverse velocity resolution $\Delta v/v$ of 0.1% can be achieved (see Sect. 3.6.1).

3.1.3 Special Designs

Magnetic Suspension. In early experiments with mechanical selectors, ball bearing problems sometimes occurred due to the use of unbalanced rotors. To avoid these problems, magnetically suspended rotors have been used [Bennewitz et al. (1972)]. In such an arrangement, the vertical axis rests on a point support and its lateral guidance is made by suitably controlled magnetic fields. The same principle, for example, is used in vacuum technology for turbopumps and in isotope separation for ultracentrifuges. The main axis of inertia of the rotor adjusts automatically parallel to the gravitational field of the Earth. The beam, therefore, must be also aligned parallel to the gravitational field. This is achieved with high accuracy (deviations $< 2 \times 10^{-4}$) using a laser beam which is reflected from a liquid surface. Static and dynamic unbalances, however, lead to disturbances of the rotation (precession and nutation) which destroy the parallelism between axis and gravitational field. The amplitudes of these disturbances increase with increasing rotational speed [Dohmann (1970)]. Hence, accurate balancing of the rotor is required.

The frictional losses of a magnetically suspended rotor are extremely low, the time constant t_0 (see (3.30)) is of the order of one to two days. The main contributions to the losses are friction due to the residual gas, due to the point support, and due to an eddy current damping caused by the magnetic suspension.

Compact Selectors and Selectors with Variable Resolution. Short selectors (with lengths ≤ 2 cm) have often been used to analyze intense target beams in scattering experiments [Buck and Pauly (1968), (1971)]. A very short selector with a resolution $R_S = 3.5$ and a transmission of 60% has been used by Buck (1969) to analyze a target beam. It consists of one slotted disk (8.5 mm thickness, R = 71.3 mm) with 250 slits, machined at an angle of 3.5° to the rotor axis. Thus, it is a short version of a "slotted cylinder" selector discussed in the next section. A rather similar selector has been described by Zatselyapin (1974).

Instead of a disk with slits tilted against the rotor axis, a disk with straight slits may also be used, if the selector axis is inclined by an angle Θ against the beam direction. The resolving power R_S and transmission $T(v)$ can be easily determined following the same geometrical considerations as those used in Sect. 3.1.1. The expressions for ($T(v)$, R_S, and $T_i(v)$) are obtained from (3.9), (3.11), and (3.14), respectively, for $\beta = 0$. Since the selector twisting is $D = Ltg\Theta$, the resolving power A as a function of Θ becomes

$$R_S = 1/\gamma = (L/a)tg\Theta. \qquad (3.33)$$

Such a selector, which has been used to produce a beam of size-selected clusters by scattering and subsequent velocity selection (see Sect. 2.8.3) [Buck and Meyer (1984), Buck et al. (1985)], has been described by Lohbrandt (1993) [Baumfalk et al. (1997)]. It consists of one disk of 19 mm diameter, having 785 slits of 0.2 mm width (the tooth width is also 0.2 mm), machined by spark erosion. This device guarantees a minimum intensity loss, and the resolution can be easily adapted to the actual requirements of the experiment. Since the path of the beam particles s is given by $s = L/\cos\Theta$, the relationship between nominal velocity v_0 and the rotor frequency ν according to (3.1) follows as

$$v_0 = 2\pi R\nu/\sin\Theta. \qquad (3.34)$$

The maximum rotational frequency is 800 Hz. According to (3.34), the maximum speed which can be selected depends on the tilting angle Θ and thus on the resolution. For the device described above, the accessible range of angles lies between $\Theta = 6°$ and $35°$, corresponding to resolving powers between $R_S = 5.2$ and $R_S = 35$.

The shortest selector described in the literature has a length of 0.3 mm [Este et al. (1975)]. Commercially available multichannel arrays (channel diameter 12.7 µm, channel length 0.3 mm) are mounted at the periphery of a single disk. The channels are parallel to the rotor axis, and the axis is inclined to the direction of the beam. Resolutions between $R_S = 3$ and $R_S = 10$ can be achieved.

3.1.4 Slotted Cylinder Velocity Selector

The first selector of this type was used by Levinstein and Crane (1946) to investigate the condensation of metals as a function of the velocity of the impinging atoms. As an example, a brief description of a device used by Miller and Kusch (1955), for precise measurements of the Maxwellian velocity distribution in beams of alkali halides, is given below. From small modifications of the Maxwellian distribution caused by dimers and trimers, Miller and Kusch were able to determine the fraction of these species in the beam.

The selector consisted of a cylinder (duralumin) of 20.4 cm length, with 700 helical slits (slit width 0.424 mm, tooth width 0.474 mm) and two straight alignment slits (see Fig. 3.5). The selector axis (steel) was attached by two flanges at both ends of the cylinder. The motor was outside the vacuum vessel, its power was transferred to a gear unit inside the vacuum using a rotary vacuum seal. The gear unit increased the rotational speed of the rotor by a factor of four. This was necessary to keep the strain on the vacuum seal as low as possible. For the same reason,

154   3. Velocity Measurement and Selection

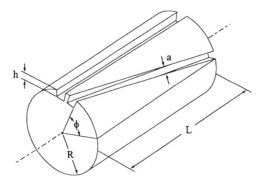

Fig. 3.5. Slotted cylinder selector [Miller and Kusch (1955)]. Length L = 204 mm, slit width a = 0.424 mm, slit depth h = 3.18 mm, radius R = 100 mm, pitch angle ϕ = 4.82°

the pitch angle ϕ was rather small to achieve velocities between 10^4 and 10^5 cm/s with rotational speeds below 100 Hz. A more recent version of this type of selector has been described by Arnold et al. (1985a).

The advantages of cylinder selectors are the complete absence of sidebands and its internal adjustment which is automatically achieved by the machining process. Disadvantages are the rather complicated machining procedure, its high weight and its high moment of inertia (about an order of magnitude higher than that of a slotted disk selector), which complicates the use of high rotational speeds, even if the present-day state of the technique would be used to rotate the selector. However, the same selector data as those described in the example above could be realized nowadays with a cylinder selector of 2 cm length, since such a rotor can be easily operated up to frequencies of 800 Hz. Since the intensity of a molecular beam decreases quadratically with distance, this would result in an intensity increase of two orders of magnitude.

3.1.5 Other Designs

Another type of rotor, often used for velocity selection of thermal neutrons, was first applied to molecular beams by Grosser et al. (1963). It consists of a horizontally operated single disk (outer diameter 19 cm, inner diameter 7.8 cm), with slightly curved radial slots. A schematic drawing is shown in Fig. 3.6. The glass–ceramic plate is framed by two metal plates, which are suspended from the axis of an induction motor. The axis of rotation is perpendicular both to the plate and to the beam direction. Thus the beam source may be either located at the center or on the outside of the disk. A disadvantage of this "slotted plate" selector is the limited space available for experiments at the center hole of the plate. This unfavorable circumstance is avoided by a rather similar device that has been described by Colgate and Imeson (1965). The spiral-shaped slots are milled into a metal plate and extend over the whole diameter of the plate. Therefore, this selector may be placed into the beam path at any position. Colgate and Imeson have also calculated transmission and resolving power of this selector. Miers et al.

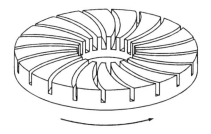

Fig. 3.6. Schematic view of a slotted plate velocity selector [Grosser et al. (1963)]. Since the path of selection is given by the radius of the plate, either the detector or the source must be located at the center hole of the plate

have used this principle to construct a selector of high resolving power ($\Delta v/v = 1.2\%$) and high transmission (38%) [Miers et al. (1988)].

Similar to the case of a slotted cylinder selector, where the spiral-shaped slots on the cylinder surface may be replaced by an appropriate arrangement of slotted disks, a slotted plate selector may be replaced by slotted concentric rings, as is shown schematically in Fig. 3.7 (Cowley et al. (1970)). In this case, a device with two rings is used. The slits in the outer ring are short; they determine the transmitted velocity interval, while the long slits in the inner ring serve to eliminate velocity sidebands. The nominal transmitted velocity v_0 is given by

$$v_0 = nL\nu, \tag{3.35}$$

where n is the number of slits, L is the distance between the two rings, and ν is the rotational frequency. The resolving power of this "slotted ring" selector depends not only on the width of the slits, but also on the width of the beam. If beam and slit widths are approximately equal, the resolving power R_S may be written as

$$R_S = 4\pi L/(na), \tag{3.36}$$

with a being the slit width.

Due to the rotor symmetry, the transmitted velocity interval is independent of the direction of rotation. This facilitates the adjustment.

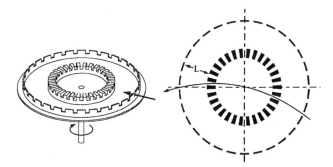

Fig. 3.7. View (*left half*) and top view (*right half*) of a slotted ring velocity selector [Cowly et al. (1970)]. The top view shows the apparent trajectory of a particle, the path of selection is the diameter of the ring. The selection is independent of the sense of rotation

In contrast to slotted disk velocity selectors, which have found widespread application in molecular beam experiments, other mechanical selectors have only been employed sporadically.

3.2 Time-of-Flight Methods

The basic principle of the time-of-flight method is extremely simple. From a measurement of the time, t, required for a particle to travel a given distance, L_0, the speed of the particle $v = L_0/t$ is determined.

In practice, this principle is realized by chopping the beam into narrow bunches of particles at the beginning of the flight path $L = 0$. These particle bunches are allowed to propagate in space to the detector at $L = L_0$, and particles of different velocities arrive at the detector at different times. Thus, the detector output yields the distribution of arrival times, which can easily be converted into the molecular speed distribution. This principle is illustrated in Fig. 3.8.

In general, chopping of the beam is performed by a rotating disk chopper with one slit at the edge of the disk, as is schematically shown at the left of Fig. 3.8. The transmitted distribution (chopper gating function) has a triangular or trapezoidal shape, as is discussed in Sect. 3.2.1. At successively later times (shown top to bottom in Fig. 3.8), this initial bunch spreads out according to the speed distribution of the beam. The detector integrates the particles arriving during a short time interval (black region under the intensity versus distance curves in Fig. 3.8) and adds this signal to a counter associated with this time interval. In this manner, the complete time-of-flight distribution is obtained from a single particle bunch. This process can be repeated until the desired signal-to-noise ratio is obtained. It must be noted, however, that the repetition rate $f_g = 1/t_g$ must be chosen such that the time between two consecutive gate pulses, t_g, is larger than the time required for the slowest beam particles to travel through the distance L_0. This condition is necessary to avoid an overlap of successive pulses. Moreover, the gate opening time, Δt, must be kept much shorter than the spread in flight times in order to obtain good resolution. Consequently, the duty factor $\eta = \Delta t/t_g$ of this "single-pulse" method will generally be small, typically of the order of 10^{-2}. This is in complete analogy to slotted disk velocity selectors, where a device of two disks with a single slit in each disk also has a transmission of this order of magnitude.

A great gain in intensity is achieved by using "multiple-pulse" methods such as the cross-correlation method, which makes use of a pseudorandom chopping sequence for which time-of-flight distributions can be extracted from the measured signal by taking the cross-correlation function of the modulation function with the measured response (see Sect. 3.2.2).

Chopping of the beam with a rotating disk with a single slit is a universally applicable method to determine the starting time of a time-of-flight measurement, but it is not the only possibility. The beam particles, for example, may be excited by electron impact into a metastable state at the beginning of their flight path

3. Velocity Measurement and Selection 157

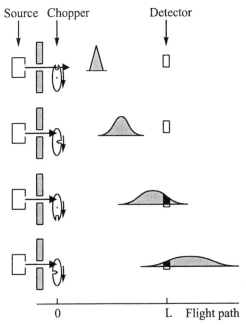

Fig. 3.8. Schematic illustration of a time-of-flight measurement [Auerbach (1988)]. The molecular beam leaving the source on the left is chopped by a rotating disk into narrow particle bunches which travel to the detector on the right. Due to the velocity spread in the beam, the particle bunches disperse in time as is indicated from top to bottom. The time-of-flight signal at which these intercept the detector is shown in black

[Clampitt and Newton (1969), Martini et al. (1987)]. Laser excitation of a high-lying Rydberg state of sufficient lifetime is another possibility. Further methods are described in Sect 3.2.3.

3.2.1 Resolution and Methods of Deconvolution

The most important factors determining the resolution of a time-of-flight measurement are (1) the finite width of the chopper gating function, and (2) the temporal resolution of the detector. In the case of a universal beam detector, for example, (see Sect. I.5.6), the latter is determined by the finite length of the ionization region.

Finite Width of the Chopper Gating Function. To investigate the influence of the finite width of the chopper gating function, we consider a rotating disk of radius R having a slit of width a, which modulates a beam of width b with the angular frequency ω. In general, the time dependence of the transmission T(t) has a trapezoidal shape, the total opening time $t_o = 2\Delta t_1 + \Delta t_2$ (see Fig. 3.9) being given by

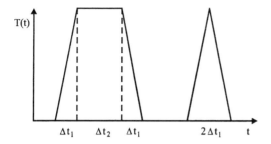

Fig. 3.9. Transmission curves of a chopper wheel. The time intervals Δt_1 and Δt_2 are given in Table 3.2

$$t_o = 2\Delta t_1 + \Delta t_2 = (a+b)/\omega R. \tag{3.37}$$

If the slit and beam width are equal (a = b), T(t) has the form of an isosceles triangle. Figure 3.9 shows the possible transmission curves T(t). The time intervals Δt_1 and Δt_2 can be expressed by the properties of the chopper and the beam (slit width a, beam width b, and angular frequency ω), as is shown in Table 3.2.

Table 3.2. Time intervals of the transmission curve

	b > a	b < a	b = a
Δt_1	$a/\omega R$	$b/\omega R$	$a/\omega R$
Δt_2	$(b-a)/\omega R$	$(a-b)/\omega R$	0

The apparent time-of-flight distribution G(t) measured at the detector is the convolution integral of the transmission function T(t) and the time-of-flight distribution g(t)

$$G(t) \propto \int_{-\infty}^{\infty} T(t-\tau)g(\tau)d\tau. \tag{3.38}$$

The time-of-flight distribution g(t)dt follows from the velocity distribution f(v)dv by the substitution v = L/t and $dv = -(L/t^2) dt$:

$$g(t)dt = -\frac{L}{t^2} f\left(\frac{L}{t}\right) dt. \tag{3.39}$$

The minus sign is generally disregarded when plotting time-of-flight distributions. It is important, however, to note that the distribution function f(v) to be used depends on the type of detector employed. As an example, we consider a triangular distribution (b = a). This may be represented by

$$T(t) = \eta \frac{\omega R}{a} t \qquad \text{for } 0 \leq t \leq \frac{a}{\omega R}, \tag{3.40a}$$

$$T(t) = \eta\left(2 - \frac{\omega R}{a}t\right) \quad \text{for } \frac{a}{\omega R} \le t \le \frac{2a}{\omega R}, \tag{3.40b}$$

$$T(t) = 0 \quad \text{for } t \le 0 \text{ and } t \ge \frac{2a}{\omega R}, \tag{3.40c}$$

where η is the particle flux of the beam. If the halfwidth of the transmission curve $T(t)$ is small compared to structures in the time-of-flight distribution $g(t)$, the convolution integral (3.38) can be approximated by the product of the distribution function g at time t with the integral transmission

$$G(t) \propto g(t) \int_{-\infty}^{\infty} T(\tau) d\tau = \eta \frac{a}{\omega R} g(t) = N_0 g(t). \tag{3.41}$$

N_0 is the number of particles within one chopper pulse. Under the above assumption, the measured apparent time-of-flight distribution $G(t)$ is proportional to the unconvoluted distribution $g(t)$.

Finite Detector Length. The limitation of the resolution due to the finite length ΔL of the ionization volume plays an important role if a universal beam detector is used. If $D(x)$ is the ionization probability at a given location x in the ion source, the time dependence of the detector signal is given by

$$\overline{G}(t) \propto \int_{L}^{L+\Delta L} D(x)G(x/v)dx. \tag{3.42}$$

$\overline{G}(t)$ is the experimentally determined time-of-flight distribution, including both the finite width of the chopper gating function and the finite length of the ion source. Instead of treating the two corrections in two steps, an effective transmission $T_{eff}(t)$ may be defined by

$$T_{eff}(t) \propto \int_{L}^{L+\Delta L} D(x)T(x/v)dx, \tag{3.43}$$

taking into account both corrections. $T_{eff}(t)$ is the time-of-flight distribution which is solely determined by the measuring device. It can be measured with a correctly monochromatic beam. Using the effective transmission in (3.38), the relationship between the measured time-of-flight distribution $\overline{G}(t)$ and the true distribution $g(t)$ becomes

$$\overline{G}(t) \propto \int_{-\infty}^{\infty} T_{eff}(t-\tau)g(\tau)d\tau. \tag{3.44}$$

160   3. Velocity Measurement and Selection

Deconvolution. If the resolution is insufficient to measure the true time-of-flight distribution g(t) of a given beam, it is necessary to determine g(t) from the measured distribution $\overline{G}(t)$ using the integral equation (3.44). Integral equations of this kind can be formally solved by appropriate mathematical methods, but serious difficulties are encountered in practice if the empirical data are subject to noise. The data, therefore, must be smoothed before deconvolution. Various methods are described in the literature [Morrison (1963), Moore (1968), Wertheim (1975)].

Another approach is based on the method of moments [Alcalay and Knuth (1969)], which, however, is also substantially limited by noise in the data.

A rather general method of obtaining the true distribution function from the measured time-of-flight data attacks the problem in reverse. In the first step, the velocity distribution is described by a parameterized form for f(v)

$$f(v) = f(v, \{a_i\}). \tag{3.45}$$

Then, using the appropriate transformations and convolutions, a calculated model of the signal $S(t,\{a_i\})$ (for the assumed values of the parameters $\{a_i\}$) is obtained, which is compared with the actual time-of-flight signal. The parameters $\{a_i\}$ are then refined and the calculation is repeated.

This iteration procedure is continued until the sum of the squares of the differences between measured and calculated data reaches a minimum.

Figure 3.10 presents an example of a deconvolution of a measured time-of-flight distribution $\overline{G}(t)$ of a highly expanded helium beam (curve 1) [Buckland et al. (1997)]. Curve 2 is the measured effective transmission $T_{eff}(t)$. It has been determined by using a very monochromatic helium beam (velocity spread $\Delta v/v$ = 0.1%) produced by diffraction from a crystal surface.

Since the widths of the measured time-of-flight distribution $\overline{G}(t)$ and of the effective transmission $T_{eff}(t)$ are of the same order of magnitude, the true time-of-flight distribution g(t) of the helium nozzle beam can only be determined by

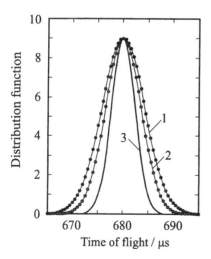

Fig. 3.10. (1) Measured time-of-flight distribution, (2) measured effective transmission, and (3) true time-of-flight distribution of a helium nozzle beam obtained by deconvolution

deconvolution. In this case, the deconvolution is especially simple, since both $\overline{G}(t)$ and $T_{eff}(t)$ can be approximated by Gaussian functions to a very high degree of accuracy. Hence, the desired function g(t) is also a Gaussian function and the integral equation (3.44) can be solved in closed form. If Δt_g is the halfwidth of g(t), Δt_G the halfwidth of $\overline{G}(t)$, and Δt_T the halfwidth of $T_{eff}(t)$, a simple integration leads to the relationship

$$\Delta t_g = \sqrt{\Delta t_G - \Delta t_T}, \qquad (3.46)$$

and thus to the desired time-of-flight distribution g(t) of the helium nozzle beam, which has a time or velocity resolution of $\Delta t/t = \Delta v/v = 0.82\%$ (see curve 3 in Fig. 3.10).

3.2.2 Cross-Correlation Method

The small duty factor of the single-pulse time-of-flight method discussed so far makes its application in experiments with low beam intensities practically impossible, since the measuring times become too large. To increase the effectiveness of time-of-flight measurements, the interdiction of overlap of consecutive pulses must be canceled, so that shutter functions with much higher duty factors can be applied. In this case, the difficulty concentrates on the extraction of the time-of-flight distribution from the measured detector signal. If Z(t) with $0 \leq Z(t) \leq 1$ is an arbitrary shutter function used for beam modulation, and if F(t) is the time-of-flight distribution within the beam, the signal function measured at the detector, designated as D(t) in the following, is essentially determined by the convolution product Z(t)*F(t)

$$D(t) = U(t) + \int Z(t - t'') F(t'') dt''. \qquad (3.47)$$

For the following considerations, the background U(t) is assumed to be time-independent (U(t) = U). The problem is now to find a shutter function Z(t) for which F(t) can be calculated from the measured signal function D(t) using (3.47). Functions Z(t), which satisfy the condition

$$\int Z(t) Z(t + \tau) dt = \delta(t), \qquad (3.48)$$

where $\delta(\tau)$ is Dirac's delta function, solve this problem. This can be easily shown by replacing t by t + t' in (3.47), multiplying both sides of this equation with Z(t'), and integrating over t'. This yields

$$\int D(t + t') Z(t') dt' = U \int Z(t') dt' + \iint Z(t + t' - t'') Z(t') F(t'') dt' dt''. \qquad (3.49)$$

Using (3.48) gives the result

$$F(t) = \int Z(t')D(t+t')dt' - U \int Z(t')dt'. \quad (3.50)$$

The integral

$$K_{D,Z}(t) = \int D(t+t')Z(t')dt' \quad (3.51)$$

is called the cross-correlation function and

$$K_{Z,Z}(t) = \int Z(t+t')Z(t')dt' \quad (3.52)$$

is termed the autocorrelation function. An example of a function satisfying (3.48) is a random function (white noise) R(t). In practice, however, such a function cannot be used for beam modulation, since the integration limits in (3.48) and (3.50) extend from $-\infty$ to $+\infty$, requiring an infinite measuring time. For practical applications, the chopper function Z(t) has to be a periodic function.

Pseudostatistical Method. Binary sequences $\{S_k\}$, which show a quasistatistical behavior resembling the white noise, are well known in mathematics. They are termed pseudorandom sequences and possess a finite number N of elements S_1, S_2, $\cdots S_k$, $\cdots S_N$, having values of 0 or 1. With the definitions

$$S(t) = S_k \quad \text{for } (k-1)\tau \leq t < k\tau \quad (3.53)$$

and

$$\begin{aligned} S(T+t) &= S(t) \\ & \quad \text{with } T = N\tau, \\ S(-t) &= S(T-t) \end{aligned} \quad (3.54)$$

a periodic, pseudorandom signal sequence of maximal length $2^k - 1$ may be constructed, which can change values only at intervals $\tau = T/N$, where T is the period of the periodic shutter function. Figure 3.11 shows a pseudorandom signal sequence of 31 elements and the corresponding autocorrelation function $K_{S,S}(t)$. Since

$$\int_0^T S(t)dt = \frac{N+1}{2}\tau \approx \frac{T}{2}, \quad (3.55)$$

the intensity loss of a beam modulated with the shutter sequence S(t) is only 50%.

The signal function S(t) is a sequence of discrete values S_k, and the detector function D(t) is also such a sequence D_k, since the detector integrates over short

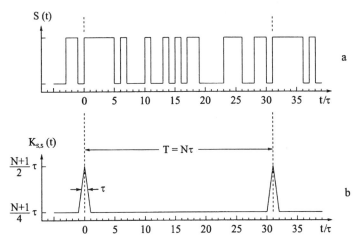

Fig. 3.11. (a) Example of a pseudorandom signal sequence of 31 elements. (b) corresponding autocorrelation function $K_{S,S}$

time intervals τ'. If τ' is the width of a channel of the multichannel memory used, we have

$$D_k = \int_{(k-1)\tau'}^{k\tau'} D(t)dt, \tag{3.56}$$

where D_k is the content of the k-th channel. Consequently, the time-of-flight distribution is also not a continuous function $F(t)$, but a sequence of some integrals over short time intervals τ'', which we designate by F_k.

For these reasons, the above considerations must be repeated using sums instead of integrals. This can be done in a completely analogous way, and the sequence $\{F_k\}$ describing the time-of-flight spectrum can be determined from the detector sequence $\{D_k\}$ according to the relation

$$F_k = \frac{4}{N+1} \sum_{i=1}^{N} S_i D_{i+k-1} - 2U_k - F_{tot}. \tag{3.57}$$

U_k is the background counted in the channel k, F_{tot} is the total signal according to

$$F_{tot} = \sum_{i=1}^{N} F_i. \tag{3.58}$$

It can be shown that the time-of-flight spectrum calculated according to (3.57) corresponds exactly to the spectrum which would have been measured with the single-pulse method.

164   3. Velocity Measurement and Selection

The cross-correlation method using pseudorandom chopping can greatly improve the signal-to-noise ratio in time-of-flight measurements. This is especially true if the time-of-flight spectrum exhibits a broad peak or several peaks of about the same amplitude. The method is not well suited in cases where the spectrum is dominated by a large peak, while the information of interest is contained in small neighboring peaks. In these cases, the statistical noise from the large peak will be distributed across the whole spectrum and decreases the signal-to-noise ratio of the smaller peaks. For a time-of-flight spectrum consisting of one narrow peak, the cross-correlation method is also inferior to the single-pulse method, especially since in this case, a large duty-factor chopper is possible, because overlap problems are reduced by the narrow width of the peak. Detailed considerations about noise and error of the cross-correlation method as well as a comparison with the single-pulse method can be found in the literature [Wilhelmi and Gompf (1970), Secrest and Meyer (1972), Comsa et al. (1981)].

3.2.3 Experimental Details

One of the most important aspects of time-of-flight experiments is the formation of very short bunches of particles. Mechanical choppers are commonly used, in most cases rotating disk choppers, since high speeds and consequently high resolution can be achieved. Oscillating devices such as tuning forks have also been used, but these are slower than rotating disks and yield, therefore, lower resolution. A rotating chopper consists of a thin disk with slits cut along the edge, which is driven by a high-speed motor. When using the single-pulse method, the narrow slits are equally spaced, whereas in the case of pseudorandom chopping the slits are cut according to a pseudorandom sequence (see Fig. 3.11). As already discussed in connection with slotted disk velocity selectors, the slits are either mechanically machined or produced by photoetching. The latter method yields smaller slit widths and a higher precision. Figure 3.12 shows an example of a disk with four equal pseudorandom sequences of 31 elements, as shown in Fig. 3.11.

The problems to be considered when constructing a rotating disk chopper are the same as those that have already been discussed in connection with mechanical velocity selectors. These include strength considerations in the choice of the material, balancing of the rotor to minimize the dynamic loading of the ball bearings, choice of suitable bearings and lubrication as well as the vacuum compatibility of these components. To operate a rotating chopper near the limit of its tearing strength, chopper disk and rotor axis should be machined from one piece of material. Furthermore, the disk must have a special thickness profile which guarantees a constant stress in any volume element of the disk independent of its radial distance. This profile d(r) is given by

$$d(r) = d_0 \exp\left(-\frac{\rho(\omega r)^2}{2\sigma}\right), \qquad (3.59)$$

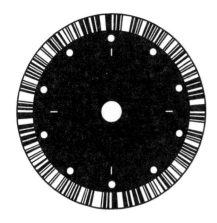

Fig. 3.12. Chopper for pseudorandom modulation (cross-correlation method). The slits were made by etching of 0.1-mm-thick stainless steel. The additional slits provide a synchronization light pulse at the start of the pseudostatistical sequence, which is repeated four times. The circular holes are used for rotational speed measurement

where σ is the breaking strength of the disk material and ρ its density. ω is the angular frequency and d_0 the thickness of the disk at $r = 0$ [Beitz and Küttner (1981)].

Motor and bearings of a rotating chopper are important for its correct operation, since high rotational speeds are generally required. For a 160-mm-diameter disk with 0.2 mm slits, for example, operation at 400 Hz is necessary to obtain a resolution of 1 µs. Both commercially available hysteresis synchronous motors and asynchronous induction motors as well as various precision ball bearings are suited for this purpose. An alternative to conventional motors and bearings is magnetic suspension, which is much easier for a chopper than for a slotted disc velocity selector because of the much lower weight of the chopper [Comsa et al. (1981), David et al. (1986)].

In this case the friction is extremely low and constant, so that phase jitters are minimized. This is of special importance in cross-correlation measurements, where phase jitter may cause background fluctuations and even ghost peaks.

Finally, the rotating disk must be synchronized with the detecting system. This can be done by using a small light source (photodiode) and a photocell to produce a synchronizing light pulse. In single-pulse experiments, this pulse may be formed by the same slit as is used to chop the molecular beam, in other cases a separate slit at an appropriate location of the disk may be used (see Fig. 3.12).

A very small device to generate extremely short particle bunches (5 to < 1 µs) at a repetition rate of 10 kHz, which is based on a commercially available spinning rotor pressure gauge [Fremerey (1985)], utilizes a magnetically suspended cylindrical rotor of 5.6 mm diameter and 45 mm length [Kuhnke et al. (1994)]. A 1 mm-diameter hole drilled perpendicular to the rotational axis permits beam passage for short time intervals. This hole is positioned off the center of the rotor, so that only one particle pulse is transmitted during one revolution of the rotor. The rotor is machined from magnetic steel and has a time constant (see (3.30)) of $t_0 = 28$ h. This is entirely due to eddy current damping of the rotor. The pulse length can be adjusted between 5 and below 1 µs by positioning a slit in front of the rotor.

Other Modulation Techniques. Instead of a mechanical chopper, a pulsed beam source can also be used for time-of-flight measurements (see Chap. I.4). Excitation of a metastable state by electron impact or the excitation of a high-lying Rydberg state have already been mentioned before. Chopping may also be achieved directly in the beam generation process, for example in beams produced by laser ablation. The physical process under investigation may also produce a pulsed beam signal which can be used for time-of-flight measurements. An example is laser desorption experiments, in which the desorbed particles leave the surface in short particle bunches, if a pulsed desorption laser is used. Detecting the desorbed particles with a second laser via multiphoton ionization and varying the delay time between desorption and detection lasers immediately yields the time-of-flight spectrum of the desorbed particles [Weide et al. (1987), Mull et al. (1992), Fischer (1994)]. Pulsed laser heating (on a time scale of the order of 10^{-8} s) of surfaces also yields a pulsed desorption, and the velocity of the desorbed molecules may be directly determined by a time-of-flight measurement [Cowin et al. (1978)]. In experiments with fast beams produced by charge exchange of an ion beam, the ion beam may be pulsed by appropriate electric fields before passing through the charge exchange cell [Gersing et al. (1973), Kita et al. (1981), Düren et al. (1986a)].

Finally, only a group of particles in a given quantum state may be modulated by depleting or enhancing its population in a molecular beam by optical means. The labeled group of molecules spreads out in time according to its velocity distribution, so that the time-of-flight spectrum can be measured by state-selectively detecting these particles at the end of the flight path with laser-induced fluorescence (time-of-flight analysis by optical population modulation) [Gaily et al. (1976), Bergmann et al. (1978), Kroon et al. (1981), Hefter and Bergmann (1988), Molenaar et al. (1998)]. This method has a higher velocity resolution than the Doppler method described in Sect. 3.3, especially for slow particles [Molenaar et al. (1998)].

A similar method has been used to measure the velocity distribution in cluster beams. If clusters of a given size are excited with a short laser pulse, they are decomposed and the measured beam intensity at the detector decreases, since the cluster fragments leave the beam due to their initial energy. From the temporal change of the detector signal the velocity distribution may be determined.

Electronics. Many schemes, both analog and digital, are available for the electronics required to measure the time distribution of the particles arriving at the detector, so that an optimum adaptation to the respective experiment is possible.

If the beam intensity is rather low, so that the average number of particles detected in one chopper pulse is typically less than one, a scheme such as that shown in Fig. 3.13 can be used [Schöttler and Toennies (1968), Dittner and Datz (1971), Gersing et al. (1973), Heindorff and Fischer (1984), Düren et al. (1986b)]. In this example, the fast beam source is opened by a gating pulse from a pulse generator, which is triggered by a quartz clock via a digital scaler 1. Its frequency v depends on the beam energy and is adjusted such that no overlap of particles from two succeeding pulses occurs. In other cases, where beam modulation is performed by a

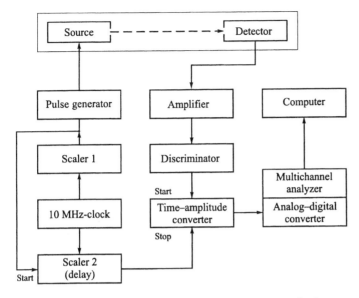

Fig. 3.13. Block diagram of a time-of-flight spectrometer for low counting rates [Gersing et al. (1973)]. For a description see text

rotating disk, the gating pulses are produced by the chopper. Synchronously with the pulse generator the scaler 2 starts, which delays the gating pulses. A particle arriving at the detector produces an output signal which, after amplification, is fed into a pulse height discriminator, which eliminates noise pulses below a given threshold and produces standard pulses of adjustable length. These standard pulses start a time-to-amplitude converter (TAC), which is stopped by the delayed trigger signal. The TAC output pulse having an amplitude which is proportional to the time between start and stop, is digitized in the fast analog-digital-converter of a multichannel analyzer and stored in the corresponding channel. Finally, the content of the memory of the multichannel analyzer can be forwarded to a minicomputer for further processing.

In contrast to this inverted time configuration method it would more straightforward (and probably easier to understand) to start the TAC with the trigger signal from the chopper and stop it with the detector signal. However, since many chopper pulses contain no particle, the TAC would be more often started than stopped. It would be necessary, therefore, to stop the TAC internally after a time corresponding to the longest observed flight time. This would drastically increase the deadtime of the time-of-flight spectrometer.

With this method, excellent time resolutions ($< 10^{-10}$ s) are achievable. Its disadvantage is the small particle rate which can be handled. High particle rates can be used if the analog system described above is replaced by a digital time-of-flight spectrometer. Instead of the time-to-amplitude converter and the multichannel analyzer a series of counters (the number of counters determines the number of channels) can be used, which is started by the trigger signal. The counters are

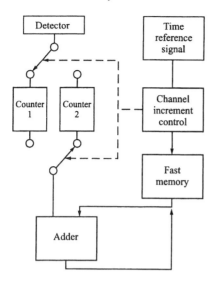

Fig. 3.14. Block diagram of a time-of-flight spectrometer for higher particle rates based on the dual scaler method [Auerbach (1988)]. For a detailed description see text

opened one after another (each for a defined time interval) registering the particles according to their arrival time. Every time when one of the eight-bit counters is near to overflow, the content of all counters is read out into a computer, where the data are gathered, analyzed and further processed [Düren et al. (1986b)].

The multiple counter arrangement discussed above may also be replaced by two counters if a device such as the one shown schematically in Fig. 3.14 is used.

Counter 1 is opened at the beginning of a time-of-flight cycle by a channel increment control unit and data are accumulated in this counter. After the channel time has elapsed the channel increment control stops counter 1, switches the input to counter 2, and starts accumulation in this unit. In the meantime, the data from counter 1 are added to the value stored at the address of the first channel in a fast memory. At the next channel increment pulse, the roles of counters 1 and 2 are interchanged, and the address in the memory is incremented by 1. This process of alternately counting and adding to memory is repeated to build up the time-of-flight spectrum. After a complete sequence the channel increment control sets the address in the memory back to zero and the whole procedure is repeated.

The deadtime of this procedure is nominally zero, since counting and adding to memory operations overlap. Further time-of-flight analyzers are described in the literature [Auerbach (1988), Huisken and Pertsch (1987)].

For very high data rates, counting techniques are not fast enough and analog detection must be used. Time-of-flight spectra may be taken with a boxcar analyzer or transient digitizer.

Velocity Selection by Time-of-Flight Methods. The time-of-flight methods discussed so far do not allow the generation of a beam of particles with uniform velocity. Instead, to perform an experiment with particles of a given velocity, the detector has to be gated in such a way that only particles of the given velocity are recorded. Thus, a temporally selected monochromatic beam is obtained. There are,

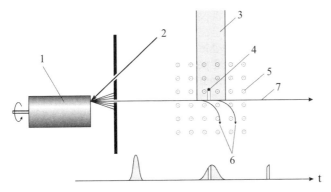

Fig. 3.15. Velocity selection by a time-of-flight method [Macler and Fajardo (1994)]. (1) rotating target, (2) ablation laser beam, (3) selection laser beam, (4) shielding wire, (5) magnetic field, (6) ions, (7) atomic beam

however, applications which require a spatially selected monochromatic beam. If it is desired, for example, to deposit atoms of a well defined velocity on a surface, which may be necessary to obtain a more homogeneous film growth, a spatially selected beam is required. A method to obtain such a selected beam on the basis of a time-of-flight method has been described by Macler and Fajardo (1994). They used pulsed beams of Al, Ga, and In, which are generated by laser ablation (see Fig. 3.15). A particle pulse disperses in time according to the speed distribution in the beam. At a given distance from the ablation target, the atom pulse is irradiated by light from a broad ArF laser beam (193 nm) (perpendicular to the atomic beam direction), which is in resonance with a strong autoionization line of the atoms.

Thus, all atoms of the pulse are ionized, except those which are in the shadow of a small wire (see Fig. 3.15). These are the particles of a well defined velocity interval around a velocity v_0, which is determined by the time difference between the ablation laser and the selection laser and by the distance between the wire and the ablation target. The ions are removed from the beam by a magnetic field which is superimposed on the ionization region.

3.2.4 Calibration

Methods to calibrate time-of-flight spectrometers are briefly discussed in the following, assuming a universal beam detector which is frequently used in molecular beam experiments. Careful investigations of all possible sources of error can be found in the literature [Beijerinck et al. (1974), Auerbach (1988)].

In a time-of-flight experiment with a universal beam detector the total time τ of a particle, starting with its passage through the chopper slit and ending with the registration of the particle in a counter, is determined. This time is composed of several contributions: The desired flight time τ_N of the neutral particles, the flight time τ_i, required for the ions to travel from the location of their formation to the conversion electrode of the multiplier, and the time τ_e, which is determined by the multiplier and the detection electronics. τ_i depends on the length of the ionization region and may also be influenced by space charge effects in the ion source [Beijerink et al. (1974)], τ_e is usually negligible if digital systems are used.

If s is the length of the path of a neutral particle between the chopper and the location of its ionization, its flight time is given by

$$\tau_N = s/v, \tag{3.60}$$

where v is the particle velocity. After ionization, the velocity of the ion is determined by the electric fields in the ion source, in the extraction area, and in the acceleration region. The velocity of an ion that has passed through the potential difference U is given by

$$v_i = \sqrt{\frac{2eU}{m}} = 1.39\sqrt{\frac{U}{M}}\, 10^6 \text{ cm/s}. \tag{3.61}$$

In the numerical part of (3.61) U is measured in V and M in atomic mass units. Without detailed knowledge of the electric fields which have been passed by the ion, its velocity v_i is always proportional to the inverse of the square root of its mass $v_i \propto M^{-1/2}$. Accordingly, its flight time τ_i scales with the given mass M as $\tau_i = kM^{1/2}$, where k is a constant determined by the operational conditions of the detector. Hence, the total flight time may be written as

$$\tau = \tau_N + \tau_i = s/v + k\sqrt{M}. \tag{3.62}$$

τ_i is an average ion flight time and s is an average ion path, since the ion source has a finite length Δs and the particles may be ionized at different locations within Δs. The finite length Δs of the ionization region determines, together with the finite width of the chopper slit, the resolving power of the time-of-flight measurement.

The average ion flight path s and the average ion flight time τ_i can be experimentally determined in a calibration measurement, using a comparison with a mechanical velocity selector. (3.62) suggests the following possibilities:

1) Calibration with beams of different velocities but equal mass. The maximum of the narrow velocity distribution of a nozzle beam of a heavy rare gas can be easily shifted within large limits by admixing a carrier gas. The exact velocity is determined by a mechanical velocity selector, which should be a symmetrical one to increase the accuracy of the velocity measurement. If the measured values of the total flight time τ are plotted against the reciprocal velocity, a straight line is obtained according to the relationship (3.62). From its slope the average flight path s can be determined, while the ordinate offset yields immediately the ion flight time τ_i for the given operation conditions of the ion source.

2) Calibration with beams of different masses but equal velocities. If beams of different particles at the same velocity (which can be adjusted by admixing a carrier gas and measured with a mechanical velocity selector) are used, the measured total flight times may be plotted against $M^{1/2}$. The slope of the resulting straight line yields the constant k and thus the ion flight time, the ordinate offset yields s/v and therefore the flight path of the neutrals. It is even simpler to use a beam of large polyatomic molecules for this calibration. During electron impact ionization, the molecules decay into various ionized fragments. Smaller molecules such as C_2H_2 and their clusters also yield a broad spectrum of fragments, which is well suited for calibration measurements [Huisken and Pertsch (1987)]. Measuring the total flight times for the different fragment masses and plotting them against the square root of their mass, $M^{1/2}$, again a straight line is obtained. Its slope yields the constant k and the ion flight time. The ordinate offset yields the flight time τ_N of the neutral molecule. If the velocity is measured simultaneously with a mechanical selector, the flight path s may be determined too.

Carrying out the measurements described above carefully, an accurate calibration of a time-of-flight spectrometer can be achieved.

3) *Measurements utilizing two different flight paths.* To determine the velocity of nozzle beams, Kresin et al (1998) used two choppers with a well defined distance Δs in the beam path. Measuring alternately the time of flight with both choppers, a time-of-flight difference Δt is obtained, which yields immediately the velocity $v = \Delta s/\Delta t$, independent of delay times within the detector.

3.3 Doppler Shift Measurements

Consider an excited atom or molecule with a velocity vector **v** in the rest frame of the observer. The frequency ω_0 (in the coordinate system of the molecule) of the light emitted from the molecule in the direction of the wave vector **k** is shifted due to the Doppler effect. For velocities $v \ll c$ (nonrelativistic case), the observer measures the emission frequency ω according to

$$\omega = \omega_0 + \mathbf{k}\mathbf{v}, \tag{3.63}$$

where **k** is the wave vector with $k = 2\pi/\lambda$ (λ = wavelength). Introducing the angle ϑ between **v** and **k** and using the relationship $\lambda v = c$ (c = velocity of light), (3.63) may be written as

$$\omega = \omega_0 \left(1 + \frac{v}{c}\cos\vartheta\right). \tag{3.64}$$

Similarly, the absorption frequency ω of an atom or molecule moving with the velocity **v** across a plane electromagnetic wave with the wave vector **k** and the angular frequency ω_L is shifted. The wave frequency ω_L in the rest frame appears in the frame of the moving molecule as $\omega' = \omega_L - \mathbf{k}\mathbf{v}$. The molecule can only absorb

172   3. Velocity Measurement and Selection

if ω' coincides with its eigen frequency ω_0 ($\omega' = \omega_0$), that means if the frequency ω_L measured in the rest frame satisfies the condition

$$\omega_L = \omega_0 + \mathbf{kv} = \omega_0(1 + (v/c)\cos\vartheta). \tag{3.65}$$

The frequency shift given by (3.65) can be used to measure the particle velocity v. This was realized already a hundred years ago and used to measure the velocity of ion beams [Stark (1905)].

3.3.1 Experimental Technique and Resolution

The principle of measuring velocities in molecular beams using the Doppler effect is illustrated in Fig. 3.16. The output of a single-mode laser is split into two beam components. One component crosses the molecular beam at right angles, the second beam component intersects the molecular beam at an angle ϑ lying between 0 and 45°.

If the frequency ω_L of the laser is scanned, a fluorescence signal is observed, the width of which is given by

$$\omega_L - \omega_0 = \omega_0 \frac{v}{c}\cos\vartheta. \tag{3.66}$$

Accordingly, a narrow fluorescence line at the frequency ω_0 is observed due to the laser beam which crosses the molecular beam at right angles. Its width is determined by the distribution of the velocity component $v_\perp$ perpendicular to the molecular beam. The fluorescence signal of the second laser beam crossing the molecular beam at the angle ϑ is determined by the distribution of the velocity component $v_\parallel$ parallel to the molecular beam. This is illustrated in Fig. 3.17, which shows the measured fluorescence signal of Na_2 molecules in a given quantum state

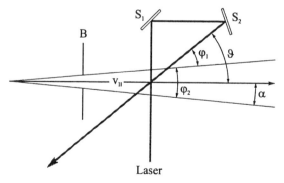

Fig. 3.16. Principle of measuring velocities utilizing the Doppler effect and definition of the quantities used to calculate that part of the resolving power which results from the divergence of the molecular beam. The velocities v_{min} and v_{max} corresponding to the angles $\varphi_1 = \vartheta - \alpha$ and $\varphi_2 = \vartheta + \alpha$ determine this contribution. B aperture, S_1 and S_2 mirrors

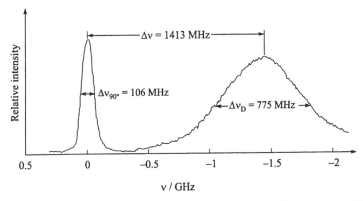

Fig. 3.17. Fluorescence signal as a function of the laser frequency for Na$_2$ molecules in a defined quantum state [Bergmann et al. (1975)]. For further details see text

(v = 0, j = 28) as a function of the laser frequency [Bergmann et al. (1975)]. From the measured frequency dependence of the fluorescence light, the measured resonance frequency ω_0, and the angle of intersection ϑ, the velocity distribution of the molecular beam may be readily determined according to (3.66).

For an accurate analysis of the fluorescence signal it must be considered that the measured intensity I(ω) is a product of the speed distribution f(v) and the excitation probability E(v), convoluted with the absorption profile L(ω,γ) (γ is the natural linewidth). Generally, the width of L(ω,γ) is negligible compared to the width of the velocity distribution f(v), so that the intensity may be written as

$$I(\omega)d\omega \propto E(\omega)f(\omega)\frac{dv}{d\omega}d\omega. \tag{3.67}$$

In the saturation limit E $\approx$ 1, the excitation probability is independent of the velocity, whereas in the other limiting case E $\ll$ 1 it is proportional to 1/v. For a two-level system (the $3^2S_{1/2}(F=2) \to 3^2P_{3/2}(F=3)$ transition in sodium atoms is an example), the number of emitted photons is proportional to the interaction time with the laser and we have E(v) $\propto$ 1/v independent of the laser intensity (see Chap. 5).

The velocity resolution $R_{SP} = v/\Delta v$ of a parallel molecular beam follows by differentiation of (3.66)

$$R_{SP} = \frac{v}{\Delta v} = \frac{\omega_L - \omega_0}{\Delta \omega_L} = \frac{\omega_0 v \cos \vartheta}{c \Delta \omega_L}. \tag{3.68}$$

Contrary to mechanical velocity selectors, the resolving power of the Doppler method is not constant but proportional to the velocity. This makes this method especially attractive for high beam velocities. The principle limit of the resolution is achieved for $\Delta\omega_L = \Gamma$, were Γ is the natural linewidth of the excited state.

A divergence of the molecular beam causes an additional contribution R_{SD} to the resolving power. If α is the half aperture angle of the beam, a fluorescence signal is observed, which begins at a velocity v_{max} and which ends at the velocity v_{min} (see Fig. 3.16). These limiting velocities can be expressed by the frequencies and the angles using (3.66) to yield

$$v_{max} = \frac{(\omega_L - \omega_0)c}{\omega_0 \cos(\vartheta + \alpha)} \quad \text{and} \quad v_{min} = \frac{(\omega_L - \omega_0)c}{\omega_0 \cos(\vartheta - \alpha)}. \tag{3.69}$$

Using the definition (3.10), the contribution of the beam divergence to the resolving power A_D can be written as

$$R_{SD} = \frac{v_{max} + v_{min}}{v_{max} - v_{min}} = \operatorname{ctg}\alpha \, \operatorname{ctg}\vartheta. \tag{3.70}$$

If this contribution cannot be neglected, the total resolution is given by

$$R_S = \frac{R_{SP} R_{SD}}{R_{SP} + R_{SD}}. \tag{3.71}$$

The resolution resulting from the finite collimation angle of the molecular beam R_{SD} is independent of the velocity. Since R_{SP} increases with increasing velocity, the maximum achievable resolution is limited by R_{SD}, which in turn is determined by the beam collimation.

A reduction of the resolution of the Doppler shift technique may be caused by closely spaced absorption lines, as in the case of hyperfine splitting.

A velocity spectrometer based on the Doppler shift technique, which can be used to measure beam velocities within large limits (corresponding to the energy interval from 10 meV up to 10 keV) has been described by Hammer et al. (1976). It has been frequently applied, in particular for measurements of the speed distributions of sputtered particles, both in their ground states and in excited states [Pellin et al. (1981), Husinsky et al. (1983), (1984)].

The Doppler shift velocity analysis of a beam of metastable particles is particularly simple, as has been demonstrated with metastable H(2S) and D(2S) atomic beams [Biraben et al. (1990)]. Irradiating the metastable atom beam collinearly with the light of a dye laser (2S–3P Balmer-α transition around 656 nm) leads to a decrease of the beam intensity at the secondary-electron detector according to the number of beam particles in resonance at the given laser frequency. Tuning the laser frequency across the atomic resonance, the absorption profile of the beam is obtained, which directly represents the speed distribution of the beam.

The Doppler shift method may also be used to produce velocity-selected beams of atoms in long-lived excited states such as high-lying Rydberg levels, if a laser is used for optical excitation. The light beam of the exciting laser crosses the atomic beam at a given angle, so that only atoms of a well defined velocity interval are excited to the desired state. This technique has been employed in experiments with the one-atom maser, which have been used, for example, to study the dynamics of the energy exchange between an atom and the radiation field of the cavity [Walther (1994)].

3.3.2 Measurements of Differential Scattering Cross Sections

An important field of application for the Doppler shift velocity analysis is the investigation of inelastic and reactive collision processes, especially if the reaction products or the inelastically scattered particles are detected by laser-induced fluorescence. In this case, the velocity analysis may also be used to determine the angular distribution of the inelastically or reactively scattered particles [Kinsey (1977), Phillips et al. (1978), Serri et al. (1981), Moskowitz et al. (1984), Park et al. (1989), Johnston et al. (1990), Taatjes et al. (1990), Brouard et al. (1994), Shafer-Ray et al. (1994), (1995), Xu et al. (1995)]. For a brief description of this technique, which has been applied in many variants, we consider as a simple example measurements of inelastic differential cross sections for fine structure transitions of K atoms in collisions with Ar atoms according to

$$K(4^2P_{3/2}) + Ar \rightarrow K(4^2P_{1/2}) + Ar + 7.16 \text{ meV}.$$

The short lifetime of the excited states (20 ns) makes a direct measurement of the angular distribution impossible.

Figure 3.18 shows a kinematic (Newton) diagram of the considered process. The velocities in the laboratory coordinate frame are designated by **v**, those in the

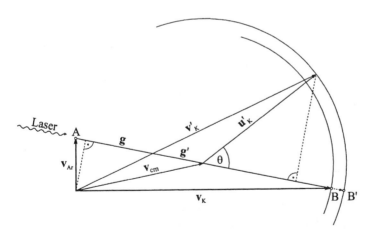

Fig. 3.18. Newton diagram for K* + Ar. The laser beam is parallel to the relative velocity **g**. The Doppler shift is proportional to the projection of the velocity v_K' on the laser beam direction. The magnitude of the relative velocity g before collision is given by the distance AB, the magnitude g' of the relative velocity after collision corresponds to the distance AB'

center-of-mass frame by **u**. The corresponding velocities after collision are **v′** and **u′**, respectively. Due to the inelasticity of the collision (7.16 meV) the radius of the Newton sphere is larger than in the elastic case. We consider a K atom, which is deflected by an angle Θ (in the center-of-mass frame). If the intersection region of the two beams is irradiated by a laser beam having the direction of the relative velocity **g**, all atomic transition lines experience a Doppler shift according to their velocity. This shift is determined by the projection v_p of the velocity vector onto the direction of the incident light (in Fig. 3.18 the fraction of g between the dashed lines) and is composed of two contributions: The projection of the center-of-mass velocity vector $\mathbf{v}_{cm}$ and the projection of the velocity of the potassium atom after collision in the center-of mass frame $\mathbf{u}_K$. According to (3.66), the frequency shift Δv due to the velocity component v_p of the considered particle is given by

$$\Delta v = \frac{v_0}{c} v_p = \frac{v_0}{c} \left[u'_K \cos\Theta + v_{cm} \cos\alpha \right]. \tag{3.72}$$

In (3.72), only the first term depends on the scattering angle Θ. Consequently, there is a unique relationship between the Doppler shift Δv and the center-of-mass angle Θ. Choosing the zero point of the frequency shift Δv arbitrarily in such a manner that we have $\Delta v = 0$ for $\Theta = 0$, we obtain (with $\lambda v_0 = c$)

$$\Delta v = \frac{u'_K}{\lambda} (\cos\Theta - 1) = \frac{m_{Ar}}{m_K + m_{Ar}} \frac{g'}{\lambda} (\cos\Theta - 1). \tag{3.73}$$

In the far right part of (3.73), u_K' has been replaced by the relative velocity g' after collision. Due to this cosine law, the scattered atoms can be selectively excited according to their scattering angle and then detected by the subsequently emitted fluorescence light. The signal observed is proportional to the differential scattering cross section per solid angle element $d\Omega = \sin\Theta d\Theta d\Phi$. Since the Doppler shift is independent of the azimuthal angle Φ, we have $d\Phi = 2\pi$. Differentiation of (3.73) yields $\sin\Theta \, d\Theta \propto d(\Delta v)$, where $d(\Delta v)$ is the spectral resolution of the detecting system, which can be considered to be constant. Consequently we have $\Delta\Omega =$ const. Thus, the Doppler profile, determined in the above manner, corresponds exactly to the differential scattering cross section if averaging processes are neglected. Presupposing a constant frequency resolution $d(\Delta v)$, the cosine law (3.73) determines the angular resolution $\delta\Theta$ of the experiment, which is given by $\delta\Theta \propto 1/\sin\Theta$. It is small in the vicinity of the poles $0°$ and $180°$, and largest at medium angles (around $90°$).

In the experiment discussed above, the potassium atoms are excited to the $4^2P_{3/2}$ state by the light of a dye laser (at a wavelength of about 766.5 nm). Simultaneously, the light of a second dye laser (581.2 nm) irradiates the scattering center, exciting those particles, which made the transition $4^2P_{3/2} \rightarrow 4^2P_{1/2}$ due to collisions with argon atoms, into the 5D state. The fluorescence intensity (6P–4S and 5P–4S) is proportional to the intensity scattered into a given angle Θ. Scanning the

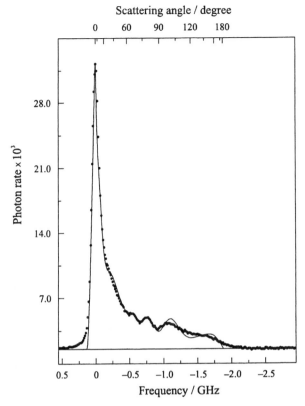

Fig. 3.19. Fluorescence intensity (differential cross section) as a function of the laser frequency (lower scale) or the angle (upper scale), respectively, for a fine structure transition in K–Ar collisions (collision energy 112 meV)). The full line is a calculated curve [Hasselbrink (1985)]

laser frequency, the differential scattering cross section is obtained. Figure 3.19 shows an example [Düren et al (1984), Hasselbrink (1985)]. The characteristic interference oscillations of atom–atom differential scattering cross sections, which are called "rainbow scattering" by analogy to optics, are clearly to be seen.

3.4 Deflection in Inhomogeneous Magnetic Fields

Atoms and molecules passing through an inhomogeneous magnetic or electric field with a gradient perpendicular to the beam trajectory, are deflected by the field, if they have an effective magnetic or electric dipole moment μ_{eff}. The deflection depends on the residence time of the particles in the field and can, therefore, be used to measure and select velocities. While inhomogeneous magnetic fields have already been used in the early days of molecular beam work to measure velocity distributions [Cohen and Ellet (1937)], and are still employed as velocity filters in present-day experiments [Este et al. (1983), Ellis et al. (1985)], electrical fields have not been used for this purpose. Hence, the following sections are restricted to magnetic deflection fields. A detailed description of inhomogeneous magnetic or electric fields, which are predominantly used for state selection of molecular beams, is given in Chap. 4. Review articles and book chapters on this subject are also available [Ramsey (1956), Kusch and Hughes (1959), Pauly and Toennies (1968), Fluendy and Lawley (1973), Reuss (1988)].

3.4.1 Two-Wire Field

The trajectory of an atom of mass m and effective magnetic moment μ_{eff}, which moves in the z direction with the velocity v through an inhomogeneous magnetic field having a gradient perpendicular to the z direction (in the y direction), is described by the equation

$$\frac{d^2 y}{dz^2} = \frac{\mu_{eff}}{mv^2} \frac{\partial H}{\partial y}. \tag{3.74}$$

The effective magnetic moment μ_{eff} of an atom is discussed in Sect. 4.2.1. In general, it depends on the coupling of the various angular momenta (electron spin, orbital angular momentum, nuclear spin), on the energy state of the particle, and on the strength of the magnetic field. For sufficiently high field strengths, where nuclear and orbital angular momenta decouple, μ_{eff} achieves a constant value. For a single angular momentum J, which couples to the external field, we obtain, in this limit, for the magnetic moment

$$\mu = -m_J g_J \mu_0, \tag{3.75}$$

where m_J is the magnetic quantum number, g_J is the ratio of the magnetic moment (in Bohr magnetons) and the angular momentum (in units of $h/2\pi$). μ_0 is the Bohr magneton. If a two-wire field (see Sect. 4.2.2) is used for deflection, which is usually generated by an electromagnet with pole pieces shaped according to its equipotential surfaces, the field gradient is given by (see Fig. 4.4 for the definition of a, r_1, and r_2)

$$\frac{\partial H}{\partial y} = -H_0 \frac{2a^2}{r_1^3 r_2^3}\left(r_1^2 + r_2^2\right) y = -2H_0 q(a, x, y), \tag{3.76}$$

with

$$r_1^2 = y^2 + (a-x)^2 \quad \text{and} \quad r_2^2 = y^2 + (a+x)^2. \tag{3.77}$$

For abbreviation, the quantity q

$$q(a,x,y) = a^2 y \frac{(r_1^2 + r_2^2)}{r_1^3 r_2^3} \qquad (3.78)$$

has been introduced in (3.76). a is a characteristic length of the field (half distance between the two wires), H_0 is the field at the point $x = 0$, $y = a$. Figure 4.4 shows a cross-sectional view of the pole pieces and defines the quantities used. If a ribbon-shaped beam with a width (dimension in the y direction) small compared to its height (dimension in the x direction) is used within the field region of nearly constant gradient (see Fig. 4.4), the x dependence of the gradient (3.76) may be neglected and q(a,x,y) can be replaced by its mean value obtained by averaging over a region of distances between y_1 and y_2, determined by the beam width and the beam deflection within the field:

$$\bar{q} = \frac{2a^2}{(y_2 - y_1)} \int_{y_1}^{y_2} \frac{y}{(y^2 + a^2)^2} dy = a^2 \frac{y_2 + y_1}{(y_1^2 + a^2)(y_2^2 + a^2)}. \qquad (3.79)$$

With this result, the beam deflection y follows from (3.74)

$$y = \frac{\mu H_0}{mv^2} \bar{q} L^2 = \frac{GH_0}{v^2}, \qquad (3.80)$$

where L is the length of the field. The quantity

$$G = \mu L^2 \bar{q}/m \qquad (3.81)$$

has been introduced for abbreviation. The approximation (3.80) together with (3.79) describes the deflection of the beam within an accuracy of ±10%. But this uncertainty in the deflection angle is unimportant for the considerations given below. If, however, the exact absolute value of the beam deflection is required, (3.74) has to be solved numerically.

To illustrate velocity measurements by beam deflection in inhomogeneous magnetic fields, we consider an effusive alkali metal atom beam. In the $^2S_{1/2}$ ground state the effective dipole moment is $\mu_{eff} = 2\mu_0 m_J$ (μ_0 = Bohr magneton, m_J = magnetic quantum number with the two states $m_J = \pm 1/2$). Thus, the beam is split into two components by the magnetic field. If the magnetic field H_0 is kept constant, and if a flux detector (Langmuir–Taylor detector) is moved along the deflection coordinate y, the Maxwellian velocity distribution, transformed to the deflection coordinate, can be measured. The Maxwellian velocity distribution is modified by the (see Table 3.1) flux detector according to

$$f(x)dx = 2x^3 \exp(-x^2) dx, \qquad (3.82)$$

180   3. Velocity Measurement and Selection

with $x = v/v_w$ (v_w = most probable velocity in the source). Using the abbreviations $y_w = (G/v_w^2)H_0$ and $\xi = y/y_w$, the measured distribution as a function of the reduced deflection ξ becomes

$$f(\xi)d\xi = \xi^{-3}\exp(-1/\xi)d\xi. \tag{3.83}$$

Moving the detector along the y direction, two distributions according to (3.83) are observed, provided that the width of the detector is small compared to the deflection y. These two distributions correspond to the two m_J components of the beam. If the detector width is not small compared to the deflection, (3.83) must be convoluted with the width of the detector.

The maximum of the distribution function (3.83) is achieved for $\xi_{max} = 1/3$, and the corresponding deflection is $y_{max} = GH_0/3v_w^2$. If both the source temperature of the effusive beam and the field strength H_0 are known, the averaged gradient $\bar{q}$ can be experimentally determined from the position of the maximum of the observed distribution. Since $y \propto 1/v^2$, we have $v/\Delta v = 2y/\Delta y$, and the velocity resolution becomes

$$R_s = \frac{v}{\Delta v} = \frac{2y}{\Delta y} = \frac{2y}{b+d} = \frac{2GH_0}{v^2(b+d)} = \frac{\mu L^2 \bar{q} H_0}{mv^2(b+d)}. \tag{3.84}$$

b is the halfwidth of the beam profile, d is the width of the detector (d < b). The resolution depends on the respective velocity and is largest for small velocities.

The inhomogeneous magnetic field can be used as a velocity selector, if a small slit (of width s) is placed at a given position y_0 at the end of the deflecting field. At a given field strength, only particles of a well defined velocity interval can pass through the slit. By varying the field strength H_0, the velocity distribution is obtained. If (3.80) is rewritten in the form

$$H_0 = \frac{y_0 v^2}{G}, \quad H_{0w} = \frac{y_0 v_w^2}{G}, \tag{3.85}$$

and if we use the reduced field strength $\zeta = H_0/H_{0w}$, the velocity distribution (3.82) as a function of the reduced field strength ζ is given by

$$f(\zeta)d\zeta = \zeta e^{-\zeta}d\zeta. \tag{3.86}$$

The maximum of this distribution is achieved for $\zeta_{max} = 1$ at the field strength H_{0w} which corresponds to the most probable velocity. The transmission of such a selector can be easily derived from simple geometrical considerations, similar to the case of a slotted disk velocity selector. For a parallel beam with a width b which is larger than or equal to the slit width s, we obtain

$$T(v) = \eta \frac{s}{b}\left[1 - \frac{GH_0}{s}\left(\frac{1}{v^2} - \frac{1}{v_1^2}\right)\right] \quad v_{min} \leq v \leq v_1,$$

$$T(v) = \eta \frac{s}{b} \quad\quad\quad\quad\quad\quad\quad\quad\quad v_1 \leq v \leq v_2, \quad\quad (3.87)$$

$$T(v) = \eta \frac{s}{b}\left[1 - \frac{GH_0}{s}\left(\frac{1}{v_2^2} - \frac{1}{v^2}\right)\right] \quad v_2 \leq v \leq v_{max},$$

$$T(v) = 0 \quad\quad\quad\quad\quad\quad\quad\quad\quad v > v_{max} \text{ and } v < v_{min}.$$

The velocities v_{max}, v_{min}, v_1, and v_2 are given by

$$v_{max} = \frac{v_0}{\sqrt{1-(b+s)v_0^2/(2GH_0)}}, \quad v_{min} = \frac{v_0}{\sqrt{1+(b+s)v_0^2/(2GH_0)}},$$

$$v_1 = \frac{v_0}{\sqrt{1+(b-s)v_0^2/(2GH_0)}}, \quad v_2 = \frac{v_0}{\sqrt{1-(b-s)v_0^2/(2GH_0)}}. \quad (3.88)$$

The factor η is given by $\eta = 1/Z$, where Z is the degeneration of the considered state. According to (3.10), the resolving power becomes

$$R_s = \frac{1+\sqrt{1-\left((b+s)v_0^2/(2GH_0)\right)^2}}{(b+s)v_0^2/(2GH_0)}. \quad (3.89)$$

This reduces for $(b+s)v_0^2/2GH_0 \ll 1$ to

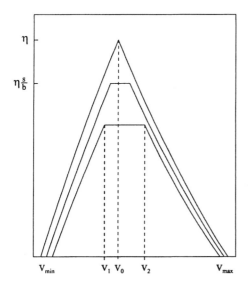

Fig. 3.20. Transmission curves of a magnetic velocity filter (two-wire field) according to (3.87) and different ratios of slit-to-beam width (s/b = 1, 0.8, and 0.6). Definition of the quantities v_0, v_1, v_2, v_{min}, and v_{max} used in (3.87) and (3.88)

182   3. Velocity Measurement and Selection

$$R_s = \frac{2H_0}{\Delta H_0} = \frac{4GH_0}{(b+s)v_0^2} = 2\frac{y_0}{b+s}. \tag{3.90}$$

For a given position y_0 of the slit and thus for a fixed nominal velocity v_0, the resolving power is constant under the above assumptions.

Figure 3.20 shows a few transmission curves $T(v)$. For $s = b$, the transmission curve has a triangular shape, the sides of the triangle are weakly curved legs of a hyperbola. For slit widths $s < b$, the transmission curves are of trapezoidal shape. The transmission, referred to the intensity of one deflected beam component, is reduced by s/b. Figure 3.21 shows a comparison between a Monte–Carlo simulation of the transmission based on the exact equation of motion (3.74) and the transmission calculated for a constant field gradient. The deviations between the two transmission curves are rather insignificant and this shows that the assumption of a constant field gradient is a very useful approximation.

Figure 3.22 shows the result of a Monte–Carlo simulation of the transmission curve of a divergent beam, calculated under the same conditions as the curve shown in Fig. 3.21. The aperture angle of the beam is twice as large as the angle, under which the resolution defining slit can be seen from the field entrance. To a first approximation, the transmission may be described by (3.87), if the beam width b is replaced by the halfwidth of the trapezoidal beam profile in the plane of the slit (full curve in Fig. 3.22).

If the beam splits into more than two components, each component yields the velocity spectrum of the beam if the magnetic field is changed from low to high values. For a beam with Maxwellian velocity distribution, the distributions of the different beam components overlap and a measurement of the velocity distribution

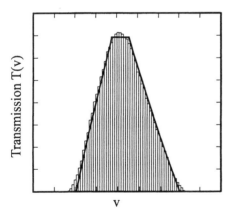
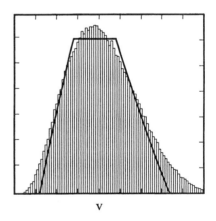

Fig. 3.21. Monte–Carlo simulation of a transmission curve (histogram) using the exact equation of motion (3.74) and transmission curve according to (3.87) (*full line*)

Fig. 3.22. Monte–Carlo simulation of the transmission for a divergent beam (histogram). Approximately, the transmission curve may be described by (3.87), if the beam width b is replaced by the halfwidth of the trapezoidal profile of the divergent beam (*full curve*)

is impossible. In the case of a beam with a narrow velocity spread, such as a nozzle beam, for instance, the different beam components can be separated. As an example, Fig. 3.23 shows the magnetic-field-induced splitting of a Ga atom beam [Toschek (1961)]. The electronic ground state of the Ga atom is a $^2P_{1/2}$ state. Since the second term of the doublet $^2P_{3/2}$ lies energetically only about 0.1 eV higher, this state is thermally populated by about 50% of the atoms at the temperatures required for evaporation (1500–1600 K). In the magnetic field, the two states split into the following components

$$^2P_{1/2}: \quad m_J = \pm\frac{1}{2}; \quad \mu = \pm\frac{1}{3}\mu_0,$$

$$^2P_{3/2}: \quad m_J = \pm\frac{1}{2}; \quad \mu = \pm\frac{2}{3}\mu_0, \quad (3.91)$$

$$^2P_{3/2}: \quad m_J = \pm\frac{3}{2}; \quad \mu = \pm 2\mu_0.$$

Since the resolution (see (3.90)) is constant and the velocity distributions are measured at different field strengths, the width of the distributions increases proportional to H_0, as can be seen from Fig. 3.23. The edges of the triangular transmission curve of the Fizeau selector ($R_S = 6.8$) are rounded by the convolution with the transmission curve of the magnetic velocity filter.

A more recent example of velocity measurements with a two-wire field has been described by Jaduszliwer and Chan (1994). They investigated velocity distributions in beams produced in gas-discharge sources.

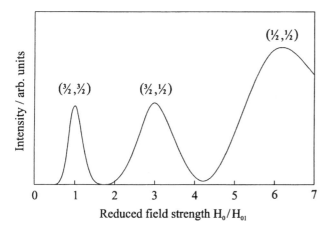

Fig. 3.23. Measured intensity (arbitrary units) of a mechanically velocity selected gallium atom beam as a function of the relative field strength H_0/H_{01} of the two-wire field [Toschek (1961)] (H_{01} is the field strength at which the (3/2, 3/2) state achieves its maximum). Since the magnetic moments scale as 1:2:6, the corresponding magnetic fields scale as 6:3:1. The intensity ratios of the (J, m_J) states I(3/2, 3/2) : I(3/2, 1/2) : I(1/2, 1/2) are 1:1:2, in accordance with the assumption of equal population of the two states

184   3. Velocity Measurement and Selection

Apart from the state-specific deflection of an atomic beam within a transversal, inhomogeneous magnetic field, a longitudinal, inhomogeneous magnetic field also causes a velocity change of atoms in different magnetic substates. This is a consequence of the decelerating or accelerating force according to their effective magnetic moment. For thermal energy beams, the relative velocity change is extremely small and thus negligible. For slow atoms, however, a substantial relative velocity change is possible, so that the magnetic substates can be separated from each other using a subsequent velocity selection [Maréchal et al. (1998)].

3.4.2 Magnetic Hexapole Fields

Two-pole fields are best suited for the geometry of a ribbon-shaped beam having a height (extension in the x direction) which is large compared to its width (extension in the y direction). For cylinder-symmetric beams, however, the best adaptation is obtained with cylinder-symmetric fields. Furthermore, these fields have focusing properties, resulting in a considerable intensity gain.

As is discussed in Sect. 4.2.5 in detail, atoms, which have a constant magnetic dipole moment μ, can be focused in a magnetic hexapole field. All atoms of the same velocity, which leave the field axis at a given point ($r = 0$, $z = -z_a$), return to the axis at a point ($r = 0$, $z = L+z_b$), independent of the angle α between their initial velocity vector v and the field axis (see Fig. 3.24). The relationship between z_a, z_b, L, and the wave number k is described by (4.68), which may be rewritten as

$$z_b = \frac{kz_a + \text{tg}(kL)}{k^2 z_a \text{tg}(kL) - k} \quad \text{with} \quad k = \frac{1}{r_0 v}\sqrt{\frac{2\mu H_0}{m}}, \qquad (3.92)$$

where r_0 is the smallest distance between the axis and the pole pieces (field radius). H_0 is the field strength at r_0, L is the length of the field, and m is the mass of the

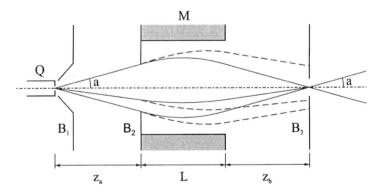

Fig. 3.24. Cross-sectional view of a magnetic hexapole field. Solid lines correspond to trajectories of particles which have the nominal velocity, so that they are focused. Dashed lines correspond to non-focused velocities. Q source, M magnetic field, B_1, B_2, and B_3 apertures [Bassi et al. (1976)]. For further details see text

atoms. To illustrate the performance of such a field as a velocity filter, Fig. 3.24 shows a cross-sectional view of a hexapole field together with some particle trajectories. To make it clearer, the lateral dimensions are largely exaggerated [Bassi et al. (1976), Caracciolo et al. (1980)].

The relevant dimensions and characteristic data of this device, which has been used as a velocity filter for H atoms, are summarized in Table 3.3. Atoms, effusing from the source Q with a maximum divergence angle α, which is determined by the apertures B_1 and B_2, enter the field and are focused according to their velocity to different points along the axis. Only atoms with a velocity within a certain velocity interval Δv around a given nominal velocity v_0 can pass through the narrow aperture B_3. Slow particles, which experience several half-periods of the radial oscillation in the field, may also be focused to the image point $z = z_b$. The corresponding velocities can be easily calculated from (3.92) (see Table 3.4). These "sidebands" and their elimination are discussed below.

Since the field exerts a harmonic force on the atoms, particles moving along the field axis experience no force and can, therefore, reach the detector independent of their state and their velocity. To block these particles, a small central obstacle is usually placed on the beam axis in the middle of the field.

The transmission curve of a hexapole velocity filters follows from simple geometrical considerations (see Fig. 3.25). If the exit aperture B_3 of radius R_B is situated at the location $z = z_b(v_0)$, all atoms having a velocity which is larger than the nominal velocity v_0 will pass through the aperture, as long as their focus at $z = z_2$ obeys the relationship

$$\frac{R_B}{z_2 - z_b} \geq \frac{r_{emax}}{z_{max}}, \tag{3.93}$$

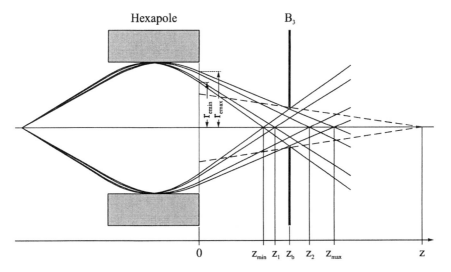

Fig. 3.25. Definitions used to calculate the velocity resolution of a magnetic hexapole filter

with (see Sect. 4.2.5)

$$\frac{r_{emax}}{z_{max}} = tg\Theta_1 = \frac{k_{max}r_0}{\sqrt{1+z_{max}^2 k_{max}^2}} \approx \frac{k_0 r_0}{\sqrt{1+z_b^2 k_0^2}}. \quad (3.94)$$

$2\Theta_1$ is the largest opening angle according to (4.82), under which the atoms with the velocity v_{max} (corresponding to the wave number k_{max}) converge at the focus at $z = z_{max}$. Similarly, all atoms having a velocity which is smaller than the nominal velocity v_0 will pass through B_3, if their focus at $z = z_1$ satisfies the inequality

$$\frac{R_B}{z_b - z_1} \leq \frac{r_{emin}}{z_{min}}, \quad (3.95)$$

with

$$\frac{r_{emin}}{z_{min}} = tg\Theta_2 = \frac{k_{min}r_0}{\sqrt{1+z_{min}^2 k_{min}^2}} \approx \frac{k_0 r_0}{\sqrt{1+z_b^2 k_0^2}}. \quad (3.96)$$

$2\Theta_2$ is the largest opening angle under which the atoms with the velocity v_{min} (k_{min}) converge at the focus at $z = z_{min}$. As indicated in (3.94) and (3.96), we neglect in the following discussion the small differences between z_{max}, z_{min}, and z_b as well as between k_{max}, k_{min}, and k_0, the nominal wave number corresponding to v_0. All atoms which are accepted and focused by the field, having velocities which lead to image points between z_{min} and z_{max}, pass through the velocity filter. For larger and smaller velocities, the transmission is determined by the aperture B_3 located at z_b and thus proportional to the solid angle, under which the aperture is seen from a given axis point at z. This yields (using (3.93) to (3.96))

$$T = \frac{\pi R_B^2}{(z-z_b)^2} = \frac{(z_{max} - z_b)^2}{(z-z_b)^2} \frac{\pi k_0 r_0}{\sqrt{1+z_b^2 k_0^2}} \quad \text{for} \quad z \geq z_{max},$$

$$T = \frac{\pi k_0 r_0}{\sqrt{1+z_b^2 k_0^2}} \quad \text{for} \quad z_{min} \leq z \leq z_{max}, \quad (3.97)$$

$$T = \frac{\pi R_B^2}{(z_b - z)^2} = \frac{(z_b - z_{min})^2}{(z_b - z)^2} \frac{\pi k_0 r_0}{\sqrt{1+z_b^2 k_0^2}} \quad \text{for} \quad z \leq z_{min}.$$

Equation (3.97) holds without a central obstacle. With a central obstacle, the transmitted intensity vanishes completely for distances z beyond or below a critical distance. At the critical distance, the orifice in the aperture B_3 is just covered by the central obstacle (if the curvature of the trajectory due to the field is taken into account).

Table 3.3. Characteristic data of a velocity filter for H atoms

Quantity	Numerical value
Field length L	75 mm
z_a	75 mm
z_b	75 mm
Field strength H_0	7800 Oe
Nominal speed v_0	3444.3 m/s
Field radius r_0	1.55 mm
Resolution A	14.6
Radius of B_1	0.14 mm
Radius of B_2	1.20 mm
Radius of B_3	0.14 mm
Transmission η	~ 50%

To express the transmission curve by velocities rather than by distances and to calculate the velocity resolution $R_S = v/\Delta v$ of a hexapole field, (3.92), which relates distances and velocities, must be reversed. This is not possible in explicit form. However, if we restrict our considerations to velocities not too far from the nominal velocity v_0, $z(v)$ may be expanded in the following way

$$z(v) - z_b = (v - v_0)\left.\frac{dz}{dv}\right|_{v=v_0} = (v - v_0)\frac{k_0}{v_0}\left.\frac{dz}{dk}\right|_{k=k_0}. \tag{3.98}$$

Referring the transmission to the beam intensity which is accepted and focused by the field, and using (3.93) to (3.98), the transmission T(v) is approximately given by

$$T(v) = \eta \frac{(v_0 - v_{min})^2}{(v_0 - v)^2} \quad \text{for} \quad v \leq v_{min},$$

$$T(v) = \eta \quad \text{for} \quad v_{min} \leq v \leq v_{max}, \tag{3.99}$$

$$T(v) = \eta \frac{(v_{max} - v_0)^2}{(v - v_0)^2} \quad \text{for} \quad v \geq v_{max}.$$

η is the fraction of beam particles (m_J components) which is focused by the field. The halfwidth of the transmission curve $\Delta v_{1/2}$ is

$$\Delta v_{1/2} = \sqrt{2}(v_{max} - v_{min}). \tag{3.100}$$

According to (3.97) we find

$$z_{max} - z_b = \left(\frac{R_B}{r_0}\right)\frac{\sqrt{1+k_0^2 z_b^2}}{k_0} \quad \text{and} \quad z_b - z_{min} = \left(\frac{R_B}{r_0}\right)\frac{\sqrt{1+k_0^2 z_b^2}}{k_0}, \tag{3.101}$$

and with (3.98) we obtain

$$v_0 - v_{min} = v_{max} - v_0 = \frac{R_B}{r_0} v_0 \frac{\sqrt{1+z_b^2 k_0^2}}{k_0^2 |dz/dk|_{k=k_0}}, \tag{3.102}$$

with

$$\left|\frac{dz}{dk}\right|_{k=k_0} = \frac{Lk^2 z_a^2 z_b^2 + Lz_a^2 + Lz_b^2 + z_a z_b(z_a + z_b) + (z_a + z_b + L)/k^2}{kz_a^2 + 1/k}. \tag{3.103}$$

The integral transmission $T_i(v_0)$ may be written as

$$T_i(v_0) = \int_{-\infty}^{\infty} T(u)du = 2\eta(v_{max} - v_{min}) = 2\eta \frac{R_B}{r_0} v_0 \frac{\sqrt{1+k_0^2 z_b^2}}{k_0^2 |dz/dk|_{k=k_0}}. \tag{3.104}$$

The resolving power R_S follows from (3.100) and (3.102)

$$R_S = \frac{v}{\Delta v} = \left(\frac{r_0}{R_B}\right)\frac{k_0^2 |dz/dk|_{k=k_0}}{2\sqrt{2}\sqrt{1+z_b^2 k_0^2}} = \frac{r_0}{R_B} F(k_0, z_a, z_b, L), \tag{3.105}$$

where $F(k_0, z_a, z_b, L) = F$ can be written (using 3.92) as

$$F = \frac{k_0^2 \sin k_0 L \left(Lk_0^2 z_a^2 z_b^2 + Lz_a^2 + Lz_b^2 + z_a z_b(z_a + z_b) + (z_a + z_b + L)/k_0^2\right)}{2\sqrt{2}\,(z_a + z_b)\sqrt{1+z_a^2 k_0^2}}. \tag{3.106}$$

When discussing (3.106), it should be noted that for a given velocity filter (z_a, z_b, L, and R_B fixed) the wave number $k_0 = \omega/v_0$ also has a constant value, given by (3.92). Changing the magnetic field changes the transmitted nominal velocity in such a way that k maintains its value. Consequently, the resolving power according to (3.105) and (3.106) is constant. The integral transmission $T_i(v)$ is proportional to v (see (3.104)). Accordingly, a measured velocity distribution is modified by a factor v, similar to the case of a slotted disk velocity selector.

Figure 3.26 shows calculated transmission curves of a hexapole velocity filter with a resolving power $R_S = 20$ ($\Delta v/v = 5\%$) with and without central obstacles of

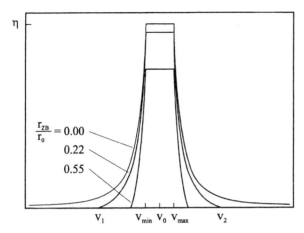

Fig. 3.26. Transmission curve of a hexapole filter for different ratios of r_{ZB}/r_0. Between v_{min} and v_{max} the transmission is constant

different diameter. If the velocities at which the transmission is $\eta/2$ are designated by v_1 and v_2, we have

$$T_i(\Delta v_{1/2}) = \int_{v_1}^{v_2} T(u)du = \eta\left(2 - \frac{1}{\sqrt{2}}\right)(v_{max} - v_{min}). \tag{3.107}$$

Thus the fraction of atoms having velocities within the halfwidth of the transmission curve, is given by $1-1/(2)^{3/2}$ (~ 65%). The relatively high fraction of atoms having higher and lower velocities is due to the wings of the transmission curve. It can be substantially reduced by a central obstacle (radius r_{co}) at $z = L/2$, modifying the transmission (3.99) in the following manner

$$T(v) = 0 \qquad\qquad v \leq v_0 - \frac{r_0}{r_{co}\sqrt{f(k)}}(v_0 - v_{min}),$$

$$T(v) = \eta\left(\frac{(v_0 - v_{min})^2}{(v_0 - v)^2} - \frac{r_{co}^2}{r_0^2}f(k)\right) \qquad v \leq v_{min},$$

$$T(v) = \eta\left(1 - \frac{r_{co}^2}{r_0^2}f(k)\right) \qquad v_{min} \leq v \leq v_{max}, \tag{3.108}$$

$$T(v) = \eta\left(\frac{(v_{max} - v_0)^2}{(v - v_0)^2} - \frac{r_{co}^2}{r_0^2}f(k)\right) \qquad v \geq v_{max},$$

$$T(v) = 0 \qquad\qquad v \geq v_0 + \frac{r_0}{r_{co}\sqrt{f(k)}}(v_{max} - v_0),$$

with

$$f(k) = \frac{(1+z_a^2 k^2)}{kz_a(\sin(kL/2) + kz_a \cos(kL/2))^2}.$$ (3.109)

f(k) is a function which depends only weakly on k in the vicinity of the nominal velocity, so that it may be replaced by its value at the argument k_0 (for a symmetric arrangement with $z_a = z_b$ we have $f(k_0) = 1$). Accordingly, the central obstacle reduces the transmission by a constant amount. As a consequence, the transmission for velocities outside an interval which is determined by the radius of the central obstacle, vanishes (see Fig. 3.26). The reduction of the transmission causes a corresponding reduction of the halfwidth of the transmission curve, which can be easily calculated using (3.108).

As has already been mentioned at the beginning of this section, the relationship (3.92) is not only satisfied for wave numbers k_0 corresponding to the nominal velocity v_0, but also for certain wave numbers k_n (n = 1, 2 ...), which correspond to smaller velocities. As an example, Table 3.4 shows the characteristic properties of the sidebands n = 1–7 of the symmetric hexapole filter described above (see Fig. 3.24 and Table 3.3).

Table 3.4. Relative integral transmission (3.102), sidebands and velocities of a symmetric velocity filter according to Table 3.3

n	Int. transmission $T_i(v_n)/T_i(v_0)\%$	k_n/k_0	v_n/v_0
0	100.0000	1.0000	1.0000
1	8.3547	2.8114	0.3557
2	1.6182	5.0398	0.1984
3	0.5323	7.3720	0.1356
4	0.2335	9.7382	0.1027
5	0.1218	12.1193	0.0825
6	0.0712	14.5080	0.0689
7	0.0451	16.9011	0.0592

The transmission of the sidebands as well as the ratio v_n/v_0 decreases drastically with increasing n. If, in a symmetric filter, a central obstacle is used, all sidebands k_n corresponding to odd values of n are eliminated. Thus, in the example discussed here, the first sideband n = 2 has a relative transmission (with respect to n = 0) of 1.6% at a velocity of 0.1984 v_0. It depends on the actual experiment, as to whether it is necessary to eliminate this sideband by a second obstacle placed at the zero passage of the trajectory. The remaining higher sidebands with n ≥ 4 are negligible in most cases. In principle, all sidebands can be eliminated by replacing the central obstacle by a rod-shaped obstacle, with a length ranging from the middle of the field L/2 to the zero passage of the trajectory belonging to n = 2

[Ebinghaus et al. (1974)]. The zero passages of the trajectories follow from (4.64) for r = 0, yielding the relationship

$$z_{0v}^{(n)} = -\frac{1}{k_n}\left(\text{arctg}(k_n z_a) - v\pi\right) \quad \text{with} \quad v = 1, 2, \cdots, n. \tag{3.110}$$

Magnetic hexapole velocity filters have often been used to select H atom beams. These beams must be generated by dissociation of molecular hydrogen (see Sect. I.4.7) and have broad velocity distributions. The generation of nozzle beams is not possible due to the recombination reaction at higher pressures [Bassi et al. (1976), Este et al. (1978), (1983), Ellis et al. (1985)].

Technical details concerning the practical realization of inhomogeneous multipole fields are described in Chap. 4.

3.5 Deflection by Gravity

Estermann et al. (1947b) used the free fall of atoms to measure the velocity distribution in thermal energy alkali metal atom beams. It was the first experimental observation of the heavy (or gravitational) mass of atoms. A narrow, ribbon-shaped atomic beam, having a height (dimension in the direction of the Earth's gravitational field) small compared to its width, was detected after a path of 2 m with a Langmuir–Taylor detector, which was moved perpendicular to the beam direction. If the height of fall is designated by h, it can be expressed by the acceleration due to gravity g, the length of atom path L, and the atom velocity v according to

$$h = gL^2/(2v^2). \tag{3.111}$$

Moving the narrow detector wire (20 µm ∅) in the vertical direction, the velocity distribution of the beam, transformed to the coordinate h, is obtained. Since the deflection is proportional to $1/v^2$, similar to the case of an inhomogeneous field, the distribution, measured with the surface ionization detector, may be written as

$$f(\xi) = \xi^{-3} e^{-1/\xi} d\xi, \tag{3.112}$$

if we use a reduced height $\xi = h/h_w$ with $h_w = gL^2/2v_w^2$, where v_w is the most probable velocity in the source. The maximum of this distribution is found at

$$h_{max} = \frac{1}{3}h_w = \frac{gL^2}{6v_w^2} = 9.829 \times 10^{-7} \frac{L^2 M}{T}. \tag{3.113}$$

In the numerical part of (3.113), L is to be measured in cm, T in K, and M is the molecular weight.

For a path length L = 2 m, the deflection is extremely small (about 0.1 mm for thermal energy cesium atoms of a most probable velocity of 200 m/s), and its measurement requires a high angular resolution of the detector. Nevertheless, the first measurements of the velocity dependence of the mean free path have been performed using this method of velocity selection [Estermann et al. (1947a)].

Several years later, an attempt to separate the slowest atoms from the rest of an atomic beam by means of the Earth's gravitational field, failed [Zacharias (1954)]. The basic idea of this experiment was simple. In a cesium atom beam, moving 10 m upwards against the Earth's gravitational field, the slowest particles of the thermal velocity distribution should reach their turning point. These slow particles were intended to establish a better time standard. Thus, the experiment was a precursor of an atomic fountain, which is easily realizable nowadays with slow atom beams. However, independent of the intermolecular interaction, the effective cross section of an atom (of velocity v), which moves through a gas (with the most probable velocity v_w) increases proportional to $1/v$ (see (I.2.36) and (I.2.39)), leading to an increasing number of collisions with residual gas molecules with decreasing atom velocity v even under ultrahigh vacuum conditions in the vacuum chamber. Furthermore, collisions with faster beam atoms overtaking the slow ones also increase the number of scattering events. Consequently, the mean free path of the slow particles tends to zero with decreasing velocity v, and no atoms with a turning point after 10 m flight path were observed.

Since methods of laser cooling have been developed and can be used to decelerate atomic beams to arbitrarily small velocities (see Chap. 5), the deflection by

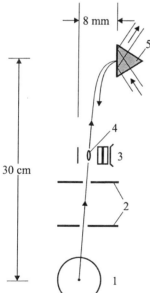

Fig. 3.27. Surface scattering with slow atoms. (1) magneto-optical atom trap, (2) slits for speed selection by gravity, (3) microchannel plate, (4) ionization laser, (5) glass surface (atom mirror) [Kasevich et al. (1991)]

gravity is a simple and effective method for velocity selection. As an example, Fig. 3.27 shows an experimental setup which has been used to determine the attractive van der Waals/Casimir potential for the interaction of atoms with solid surfaces [Kasevich et al. (1991)]. Slowly moving atoms from an atomic fountain are launched on a trajectory that causes them to collide with a glass surface as they are beginning to turn around due to gravity. Velocity selection is achieved by the apertures and by the dimensions of the ionizing laser beam used to detect the bounced atoms. The incident atoms experience an attractive van der Waals potential, the Casimir potential due to radiation retardation effects, and a repulsive potential caused by the evanescent field from a totally internally reflected beam of light in a prism. Measuring the reflection as a function of the height of the repulsive portion of the surface potential, information on the van der Waals/Casimir attractive potential can be obtained.

Velocity selection by gravity has also been used to measure state-to-state differential crossed-beam scattering cross sections at μK collision energies [Legere and Gibble (1998)]. Two laser-cooled balls of Cs atoms are launched vertically from a double magnetooptical trap in rapid succession. The atoms collide near the top of the fountain (see Chap. 5) and, by varying the delay between the launches, the collision energy can be scanned from 19 to 150 μK. The differential cross section is measured by detecting the vertical velocity component of the atoms after scattering according to the principle described in Sect. 3.3.2.

3.6 Determination of the de Broglie Wavelength

Soon after the first experiments, in which the diffraction of atoms from crystal gratings was observed [Stern (1929), Knauer and Stern (1929), Estermann and Stern (1930)], diffraction was used for velocity selection [Estermann et al. (1931)]. In an experiment an effusive helium beam was successively diffracted from two crystal surfaces, the first crystal served as "monochromator", in order to achieve a better angular resolution in the second diffraction process.

While the diffraction from crystals is more or less restricted to helium atom beams, the recent progress in nanotechnology allows the fabrication of transmission gratings with periods down to 100 nm, which can be universally applied. These enable velocity measurements for molecular beams with an unrivaled precision, as is shown in the following sections.

3.6.1 Diffraction from Crystal Surfaces

Due to a number of requirements which must be satisfied in a diffraction experiment, both for the beam particles and the surface [Boato (1992)], the diffraction from crystal surfaces is best suited for helium beams. But even in this case, the intensity in a single diffraction maximum is only about 10^{-2} to 10^{-3} of the

3. Velocity Measurement and Selection

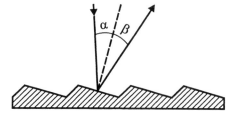

Fig. 3.28. Echelle grating. The largest fraction of a wave incident at an angle α is specularly reflected under the angle β. By an appropriate choice of the angle of incidence α, a desired order of diffraction can be observed in this direction

intensity of the incident beam. This fraction is further reduced by selecting a small interval of wavelengths.

On the other hand, extremely high velocity resolutions (e.g. $R_S = 1000$, $\Delta v/v = 0.1\%$) are achievable, which may be used for calibration purposes (see Sect. 3.2.4) [Buckland et al. (1997)].

A substantial increase of the reflected intensity would be achievable by nanoscaled echelle reflection gratings (see Fig. 3.28), which diffract a wave predominantly into one order [Hecht (1987)]. They consist of flat terraces with an almost perfect reflectivity, separated by steps with a height which is much smaller than the width of the steps. Nanofabrication of stepped gratings in the 1 nm range, however, exceeds the capabilities of current technologies. But nature provides a class of crystal surfaces, which exhibit the desired regular stepped structure [Comsa et al. (1979), Hahn et al. (1994)]. These surfaces may be directly used as nanoscaled echelle gratings [Schief et al. (1996)]. An example is the vicinal Pt(997) surface, consisting of (111) terraces about 20 Å wide, separated by monatomic steps (height 2.27 Å). This surface has both the high specular reflectivity (for helium) of the Pt(111) surface [Poelsema et al. (1982)] and the desired step structure. By an appropriate choice of the angle of incidence, it may be arranged that the total reflected intensity is concentrated in one diffraction order. Up to 20% of the incident intensity of a helium beam ($\Delta v/v = 1\%$) with a wavelength of $\lambda = 0.96$ Å has, for example, been observed in the diffraction order –4 [Schief et al. (1996)]. This is at least an order of magnitude more than is achievable with other crystal surfaces.

The surface is kept at a temperature of 750 K in a ultrahigh vacuum chamber, with an oxygen partial pressure of 5×10^{-9} mbar and a hydrogen partial pressure of 5×10^{-7} mbar. These gases provide a permanent autocatalytic cleaning of the surface. Carbon impurities are oxidized to desorbing CO, and a buildup of an oxygen coverage on the surface is prevented by the hydrogen gas, which leads to the formation of desorbing water. Thus, the Pt(997) surface can be used for several weeks without any significant intensity loss and without any further treatment.

Schief et al. (1996) have used such a crystal to generate an almost monochromatic helium beam ($\Delta v/v = 0.6\%$) from a nozzle beam with a velocity spread of $\Delta v/v = 1.3\%$. The velocity selected beam has been analyzed by a second crystal. The attainable resolution is limited by the finite spread of the orientation of the crystallites on the surface. For the crystals employed in the above described experiment, the angular range of this spread was about 0.02°. By an appropriate treatment of the crystal, this value may be reduced to values of the order of 0.003°

[Sandy et al. (1992)]. With such crystals, velocity resolutions between 1000 and 10 000 seem to be feasible.

3.6.2 Diffraction from Transmission Gratings

Recent progress in the fabrication of nanostructures by means of electron-beam lithography have made available free-standing transmission gratings with periods down to 1000 Å [Keith et al. (1988), (1991b), Ekstrom et al. (1992)]. Another possibility to obtain transmission gratings with similar periods is given by standing laser waves (see Sect. 5.4.5).

As is discussed in detail in Sect. 5.4.5, the angle θ at which the main maximum of the n-th diffraction order is observed for normal incidence, is given by

$$\sin\theta = n\frac{\lambda}{d} = \frac{nh}{mvd} = n\frac{3.9614}{dMv} \text{ mrad.} \qquad (3.114)$$

In the numerical part of this equation, the period d of the grating is measured in cm, the velocity v in cm/s, and the mass M in atomic units. For thermal energy beams the angles of diffraction are small and sinθ can be replaced by its argument θ in (3.114). Thus, the angle of diffraction is inversely proportional to the velocity.

Numerous diffraction experiments have already been performed with such diffraction gratings using beams of He, H_2, D_2, and clusters of these gases as well as Ne, Ar, Kr, Na atoms and Na_2 molecules (see Chap. 5), investigating different physical phenomena [Chapman et al. (1995), Schmiedmayer et al. (1997), Schöllkopf and Toennies (1994), (1996)]. Even the diffraction of C_{60} molecules has been observed recently [Arndt et al. (1999)]. An accurate determination of velocities and velocity distributions has been performed with nozzle beams from Na and Na_2 [Chapman et al. (1995), Schmiedmayer et al. (1997)]. From the observed diffraction pattern it was possible to determine beam velocities with an accuracy of 0.1%, a precision which is unattainable with other methods, except the diffraction from crystal gratings discussed in Sect. 3.6.1. From the broadening of the diffraction maxima the velocity distribution in the beams has been determined. Even in beams consisting of a mixture of sodium atoms and sodium dimers, it was possible to determine the velocity and the velocity distributions for both beam components. Consequently, the velocity slip between atoms and dimers, which lies — depending on the expansion conditions — between 1 and 3.5%, could be measured.

3.7 Beam Deflection by Photon Recoil

If an atomic beam is crossed by a laser beam perpendicular to the flight direction of its atoms, and if the laser frequency ν is in resonance with an atomic transition, the beam is deflected due to the photon recoil. This deflection is inversely proportional to the square of the atom velocity and can therefore be used for velocity measurement and selection. Since this method is especially useful for slow atoms, which are produced by laser deceleration, this method is discussed in Chap. 5.

4. State Selection

With the fundamental experiment on space-quantization of atoms in magnetic fields [Gerlach and Stern (1924), (1925)] it was demonstrated that atoms in different quantum states (characterized by their magnetic quantum number) can be spatially separated by inhomogeneous magnetic fields and that such fields are well suited for state selection.

The disadvantage of the initially used deflection fields (designated as Stern–Gerlach fields in the following), which had a strongly varying gradient over the cross section of the atomic beam and the beam deflection, was soon eliminated by the introduction of the "two-wire field" [Rabi et al. (1934)]. This field is generated by two straight, parallel wires at a distance $2a$, carrying equal currents in opposite directions. The maximum achievable field strengths and field gradients are, however, severely limited by the maximum current I which can be sent through the wires consistent with adequate cooling. Currents between 400 and 800 A have been used [Hamilton (1939), Prodell and Kusch (1952)]. Since superconductive magnets were not available at that time, a practical solution of this problem was to simulate the two-wire field with an iron magnet whose pole faces are cylindrical surfaces which coincide with the equipotential surfaces of the two-wire system [Millman et al. (1938)]. In the following, we will call this field the "Rabi-field", to have a name that distinguishes this field from other two-pole fields, particularly from its orthogonal field, in which the roles of field and equipotential surfaces are interchanged.

These fields were the basis of the atomic and molecular beam spectroscopy (atomic beam resonance method), which was introduced in 1938 by Rabi and co-workers [Rabi et al. (1938)] and which has been used to detect and measure energy differences of atoms and molecules at different magnetic fields. From these measurements, important properties of the atoms, but in particular also of the atomic nuclei (spin, magnetic dipole moment, electric quadrupole moment, and magnetic octupole moment) have been determined. Our present knowledge about these properties has been gathered over more than forty years by measurements with this atomic beam resonance method. Numerous review articles and books on this subject are available [Ramsey (1956), Kusch and Hughes (1959), Radford (1967), Childs (1972), Zorn and English (1973), English and Zorn (1974), Penselin (1978), Legon (1983), Dyke (1984), Legon and Millen (1986), Muenter (1992)]. Further applications of Rabi fields have been measurements of electric dipole polarizabilities of atoms and molecules, in which the magnetic deflection has often been used to compensate the deflection by the inhomogeneous electric field to achieve higher accuracy [Miller and Bederson (1977), Tarnovsky et al. (1993)]. The first scattering experiments with atoms in different Zeeman states were also performed using these fields [Berkling et al. (1962)]. In subsequent scattering experiments with magnetic Rabi-fields for state selection and for generation of polarized H_2 beams, the anisotropy of a series of atom–molecules and molecule–molecule interactions has been determined [Zandee and Reuss (1977)]. In present-day experiments, these fields are often used for state selection in atomic beams, for example, in beams generated in discharge sources [Aquilanti et al. (1992)], or in scattering experiments for a defined variation of the relative fractions of fine structure components in a given beam [Aquilanti et al. (1988)]. Finally, magnetic deflection fields find an increasing application in investigations of the magnetic properties of clusters [de Heer et al. (1990), Billas et al. (1994), Apsel et al. (1996)].

An important step in the development of magnetic state selectors was achieved with fields of higher symmetry (multipole fields), which are not only state selective, but also have focusing properties, resulting in a substantial intensity gain. These fields have been proposed by Vauthier (1949) and independently by Korsuski and Fogel (1951). Paul and co-workers were the first to investigate these fields experimentally [Friedburg and Paul (1951), Bennewitz and Paul (1954)]. They were first applied in atomic beam resonance experiments with beams of radioactive atoms [Lemonick and Pipkin (1954), Lemonick et al. (1955), Christensen et al. (1956)], where the intensity gain reduces the measuring time required by the radiographic detection substantially (see Sect. I.5.1.4). Further magnetic resonance experiments have been performed with paramagnetic molecules [Muenter (1992)]. Furthermore, magnetic state selectors are an essential component of the atom beam maser [Goldenberg et al. (1960), Audoin et al. (1971), Wineland and Ramsey (1972), Lainé (1992a)]. Finally, magnetic state selectors have found widespread application in the production of spin-polarized beams, which are used as the starting point in the generation of polarized electrons or ions for applications in particle accelerators or for use as polarized targets [Celotta and Pierce (1980), Baum et al. (1988), Chan et al. (1988), Schiemenz (1992), Schmelzbach (1994)].

Electrostatic state selectors, although technically easier to realize, have been developed rather late in connection with the molecular beam electric resonance method, the electrical analog of the magnetic resonance method [Hughes (1947), Trischka (1948), Gordon et al. (1954), (1955), Bennewitz et al. (1955), Schlier (1957), Trischka (1962), English and Zorn (1974), Muenter (1992)]. Numerous scattering experiments to investigate inelastic and reactive processes with molecules in defined quantum states have also been performed with electrostatic state selectors [Bennewitz et al. (1964), Toennies (1962), (1965), (1966), Beuhler and Bernstein (1969), Brooks (1976), Borkenhagen et al. (1975), (1979), Bernstein (1982), Stolte (1988), Parker and Bernstein (1989)]. Electrostatic multipole fields are also often used in investigations (spectroscopy, photolysis) which require oriented molecules [Kaesdorf et al. (1985), Gandhi and Bernstein (1988), Bernstein et al. (1989), Mastenbroek et al. (1995)].

Another field of application of electrostatic state selectors is the molecular beam maser and its use in frequency standards and in beam–maser spectroscopy [Dymanus (1992), Lainé (1992a), (1992b)]. In the latter case, also higher order multipole fields which are not discussed in this book are often used, since in beam–maser spectrometers precise focusing is usually not required. 8- or 12-pole fields may be used for states with a second order Stark effect, while 6- or 10-pole fields are used for states with a linear Stark effect. The higher repelling force of these fields results in an enlarged angle of acceptance and increases the density of the state-selected particles in the resonator. A higher angle of acceptance can also be achieved by using multipole fields with parabolic or linear-tapered cross sections [Helmer et al. (1960)]. Further devices for state selection in beam–maser spectroscopy are discussed by Dymanus (1992).

Recently, three-dimensional, inhomogeneous magnetostatic or electrostatic fields have found increasing application as traps for neutral particles. In combination with techniques to cool the stored particles, traps have become a source of new states of matter, such as, for instance, Bose–Einstein condensates of weakly interacting gases. Many of these devices were already proposed many years ago [Friedburg and Paul (1951), Heer (1963), Kügler et al. (1978), Wing (1980), Pritchard (1983), Bergeman et al. (1987), Ketterle and van Druten (1996)]. Some important trap devices are discussed in this chapter in connection with the treatment of inhomogeneous fields for state selection.

Finally, the first experiments on optical pumping provided the basis for state selection by pure optical methods [Kastler (1950)]. These techniques, which have a much wider range of application compared to inhomogeneous electric and magnetic fields, have been developed since the middle of the 1970s, taking advantage of the development of CW and pulsed lasers and refined spectroscopic methods [Bergmann (1988)]. These methods are discussed in Sect. 4.8 and illustrated by typical examples.

4.1 Potentials of Cylinder-Symmetric and Planar Fields

The potentials necessary to calculate the deflection fields described in this chapter as well as of two-dimensional, periodic fields used as atom mirrors (see Chap. 5), are derived in the following. The starting point is Poisson's equation

$$\nabla^2 G = 0. \tag{4.1}$$

The potential G is considered to be a complex function of $\zeta = x + iy = r\exp(i\varphi)$, which may be written as

$$G(x+iy) = U(x,y) + iV(x,y). \tag{4.2}$$

U and V are real functions which satisfy the Poisson equation. We restrict our considerations to the space outside of electrodes or current-carrying conductors and pay attention to avoid multiple values of the potential.

Since the devices which are used as state selectors are independent of the z direction (beam direction) (if we assume an infinite length), the potential G is also independent of z. Therefore, (4.1) obtains the form

$$\frac{1}{r}\frac{\partial}{\partial r}r\frac{\partial G}{\partial r} + \frac{1}{r^2}\frac{\partial^2 G}{\partial \varphi^2} = 0 \quad \text{or} \quad \frac{\partial^2 G}{\partial x^2} + \frac{\partial^2 G}{\partial y^2} = 0, \tag{4.3}$$

if cylindrical or Cartesian coordinates, respectively, are used. The solutions of (4.3) are two conjugate functions U(x,y) and V(x,y). It may be easily shown that the lines of constant U are orthogonal to the lines of constant V, so that the former describe the equipotentials, while the latter represent the lines of force, or vice versa. Introducing

$$G(re^{i\varphi}) = \Psi(r)\Phi(\varphi) \tag{4.4}$$

into (4.3) (in cylindrical coordinates) yields the two differential equations

$$r\frac{d}{dr}r\frac{d\Psi}{dr} = c\Psi \quad \text{and} \quad \frac{d^2\Phi}{d\varphi^2} = -c\Phi, \tag{4.5}$$

where c is the separation constant. c = 0 yields the potential of a current-carrying wire of infinite length (or, in the electric case, of a conducting rod of circular cross section of infinite length kept at a given potential), which we call a "monopole" field to distinguish it from the multipole fields which follow from c ≠ 0.

4.1.1 Monopole Field

First we consider the case c = 0. From (4.5) we obtain the solutions $\Psi \propto \ln r$ and $\Phi \propto \varphi$, if we do not account for the integration constants. Thus, according to (4.2) we obtain

$$G \propto (\ln r + i\varphi) = \ln(r\exp(i\varphi)), \tag{4.6}$$

with U = Ψ and V = Φ. In the (x,y) plane the lines U = const. are circles with the origin of the coordinate system as center (intersection of the wire or rod with the (x,y) plane). The curves of constant V are straight lines, emanating in radial directions from the origin. In the electric case the curves of constant U are the potential lines, the curves V = const. are the field lines. In the magnetic case the roles of U and V are interchanged, yielding for the field (in A/m) H = I/2πr, where I is the current through the wire, as is well known from the Biot–Savart law.

4.1.2 Multipole Fields

For $c \neq 0$ the solutions are

$$\Phi \propto \exp(\pm i\sqrt{c}\varphi) \quad \text{and} \quad \Psi \propto \exp(\pm\sqrt{c}\ln r). \tag{4.7}$$

Using the boundary condition of periodicity of Φ and finiteness of Ψ for $r = 0$ yields

$$G = Cr^n e^{in\varphi} = Cr^n(\cos(n\varphi) + i\sin(n\varphi)), \tag{4.8}$$

where $n = \sqrt{c}$ is a positive integer. Thus, the functions U and V are given by

$$U(r,\varphi) = Cr^n\cos(n\varphi) \quad \text{und} \quad V(r,\varphi) = Cr^n\sin(n\varphi). \tag{4.9}$$

The radius of curvature of the equipotential lines $U(r,\varphi)$ at those points where $\sin(n\varphi) = 0$, i.e. for $\varphi = 0, \pi/n, 2\pi/n, \cdots (2n-1)\pi/n$, is of practical importance for an approximation of the field by pole pieces (or electrodes) of circular shape. It is also the minimum radius of curvature of the potential lines. Furthermore, at these locations, the distance between two opposite pole pieces or electrodes, respectively, also has a minimum. This distance is usually termed the field diameter $2r_0$. For the minimum radius of curvature ρ_0 in these points we have

$$\rho_0 = \frac{r_0}{n-1}. \tag{4.10}$$

We return to this subject in Sect. 4.6. Of the fields which are represented by (4.9) we consider the following cases in some detail:

$n = 1$ (homogeneous field between parallel, planar electrodes):

$$U = r\cos\varphi = x \quad ; \quad V = r\sin\varphi = y. \tag{4.11}$$

In the x,y plane, the curves $U = $ const. are straight lines parallel to the y axis, the curves of constant V are straight lines parallel to the x axis. If $U = $ const. describes the potential lines, the field lines follow from $V = $ const. The homogeneous field is generated by two planar electrodes (of infinite size) or pole pieces, respectively, parallel to the (x,z) plane.

$n = 2$ (quadrupole field):

$$U = r^2\cos 2\varphi = x^2 - y^2 \quad ; \quad V = r^2\sin 2\varphi = 2xy \tag{4.12}$$

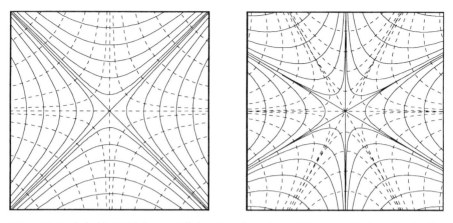

Fig. 4.1. Left half: Field lines (*full lines*) and equipotential lines (*dashed lines*) of a quadrupole field. Right half: Field lines (*full lines*) and equipotential lines (*dashed lines*) of a hexapole field

Both the lines U = const. and V = const. in the xy plane are hyperbolae (see left half of Fig. 4.1). The minimum radius of curvature ρ_0 is given by $\rho_0 = r_0$.

n = 3 (hexapole field):

$$U = r^3 \cos 3\varphi = x^3 - 3xy^2 \quad ; \quad V = r^3 \sin 3\varphi = 3yx^2 - y^3. \tag{4.13}$$

The curves of constant U and V build two systems of orthogonal hyperbolae of third order (see right half of Fig. 4.1). The radius ρ_0 is given by $\rho_0 = r_0/2$.

n = 4 (octupole field):

$$U = r^4 \cos 4\varphi = x^4 + y^4 - 6x^2y^2; \quad V = r^4 \sin 4\varphi = 4x^3y - 4xy^3, \tag{4.14}$$

with $\rho_0 = r_0/3$.

4.1.3 Two-Dimensional, Periodic Fields

In the following we consider two-dimensional field configurations (both magnetic and electric) with a periodic potential U(x,y), which is independent of the third space coordinate z (see Fig. 4.2 for a cross section in the x,y plane). Such fields have received much interest in recent years as atom mirrors, since they reflect beams of slowly moving atoms (see Chap. 5) [Opat et al. (1992), Lau et al. (1999)]. The periodicity in the x direction yields the condition

$$U(x + a, y) = U(x, y). \tag{4.15}$$

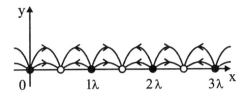

Fig. 4.2. Schematic view of a two-dimensional field of period λ which is independent of the z coordinate and generated by parallel rods, alternately at the potential +U and –U

The (magnetic or electric) field strength **F** is given by

$$\mathbf{F} = -\left(\mathbf{e}_x \frac{\partial U}{\partial x} + \mathbf{e}_y \frac{\partial U}{\partial y}\right), \tag{4.16}$$

where $\mathbf{e}_x$ and $\mathbf{e}_y$ are unit vectors in the x or y direction, respectively. In the magnetic case, such a field may be realized by an array of parallel wires (at a distance λ/2) alternately carrying current in opposite directions. Another possibility to realize an atom mirror is to write sinusoidal signals on appropriate magnetic materials such as an audio-tape or a floppy disk (see Chap. 5).

In the electric case, an array of parallel conductors may be used, kept alternately at positive or negative potentials.

The general solution of Poisson's equation (4.3) in Cartesian coordinates, taking into account the boundary conditions (field periodicity and vanishing of the field at infinity) may be written (with $\zeta = x + iy$):

$$G(\zeta) = U(x,y) + iV(x,y) = \sum_{n=0}^{\infty} A_n \exp(i2\pi n \zeta/\lambda)$$
$$= \sum_{n=0}^{\infty} A_n \exp\left(-\frac{2\pi n}{\lambda} y\right)\left(\cos\frac{2\pi n}{\lambda}x + i\sin\frac{2\pi n}{\lambda}x\right). \tag{4.17}$$

For $y = 0$, the potential $U(x,y)$ is identical with the Fourier expansion of the periodic potential. Accordingly, the contributions of higher harmonics $n > 1$ to the potential is determined by $U(x,0)$ and thus depends on the geometry of the device. If, for example, a magnetic material is magnetized according to a cosine function, only $n = 1$ occurs in the expansion (4.17). However, an unevenness or a curvature of the mirror device causes contributions of higher harmonics also in this case. For alternating potentials, defined by the relationship

$$U(x + \lambda/2, 0) = -U(x,0), \tag{4.18}$$

the expansion (4.17) has no even harmonics. If we consider only the first term ($n = 1$) in the expansion (4.17), the magnitude of the field strength F is given as

$$F = A_1 \frac{2\pi}{\lambda} \exp\left(-\frac{2\pi}{\lambda} y\right). \tag{4.19}$$

The magnitude of the field strength decreases exponentially with increasing distance from the mirror surface, similar to the field of an evanescent light wave (see Chap. 5).

4.2 Deflection in Inhomogeneous Magnetic Fields

Neutral particles having a magnetic dipole moment μ experience a force **F** in an inhomogeneous magnetic field, which is proportional to the field gradient and to the dipole moment according to

$$\mathbf{F} = \mu \, \text{grad} \, \mathbf{H}. \tag{4.20}$$

Since the effective magnetic moment of atoms or molecules μ_{eff} depends on their quantum state, such fields may be used for state selection.

4.2.1 Effective Magnetic Dipole Moment of Atoms

The effective magnetic dipole moment μ_{eff} may be expressed as

$$\mu_{\text{eff}} = -\frac{\partial W}{\partial H}, \tag{4.21}$$

where W is the energy of the particle in the magnetic field. The effective moment depends, in general, on the state of coupling of the several angular momenta, connected with the electron spin, the orbital angular momentum, the nuclear angular momentum, and (for molecules) the rotational angular momentum. Furthermore, it depends on the energy level and on the magnitude of the magnetic field. For an atom with the total angular momentum F, which couples to the external field, the effective magnetic moment is — in the limit of vanishing field — given by

$$\mu_{\text{eff}} = -M_F g_F \mu_0, \tag{4.22}$$

where M_F is the magnetic quantum number, and g_F is the ratio of the magnetic moment in units of the Bohr magneton to the angular momentum in units of $h/2\pi$. μ_0 is the Bohr magneton. The contribution due to the nuclear magnetic moment, which is about three orders of magnitude smaller, has been neglected. In this approximation, g_F may be written as

$$g_F = g_J \frac{F(F+1) + J(J+1) - I(I+1)}{2F(F+1)}, \tag{4.23}$$

with

$$g_J = \frac{3J(J+1)+S(S+1)-L(L+1)}{2J(J+1)}. \tag{4.24}$$

J is the electronic angular momentum, composed of the electronic spin S and the orbital angular momentum L. I is the nuclear angular momentum.

At high fields, where orbital angular momentum and nuclear spin decouple, the effective magnetic moment is given as

$$\mu_{\text{eff}} = -M_J g_J \mu_0. \tag{4.25}$$

As an example, we consider in the following an atom with the electronic angular momentum $J = 1/2$ and the nuclear spin $I = 3/2$. Due to the coupling of the two angular momenta the atom has the total spin F, which may accept, in the given example, the values $F = I \pm 1/2$. The corresponding magnetic quantum number M_F has $2F+1$ values from $+F$ to $-F$. At zero field the effective magnetic moment corresponding to the quantum numbers (F, M_F) is described by (4.22) to (4.25) (see Fig. 4.3). With increasing external field the electronic angular momentum becomes decoupled from the nuclear spin, and the components of the magnetic moments in the field direction, μ_{eff}, follow the dependence shown in Fig. 4.3. Except for the states with $M_F = \pm(I+1/2)$ the effective magnetic moments change with the field strength, and only at sufficiently high fields, where electronic angular momentum and nuclear spin decouple completely, do the effective magnetic moments achieve the limiting values given by (4.25). In Fig. 4.3, the effective magnetic moments are plotted as a function of a reduced field strength $x = g_J \mu_0 H / \Delta W$, where ΔW is the hyperfine structure splitting without external field. Two regions may be

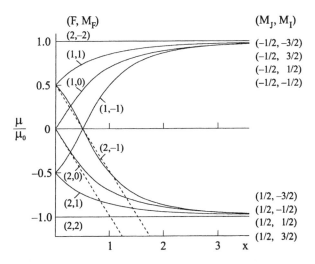

Fig. 4.3. The magnetic moments of the magnetic sublevels of the hyperfine states of an atom with $J = 1/2$, $I = 3/2$ as a function of the reduced field strength x. For a further description see text

distinguished. At low fields, some (F,M_F) states exist, for which the magnetic moment can be approximated by

$$\mu_{eff} = -\alpha H \quad \text{or} \quad \mu_{eff} = -\alpha H + \beta, \tag{4.26}$$

(dashed lines in Fig. 4.3). These states with $(F,M_F) = (2,-1)$ and $(2,0)$ can be focused in a magnetic quadrupole field. At higher fields, where the effective magnetic moment is constant, the states with $M_J = 1/2$ can be focused in a magnetic hexapole field.

4.2.2 Two-Wire Field (Rabi-Field)

The field of two straight, parallel wires at a distance 2a, extending to infinity in the z direction and carrying equal currents in opposite directions, can be easily calculated using the formulae derived in Sect. 4.1.1 [Ramsey (1956), Kusch and Hughes (1959), Reuss (1988)], by simply adding the potentials of the two wires. Using the notation defined in Fig. 4.4 together with (4.6) gives the result

$$V = C\ln(r_1/r_2) \quad \text{and} \quad U = -C(\varphi_1 - \varphi_2), \tag{4.27}$$

if the remaining integration constants are chosen in such a manner that V and U vanish at infinity. It can be easily shown that V is identical to the magnitude of the vector potential **A**. The direction of **A** is the z direction. Thus we may write

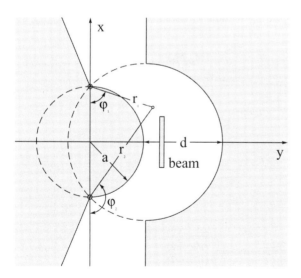

Fig. 4.4. Cross section (perpendicular to the z axis) of a two-wire field and definition of the notation used in text. The current-carrying wires intersect the drawing plane perpendicularly at the points $(x,y) = (a,0)$ and $(-a,0)$. The field is usually generated by an iron magnet with pole pieces corresponding to two equipotential surfaces. Within the rectangle, the field is nearly constant

$$\mathbf{A} = \mathbf{e}_z V, \tag{4.28}$$

where $\mathbf{e}_z$ is a unit vector in the z direction. Hence, the field $\mathbf{H}$ may be calculated either from the vector potential $\mathbf{A}$ or from the scalar potential U of the magnetic field:

$$\mathbf{H} = \nabla \times \mathbf{A} \quad \text{or} \quad \mathbf{H} = -\text{grad} U. \tag{4.29}$$

In the plane perpendicular to the wires (x,y plane), the magnetic equipotentials U = const. and the magnetic field lines V = const. are families of orthogonal circles (see Fig. 4.5). The former (full lines in Fig. 4.5) are circles whose centers lie on the y axis. They all pass through the centers of the wires. The latter (dashed lines in Fig. 4.5) are circles with centers on the x axis, asymmetrically surrounding the centers of the wires. The equipotentials are described by the equation

$$(y - a\,\text{ctg}U)^2 + x^2 = \frac{a^2}{\sin^2 U}, \tag{4.30}$$

while the lines of force are given by

$$y^2 + (x - a\,\text{ctgh}V)^2 = \frac{a^2}{\sinh^2 V}. \tag{4.31}$$

With

$$r_1^2 = y^2 + (a - x)^2, \quad r_2^2 = y^2 + (a + x)^2, \quad \text{tg}\varphi_1 = \frac{y}{x+a}, \quad \text{tg}\varphi_2 = \frac{y}{x-a}, \tag{4.32}$$

(see Fig. 4.4), (4.27) may be rewritten as

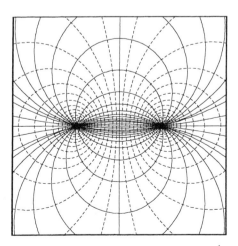

Fig. 4.5. Lines of force (*dashed lines*) and equipotentials (*full lines*) of the two-wire field. The wires are perpendicular to the drawing plane

$$V = \frac{C}{2}\ln\left(\frac{y^2+(x+a)^2}{y^2+(x-a)^2}\right), \quad U = -C\left[\text{arctg}\left(\frac{y}{x+a}\right) - \text{arctg}\left(\frac{y}{x-a}\right)\right]. \tag{4.33}$$

This yields for the field components H_x and H_y

$$H_x = -Cy\left(\frac{1}{r_1^2}-\frac{1}{r_2^2}\right), \quad H_y = C\left(\frac{x+a}{r_1^2}-\frac{x-a}{r_2^2}\right). \tag{4.34}$$

The magnitude of the field is given by

$$H = \sqrt{H_x^2 + H_y^2} = 2C\frac{a}{r_1 r_2}, \tag{4.35}$$

and the gradients in the x and y directions may be written as

$$\frac{\partial H}{\partial x} = 2Cax\frac{r_1^2+r_2^2}{r_1^3 r_2^3}; \quad \frac{\partial H}{\partial y} = -2Cay\frac{r_1^2+r_2^2}{r_1^3 r_2^3}. \tag{4.36}$$

If the field is generated by two current-carrying wires, the constant C is given by $C = 2I$, where I is the electric current. C has the dimension field strength×length and depends on the units used. In SI-units (Ampere, meter, Tesla), C must be multiplied by $4\pi \times 10^{-7}$. In "mixed" units (Ampere, cm, Oersted), the multiplication factor is $4\pi/10$. This also holds for (4.41).

As already mentioned, the field is usually generated by an iron magnet with pole pieces having the shape of equipotential surfaces [Millman et al. (1938)]. The field of the iron magnet, however, is only equivalent to the two-wire field, when the permeability of the iron in the region of the pole faces is so high that saturation does not occur. In this case, the magnitude of the field and its gradient in the gap are described by (4.35) and (4.36). The factor C can be expressed by the mean value of the field $\overline{H}$ in the gap (the length of the gap being d) along the y axis

$$\overline{H} = \frac{1}{d}\int_a^{a+d} H(x=0)dy = \frac{2C}{d}a\int_a^{a+d}\frac{dy}{y^2+a^2} = \frac{2C}{d}\left[\text{arctg}\left(1+\frac{d}{a}\right)-\frac{\pi}{4}\right]. \tag{4.37}$$

This yields

$$C = \frac{\overline{H}d}{2[\text{arctg}(1+d/a)-\pi/4]}. \tag{4.38}$$

Using Ampere's law

$$\oint H ds = NI, \tag{4.39}$$

where the integral is taken along a path through the magnet (of length L_{Fe}) and its gap (of length d). NI is the total number of ampere turns through the surface enclosed by the path. With (4.37) we obtain

$$\oint H ds = \overline{H} d + \frac{\overline{H}}{\mu_p} L_{Fe}, \tag{4.40}$$

if the field in the iron is approximated by $\overline{H}/\mu_p$ (μ_p is the permeability of the iron). Thus the constant C is given by

$$C = \frac{NI}{2(1+L_{Fe}/d\mu_p)[\arctg(1+d/a)-\pi/4]}. \tag{4.41}$$

If the permeability μ_p of the iron is high, the main contribution to the line integral is due to the field in the gap and the factor $L_{Fe}/\mu_p d$ in the denominator of (4.41) can be neglected. To abbreviate the notation, we express in the following the constant C by the magnetic field H_0 at the location x = 0, y = a. According to (4.35) we obtain $C = aH_0$. Introducing the reduced quantities

$$\xi = x/a, \quad \eta = y/a, \quad \rho_i = r_i/a \quad i = 1,2, \tag{4.42}$$

the reduced gradient follows from (4.36):

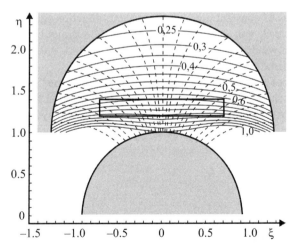

Fig. 4.6. Curves of constant gradient G_η (*full lines*) in a two-wire field. The rectangle corresponds to the interval (4.44). The numbers indicate the reduced gradients G_η according to (4.43). The dashed lines are curves of constant gradient G_ξ. G_ξ vanishes on the η axis. The other values are 0.01, 0.02, 0.04, 0.06, 0.08, 0.10, 0.14, 0.18, 0.24, 0.3, and 0.4

210   4. State Selection

$$\frac{a}{H_0}\frac{\partial H}{\partial y} = G_\eta = 2\eta\frac{\rho_1^2+\rho_2^2}{\rho_1^3\rho_2^3}, \quad \frac{a}{H_0}\frac{\partial H}{\partial x} = G_\xi = 2\xi\frac{\rho_1^2+\rho_2^2}{\rho_1^3\rho_2^3}. \tag{4.43}$$

The field gradient $\partial H/\partial y$ of the two-wire field is reasonably constant with x, as is shown by Fig. 4.6, which shows lines of constant gradient G_η (full lines) in the region between the pole faces. In the interval (see rectangle in Figs. 4.4 and 4.6)

$$1.2 \leq \eta \leq 1.4, \quad -0.7 \leq \xi \leq 0.7, \tag{4.44}$$

the maximum variation of this gradient with x (or ξ) is less than 4%.

Using a ribbon-shaped beam with a width (extension in the y direction) which is small compared to its height (extension in the x direction), a homogeneous deflection over the full height of the beam is obtained, preserving the beam profile. The gradient in the x direction varies more strongly with x, but this dependence has no influence upon the beam deflection. In many cases, therefore, the deflection can be considered as being caused by a field of constant gradient. Thus, the actual gradient may be replaced by its mean value averaged over that range of y values, which is covered by the particle trajectories (see Sect. 3.4.1). With this approximation, the deflection y(v) follows from (4.20) yielding

$$y(v) = \frac{1}{2mv^2}\mu_{\text{eff}}\overline{\frac{\partial H}{\partial y}}L^2, \tag{4.45}$$

where μ_{eff} is the effective magnetic moment, m the mass, and v the velocity of the particles. L is the length of the field. If this approximation is insufficient, the equations of motion have to be solved numerically.

Figure 4.6 shows also lines of constant gradient G_ξ in the ξ direction (dashed lines). Along the η axis we have $G_\xi = 0$, and G_ξ is very small in the vicinity of this axis. At the ends of the "beam window" shown in Fig. 4.6, we have $G_\xi = 0.14$, a value which is still small compared to G_η.

If two current-carrying wires are used to generate the field, not only the "beam window", given by (4.44) and shown in Fig. 4.4, can be used for deflection measurements, but also the region between the two wires in the vicinity of the symmetry axis (x = 0, y = 0) (see Fig. 4.7) may be used. In this region the equations (4.36) may be expanded to yield

$$\begin{aligned}\frac{\partial H}{\partial x} &= \frac{4H_0}{a^2}x\left[1+2\left(\frac{x}{a}\right)^2-4\left(\frac{y}{a}\right)^2+\cdots\right], \\ \frac{\partial H}{\partial y} &= -\frac{4H_0}{a^2}y\left[1+4\left(\frac{x}{a}\right)^2-2\left(\frac{y}{a}\right)^2+\cdots\right].\end{aligned} \tag{4.46}$$

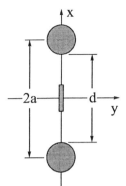

Fig. 4.7. Two-pole field with focusing in one direction (cylindrical lens)

An atom with a positive magnetic moment, starting within the field at a location $x=0$, $y=0$, $z=0$ and moving in the z direction (having a small velocity component perpendicular to the z direction), performs an harmonic oscillation in the y direction and returns to the z axis at the position

$$z_0 = \frac{\pi v a}{2} \sqrt{\frac{m}{\mu_{\text{eff}} H_0}}. \tag{4.47}$$

In the x direction the deviations from the z axis increase exponentially. Thus, a focusing action in the y direction is achieved, while a defocusing occurs in the x direction. In the case of a negative magnetic moment, the roles of x and y are interchanged. However, the image distance z_0 depends on the initial value of x, so that the beam dimension in the x direction has to be kept rather small to obtain a good focusing (for $x = 0.2a$ the image distance is about 15% smaller than for $x = 0$).

An electric two-pole field is generated by two metal rods kept at opposite potentials.

4.2.3 Quadrupole Sector Field

A two-pole configuration having a gradient $\partial H / \partial y$ which is almost independent of y, is obtained by using a quadrant of a quadrupole field [McColm (1966)]. It consists of a pole face of hyperbolic cross section (which may be approximated by a pole face of circular cross section), and a pole face build by the corresponding asymptotic planes (see Fig. 4.8). This field is often used in modern cluster experiments [Bucher et al. (1991a), (1991b), Douglass et al. (1993)].

Using (4.12) and the coordinates and notations of Fig. 4.8, the magnetic potential outside the iron is determined by

$$U(x,y) \propto \frac{y^2 - x^2}{4a^2}, \tag{4.48}$$

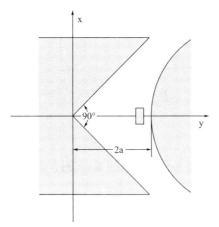

Fig. 4.8. Quadrupole sector field. In the rectangular beam cross section (according to the interval (4.52)), the gradient is constant to better than 1% (see Fig. 4.9)

yielding for the magnitude of the magnetic field

$$H = H_0 \frac{\sqrt{x^2 + y^2}}{2a}. \tag{4.49}$$

H_0 is the field at $x = 0$, $y = 2a$ (see Fig. 4.8). The gradients in the x and y directions are

$$\frac{\partial H}{\partial x} = \frac{H_0 x}{2a\sqrt{x^2 + y^2}}, \quad \frac{\partial H}{\partial y} = \frac{H_0 y}{2a\sqrt{x^2 + y^2}}, \tag{4.50}$$

and the reduced gradients (see (4.43), can be written as

$$\frac{a}{H_0} \frac{\partial H}{\partial x} = G_\xi = \frac{\xi}{2\sqrt{\xi^2 + \eta^2}}, \quad \frac{a}{H_0} \frac{\partial H}{\partial y} = G_\eta = \frac{\eta}{2\sqrt{\xi^2 + \eta^2}}. \tag{4.51}$$

In the ξ, η plane the curves of constant gradient G_η are straight lines through the origin of the coordinate system (see Fig. 4.9). Between the pole pieces, G_η changes from 0.35 on the asymptotes and 0.5 on the η axis. In the interval (rectangle in Figs. 4.8 and 4.9)

$$1.5 \leq \eta \leq 1.8, \quad -0.2 \leq \xi \leq 0.2, \tag{4.52}$$

G_η is constant to better than 1%. This field, therefore, is an ideal deflection field with constant gradient for beams with circular or rectangular cross section (beam height (extension in the x direction) ≥ beam width (extension in the y direction). But it is also suited for ribbon-shaped beams which have been discussed in Sect. 4.2.2, since the x dependence of the gradient G_η is not larger than in the case of a two-wire field, as long as the beam height h satisfies the relationship $h \leq a$.

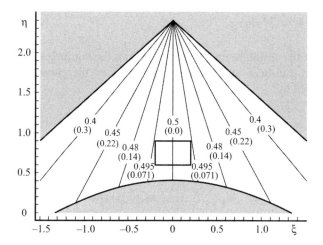

Fig. 4.9. Lines of constant gradient G_η in a quadrupole sector field. The rectangle corresponds to the interval defined in (4.52). The numbers give the reduced gradient G_η, the values in brackets refer to G_ξ. Both gradients have equal values on the pole face planes ($G_\xi = G_\eta = 0.35$). For further details see text

The curves of constant gradient G_ξ are (in the ξ,η plane) also straight lines through the coordinate origin. G_ξ vanishes on the η axis ($\xi = 0$) and is very small in its vicinity compared to G_η. On the pole face planes we have $G_\xi = G_\eta = 0.35$.

4.2.4 Two-Pole Field

For completeness, the magnetic field, which is orthogonal to the two-wire field shown in Fig. 4.7, is also mentioned here. It is obtained by interchanging the lines of force with the equipotentials (interchanging of V and U in (4.33)) and can be realized by a magnet with two opposite pole faces of half-circular cross section. The magnitude of this field and its gradient are described by (4.35) and (4.36), and it has the same focusing properties as the field shown in Fig. 4.7. We will return to this field when discussing electrostatic two-pole fields (Sect. 4.3.2).

4.2.5 Multipole Fields

The two-wire field is well adapted to the geometry of a narrow parallel beam with a width small compared to its height. For beams of circular cross sections, cylinder-symmetric fields are much more appropriate. Furthermore, these fields provide radial focusing connected with a considerable increase in intensity.

Paul and co-workers [Friedburg and Paul (1951), Friedburg (1952), Bennewitz and Paul (1954), Bennewitz et al. (1955)] were the first to demonstrate experimentally that these fields are quite analogous to optical lenses. According to (4.9), multipole fields may be described by the functions

$$U = Cr^n \cos n\varphi \quad \text{and} \quad V = Cr^n \sin n\varphi. \tag{4.53}$$

If, in a plane perpendicular to the beam direction (z axis), the equipotentials are described by U = const. the lines of force are represented by the curves of constant V. The Field is given by

$$\mathbf{H} = -\text{grad}\, U, \tag{4.54}$$

with the components (in cylindrical coordinates)

$$H_r = -\frac{\partial U}{\partial r} = -nCr^{n-1} \cos n\varphi \quad \text{and} \quad H_\varphi = -\frac{1}{r}\frac{\partial U}{\partial \varphi} = nCr^{n-1} \sin n\varphi, \tag{4.55}$$

and the magnitude

$$H = nCr^{n-1}. \tag{4.56}$$

The constant C may be expressed by the field strength H_0 at $r = r_0$, where r_0 is the field radius (see Sect. 4.1.2):

$$C = \frac{H_0}{nr_0^{n-1}}. \tag{4.57}$$

The field gradient in the radial direction is given by

$$\frac{\partial H}{\partial r} = n(n-1)Cr^{n-2} = \frac{(n-1)}{r_0^{n-1}} H_0 r^{n-2}. \tag{4.58}$$

Hexapole Field. The force experienced by particles of constant effective magnetic moment, μ, in a multipole field, is proportional to the distance from the axis in a hexapole field (n = 3, see (4.58)). Due to this force, particles with an appropriate sign of the magnetic moment undergo a harmonic radial oscillation when traveling through the field. All particles of equal velocity (and mass), leaving the axis at a given point (r = 0, z = 0) within the field, return to the axis at a point (r = 0, z = z_0) independent of their starting angle. This follows from the equations of motion:

$$\frac{d^2 r}{dt^2} = -\frac{2\mu H_0}{mr_0^2} r, \quad \frac{d^2 z}{dt^2} = 0. \tag{4.59}$$

Since no force acts in the z direction, the particles move with constant velocity v in this direction. Their path is given by z = vt, if the motion starts at t = 0 at the location z = 0. The general solution of (4.59) may be written in the form

$$r = c_1 \sin\omega t + c_2 \cos\omega t \quad \text{with} \quad \omega = \frac{1}{r_0}\sqrt{\frac{2\mu H_0}{m}} = \frac{105.303}{r_0}\sqrt{\frac{\mu H_0}{M}} \text{ s}^{-1}. \quad (4.60)$$

The units to be used in the numerical part of (4.60) are the Bohr magneton for μ, Oe for H_0, and atomic mass units for M. If the radial motion starts at $t = 0$ and $r = 0$ with the radial velocity component v_r, these initial conditions determine the integration constants in (4.60): $c_1 = v_r/\omega$ and $c_2 = 0$. This yields the final result

$$r = \frac{v_r}{\omega}\sin\omega t. \quad (4.61)$$

After the time $t = \pi/\omega$, the particles have finished a half-period of their oscillation and return to the axis at $z = z_0$. For the "image point", which depends on the velocity v_0 of the particles, we find

$$z_0 = \pi\frac{v_0}{\omega} = \pi r_0 v_0\sqrt{\frac{m}{2\mu H_0}} = 2.984 \times r_0 v_0\sqrt{\frac{M}{\mu H_0}} \text{ cm}. \quad (4.62)$$

r_0 is the field radius, and H_0 is the field at this point. The units to be used in the numerical part of (4.62) are cm for r_0, m/s for v, the Bohr magneton for μ, Oe for H_0, and atomic mass units for M.

Particles with velocities $v = v/n$ ($k_n = n\pi/z_0$ with n being an integer), perform n half-periods of oscillation in the field on their path z_0 and cross the field axis also at $z = z_0$ (velocity sidebands, see Chap. 3).

Imaging by a hexapole field has a large "chromatic" aberration: The focus length is proportional to the particle velocity. Therefore, a velocity selector has to be used when working with effusive beams.

Since the field exerts an harmonic force on the atoms, particles moving along the field axis experience no force and can pass the filter independent of their state and their velocity. To block these particles, a small central obstacle is usually placed on the beam axis in the middle of the field (see Fig. 4.10). This central obstacle also blocks those slow particles which experience an even number of half-periods of their radial oscillation during their passage through the field. Since the transmission of the state selector decreases drastically with increasing n, as is discussed in detail in Sect. 3.4.2 (see also Table 3.4), the contributions of sidebands corresponding to odd values of n (n = 3, 5, ...) are usually negligible.

In the case where starting and image points of the particles are situated outside the field, as is shown in Fig 4.10, the trajectory of a particle consists of three parts, which have to be joined continuously and with continuous derivative. If the zero of the z axis is located at the field entrance (see Fig. 4.10), the solution of the equation of motion in the field-free space in front of the field is given by

216   4. State Selection

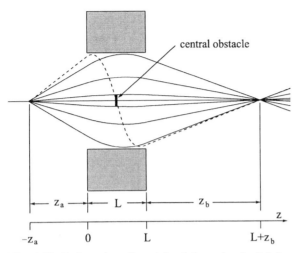

Fig. 4.10. Trajectories of particles (of equal velocity) in a magnetic hexapole field and definition of the geometry used to derive the lens equation. The dashed trajectory corresponds to the first velocity sideband. The drawing is vastly exaggerated in the lateral direction

$$r = \frac{r_a}{z_a} z + r_a, \quad \frac{dr}{dz} = \frac{r_a}{z_a} \quad \text{for} \quad -z_a \leq z \leq 0. \tag{4.63}$$

The solution inside the field is

$$r = \frac{r_a}{kz_a} \sin kz + r_a \cos kz, \quad \frac{dr}{dz} = \frac{r_a}{z_a} \cos kz - kr_a \sin kz, \quad \text{for} \quad 0 \leq z \leq L, \tag{4.64}$$

where we have used the wave number k

$$k = \frac{\omega}{v}. \tag{4.65}$$

ω is determined by (4.60). The continuity of (4.63) and (4.64) at the field entrance $z = 0$ has also been taken into account. Between the field exit at $z = L$, $r = r_e$ and the image point at $z = L + z_b$, $r = 0$, the motion of the particles is described by

$$r = -\frac{r_e}{z_b} z + \frac{r_e}{z_b}(L + z_b), \quad \left(\frac{dr}{dz}\right)_{r=r_e} = -\frac{r_e}{z_b} \quad \text{for} \quad L \leq z \leq L + z_b. \tag{4.66}$$

The continuity at $z = L$ yields the equations

$$r_e = \frac{r_a}{kz_a}\sin kL + r_a \cos kL, \quad \left(\frac{dr}{dz}\right)_{r=r_e} = -\frac{r_e}{z_b} = \frac{r_a}{z_a}\cos kL - kr_a \sin kL. \quad (4.67)$$

Equation (4.67) is a relationship between the "object distance" z_a, the "image distance" z_b, and the length L of the field, which may be expressed in the following form

$$\frac{1}{z_a} + \frac{1}{z_b} = \text{tg}(Lk)\left(k - \frac{1}{kz_a z_b}\right). \quad (4.68)$$

A particle entering the field parallel to the field axis ($z_a \to \infty$) crosses the axis in the focal point P'. Its distance F' from the end of the field is given by

$$F' = \frac{1}{k}\text{ctg}(kL). \quad (4.69)$$

On the other hand, a particle starting at the focus P leaves the field exit parallel to the z axis ($z_b \to \infty$). Correspondingly, the focal length on the object side, referred to the field entrance, is given by

$$F = \frac{1}{k}\text{ctg}(kL). \quad (4.70)$$

If the distances z_a, z_b, F, and F' are not referred to the field entrance and the field exit, but to the principle planes H and H', which have the distances g and g' from the field entrance or exit, respectively (see Fig. 4.11), (4.68) can be rewritten in the form of the optical lens equation

$$\frac{1}{a} + \frac{1}{b} = \frac{1}{f}, \quad (4.71)$$

with $a = z_a + g$, $b = z_b + g'$, $f = F + g$, $f' = F' + g'$. The focal lengths and the location of the principle planes follow, with (4.68):

$$f = f' = \frac{1}{k}\left[\text{ctg}(kL) + \text{tg}\left(\frac{kL}{2}\right)\right], \quad (4.72)$$

$$g = g' = \frac{1}{k}\text{tg}\left(\frac{kL}{2}\right). \quad (4.73)$$

These equations are valid for both magnetic and electrostatic multipole fields, provided that the magnetic or electric dipole moments lead to an harmonic os-

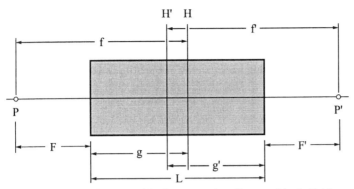

Fig. 4.11. Nomenclature used in the text to describe a multipole field as a thick lens

cillation in the field. Therefore, multipole fields can be treated as thick lenses, and their imaging properties can be determined by applying the well known methods of geometrical optics. Correspondingly, a refractive index can also be defined by

$$n = \sqrt{1 + \mu \frac{H_0}{W}}, \qquad (4.74)$$

where W is the total energy of the particles. This expression also holds for electrostatic fields, if the magnetic field and the magnetic moment are replaced by the electric field and the electric dipole moment. For thermal energy beams, the index of refraction defined by (4.74) is very close to unity. Using, for example, for the total energy W the thermal energy kT at a temperature T = 1000 K, a magnetic field of 5000 Oe, and a dipole moment of one Bohr magneton, the index of refraction is $n - 1 = 1.7 \times 10^{-4}$. For a magnetic dipole moment of one nuclear magneton, the refractive index is $n - 1 = 9 \times 10^{-8}$. However, for slow atoms produced by laser cooling (see Chap. 5), the refractive index can achieve large values. For example, a rubidium beam with a velocity of 100 m/s and a field of 5000 Oe has an index of refraction of n = 8.1. Thus, very short multipole fields can be used for focusing [Kaenders et al. (1996)]. The velocity sidebands of devices having their focuses outside of the field are no longer integer multiples of the wave number, but must be calculated for a given setup from (4.68) (see Sect. 3.4.2). The half angle of acceptance Θ_a of the field for a diverging beam starting at $z = -z_a$ is

$$tg\Theta_a = \frac{r_{am}}{z_a}, \qquad (4.75)$$

where r_{am} is the maximum distance from the axis, at which these particles may enter the field without touching the pole pieces during their passage through the field. r_{am} follows from (4.64) by assuming $dr/dz = 0$ and $r = r_0$ at a given value

$z = z_m$. From the two equations (4.64) both z_m and r_a can be determined, yielding for the half angle of acceptance Θ_a

$$\text{tg}\Theta_a = \frac{k r_0}{\sqrt{1 + z_a^2 k^2}}. \tag{4.76}$$

The half angle Θ_b under which the particle trajectories converge at the image point can be determined similarly with the result

$$\text{tg}\Theta_b = \frac{k r_0}{\sqrt{1 + z_b^2 k^2}}. \tag{4.77}$$

Finally, the half-opening angle Θ_{co}, which is covered by the central obstacle of radius r_{co}, is given by

$$\text{tg}\Theta_{co} = \frac{k r_{co} \sqrt{1 + \text{tg}^2(kL/2)}}{z_a k + \text{tg}(kL/2)}. \tag{4.78}$$

For completeness it should be pointed out here that the equations of motion in the different regions (4.63), (4.64), and (4.67) can be written in matrix form. By multiplication of the matrices corresponding to the different regions, a matrix is obtained which connects the initial vector $\{r, dr/dz\}$ with the final vector $\{r_f, dr_f/dz\}$. This method is of particular advantage when calculating the imaging properties of several multipole fields connected in series [Septier (1961), Berg et al. (1965)].

Quadrupole Field. As pointed out in Sect. 4.2.1, atoms in relatively weak magnetic fields may exhibit (F, M_F) states, for which the magnetic dipole moment can be approximated by

$$\mu_{eff} = -\alpha H \quad \text{or} \quad \mu_{eff} = -\alpha H + \beta, \tag{4.79}$$

where α and β are constants. Hence

$$\frac{d^2 r}{dz^2} + \frac{\alpha H_0^2}{mv^2 r_0^2} r = \frac{\beta H_0}{r_0 mv^2} \tag{4.80}$$

is the equation of motion in this case. Its general solution is

$$r = c_1 \sin kz + c_2 \cos kz + \frac{\beta}{\alpha} \frac{r_0}{H_0} \quad \text{with} \tag{4.81}$$

220   4. State Selection

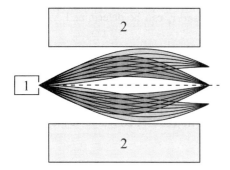

Fig. 4.12. Focusing in a magnetic quadrupole field [Bennewitz et al. (1956)]. (1) source, (2) pole pieces of the magnetic quadrupole field. For details see text

$$k = \frac{H_0}{r_0 v} \sqrt{\frac{\alpha}{m}}. \tag{4.82}$$

The integration constants c_1 and c_2 are determined by the initial conditions. For a given wave number, the imaging properties can be calculated from (4.68). For $\beta \neq 0$, a constant force is superimposed on the harmonic force of the field. Due to this constant force, the particles are focused onto a ring of radius R in the image plane (see Fig. 4.12). The radius R is given by:

$$R = \alpha\, r_0 / \beta H_0. \tag{4.83}$$

For the example discussed in Sect. 4.2.1 (see Fig. 4.3) of atoms with the electronic angular momentum $J = 1/2$ and the nuclear spin $I = 3/2$ all atoms with $M_F = 0, -1, \cdots, -(F-1)$ are focused. The number of rings yields the nuclear spin, if the central component ($M_F = 0$) is counted as 1/2. In the absence of a central component the nuclear spin is integer. From the field strength required to obtain a sharp image of the rings, the constant of the magnetic hyperfine-structure coupling may be determined. This method has been applied to some radioactive isotopes, using a radiographic detection method [Bennewitz et al. (1956)].

4.3 Deflection in Inhomogeneous Electrostatic Fields

As in the magnetic case, molecules having an electric dipole moment μ experience a force **F** in an inhomogeneous electrostatic field, which is given by

$$\mathbf{F} = \mu\, \text{grad} E. \tag{4.84}$$

As is shown below, the effective electric moment of a molecule μ_{eff} depends upon the permanent dipole moment μ of the molecule and upon the motion of the nuclei, i.e. upon the molecular rotation and the corresponding quantum numbers. Consequently, these fields may be used for state selection.

4.3.1 Effective Electric Dipole Moment of Molecules

The effective electric dipole moment μ_{eff} of a molecule is given by

$$\mu_{\text{eff}} = -\frac{\partial W}{\partial E}, \tag{4.85}$$

where μ_{eff} depends upon the permanent dipole moment μ of the molecule and its rotational quantum numbers. In perturbation theory, the energy W of a molecule in an electric field is expanded in powers of the electric field E

$$W = W^{(0)} + \lambda_1 E + \lambda_2 E^2 + \cdots, \tag{4.86}$$

where $W^{(0)}$ is the energy of the unperturbed rotator at vanishing field. λ_1 and λ_2 are the Stark effect coefficients of first and second order. While classically the energy W of an electric dipole moment μ in an electric field E can be described by

$$W = W^{(0)} + \mu E \cos\vartheta, \tag{4.87}$$

where ϑ is the angle between the two vectors μ and **E**, in quantum mechanics, $\cos\vartheta$ has to be replaced by the expectation value $\langle \cos\vartheta \rangle$ of the cosine of the angle of orientation of the dipole moment with respect to the field. This yields

$$W = W^{(0)} + \mu E \langle \cos\vartheta \rangle, \tag{4.88}$$

and the effective electric dipole moment μ_{eff} becomes

$$\mu_{\text{eff}} = -\frac{\partial W}{\partial E} = \mu \langle \cos\vartheta \rangle = -\lambda_1 - 2\lambda_2 E + \cdots. \tag{4.89}$$

The force F on a molecule in a field with a gradient in the y direction is, according to the expansion (4.86):

$$F = -\lambda_1 \frac{\partial E}{\partial y} - 2\lambda_2 E \frac{\partial E}{\partial y} + \cdots. \tag{4.90}$$

Knowing the Stark coefficients λ_1 and λ_2 (see, for instance, Table 4.1) and the field gradient (see Sect. 4.3.2), the deflection in an electric Rabi or two-pole field may be readily calculated.

Table 4.1 shows the first nonvanishing coefficients λ_1 and λ_2 of the expansion (4.86) for some types of molecules, as well as possible deflection fields and examples of typical molecules. μ is the permanent electric dipole moment of the molecule, and J, M, and K are the rotational quantum numbers determining the

4. State Selection

Table 4.1. Stark coefficients λ_1 and λ_2 for some molecules

Molecule	λ_1 or λ_2	Examples	Field
Linear ($^1\Sigma$) (no bending)	$\lambda_1 = 0$ $\lambda_2 = -\dfrac{\mu^2}{2hcB_e}\left[\dfrac{J(J+1)-3M^2}{J(J+1)(2J-1)(2J+3)}\right]$	NaCl, CsF, TlF	Two-pole Rabi Quadrupole
Symmetric top	$\lambda_1 = \mu\dfrac{KM}{J(J+1)}$	CH$_3$F, CH$_3$I, CF$_3$H	Rabi Hexapole
Linear (Π) (no bending)	$\lambda_1 = \mu\dfrac{\Omega M}{J(J+1)}$	NO, OH	Hexapole
Linear ($^1\Sigma$) (with bending)	$\lambda_1 = \mu\dfrac{\ell M}{J(J+1)}$	N$_2$O, LiCN	Hexapole

angular momentum J and its projections on a space-fixed axis (M) and the molecular axis (K). ℓ is the vibrational angular momentum projected on the molecular axis, Ω is the electronic angular momentum, and B_e is the rotational constant according to (I.2.56).

A characteristic value for the effective electric dipole moment of a polar molecule in a low rotational state is of the order of 1 debye ($= 10^{-18}$ esu), while the characteristic value for the magnetic moment of an atom is of the order of 1 Bohr magneton ($\approx 10^{-20}$ erg G^{-1}). Comparing the deflection of a molecule obtained in an electric field with that of an atom obtained in a magnetic field, assuming that both fields have the same geometry, equal deflection will be achieved if E (esu) = 0.01 H (in G). A typical strong magnetic deflecting field is 10 000 G and the corresponding electric field would be 100 esu or 30 000 V/cm.

4.3.2 Rabi and Two-Pole Fields

The inhomogeneous electrostatic fields that have been used in the first electric resonance experiments were the electrical analog to the two-wire field. Such a field is generated by two cylindrical electrodes (see Fig. 4.21) being identical with two equipotential surfaces. Accordingly, the formulas derived in Sect. 4.2.2 can be used if the magnetic field H is replaced by the electric field E. In particular, the electric field and its gradient in the y direction may be written as

$$E = \sqrt{E_x^2 + E_y^2} = 2C\frac{a}{r_1 r_2}, \quad \frac{\partial E}{\partial y} = -2Cay\frac{r_1^2 + r_2^2}{r_1^3 r_2^3}. \tag{4.91}$$

The constant C can be expressed by the potential difference $V_0 = V_1 - V_2$ applied to the two electrodes. For the potential V_1 of the convex electrode (the point $x = 0$, $y = a$ belongs to this surface) and of the concave electrode V_2 (containing the point $(x = 0, y = a + d)$, respectively, follows according to (4.33):

$$V_1 = -C\,\text{arctg}(1) = -C\pi/4, \quad V_2 = C\,\text{arctg}(1 + d/a). \tag{4.92}$$

This yields for the constant C

$$C = V_0\left[\text{arctg}(1 + d/a) - \pi/4\right]^{-1}. \tag{4.93}$$

As in the magnetic case, the role of the equipotentials and the lines of force may be interchanged. The corresponding field can be realized by two oppositely charged parallel rods, whose cross section in the (x,y) plane corresponds to two circles of equal potential determined by (4.31). In the following, this field is designated as a two-pole field. Its potential U is represented by (4.33):

$$U = \frac{C}{2}\ln\left(\frac{y^2 + (x+a)^2}{y^2 + (x-a)^2}\right), \tag{4.94}$$

yielding the following components E_x and E_y of the electric field E

$$E_x = -\frac{\partial U}{\partial x} = -C\left(\frac{x+a}{r_1^2} - \frac{x-a}{r_2^2}\right), \quad E_y = -\frac{\partial U}{\partial y} = -Cy\left(\frac{1}{r_1^2} - \frac{1}{r_2^2}\right). \tag{4.95}$$

r_1 and r_2 are defined in (4.32). The magnitude of the field E and its gradient in the y direction are given by (4.91). The constant C can be expressed by the potential difference U_0 between the two rods at distance d (see Fig. 4.7)

$$C = U_0\left[\ln(a + (d/2)) - \ln(a - (d/2))\right]^{-1}. \tag{4.96}$$

If the beam passes through the region between the two rods (in the vicinity of $x = 0$, $y = 0$), the field can be expanded at the origin of the coordinate system in the following manner:

$$\frac{\partial E}{\partial x} = \frac{4C}{a^3}x\left[1 + 2\left(\frac{x}{a}\right)^2 - 4\left(\frac{y}{a}\right)^2 + \cdots\right],$$

$$\frac{\partial E}{\partial y} = -\frac{4C}{a^3}y\left[1 + 4\left(\frac{x}{a}\right)^2 - 2\left(\frac{y}{a}\right)^2 + \cdots\right]. \tag{4.97}$$

C is defined by (4.96). The focusing properties in the y direction (and defocusing in x direction) for particles having a positive dipole moment have already been pointed out for magnetic two-pole fields (Sect. 4.2.2). Using a series of two-pole fields, which are alternately turned around their axis by 90°, field portions with y focusing and x defocusing alternate with those yielding x focusing and y defocusing. Such an array may be used to obtain beam focusing in both directions (alternating gradient or "strong" focusing). By applying time-dependent fields an acceleration or deceleration of molecules becomes possible [Auerbach et al. (1966), Wolfgang (1968), Kakati and Lainé (1967), (1971)]. Recently, this has been successfully demonstrated experimentally [Bethlem et al. (1999)].

4.3.3 Electrostatic Multipole Fields

The formulas derived in Sect. 4.2.5 for magnetic multipole fields also hold in the electrostatic case. Thus, the magnitude of the electric field and the gradient in the radial direction may be written as

$$E = nCr^{n-1} \quad \text{and} \quad \frac{\partial E}{\partial r} = n(n-1)Cr^{n-2}. \tag{4.98}$$

With

$$\frac{V_0}{2} = \int_0^{r_0} E dr = C r_0^n, \tag{4.99}$$

the constant C can be expressed in terms of the applied voltage V_0 (voltage between two opposite electrodes) and the field radius r_0

$$C = \frac{V_0}{2 r_0^n}. \tag{4.100}$$

It is straightforward, therefore, to calculate the motion of molecules with linear Stark effect in an electrostatic hexapole field and of molecules with quadratic Stark effect in a quadrupole field. The wave number k describing the harmonic motion of a molecule with *linear Stark effect* in a hexapole field is given by

$$k = \frac{1}{r_0 v} \sqrt{\frac{3 \mu_{eff} V_0}{m r_0}}. \tag{4.101}$$

It should be noted that the only states that can be focused are those for which $\mu_{eff} < 0$. The length of the path z_0 for a single-loop trajectory in the field follows in complete analogy to the magnetic case:

$$z_0 = \frac{\pi}{k} = \pi r_0 v \sqrt{\frac{mr_0}{3\mu_{eff}V_0}}. \tag{4.102}$$

Considering a symmetric top molecule as an example, we find

$$z_0 = \pi r_0 v \sqrt{\frac{mr_0}{3\mu V_0 f(J,K,M_J)}} = 4.047\, r_0 v \sqrt{\frac{r_0 M}{\mu V_0}} \frac{1}{\sqrt{f(J,K,M_J)}} \text{ cm}, \tag{4.103}$$

with

$$f(J,K,M_J) = \frac{KM_J}{J(J+1)}. \tag{4.104}$$

On the right hand side of (4.103) μ is in debye, V_0 in V, r_0 in cm, v in m/s. M is the molecular weight. At a field radius of 0.5 cm, a length of the field of about 1 m is necessary for focusing CH_3I molecules in low-lying quantum states. The wave number k of the harmonic motion of a molecule with *quadratic Stark effect* in a quadrupole field is determined by

$$k = \frac{\mu V_0}{r_0^2 v} \sqrt{\frac{f(J,M_J)}{hcBm}} \quad \text{with} \quad f(J,M_J) = \frac{J(J+1) - 3M_J^2}{J(J+1)(2J-1)(2J+3)}. \tag{4.105}$$

States with $f(J, M_J) > 0$ are focused. For the focus length z_0 the following result is obtained

$$z_0 = \frac{\pi v r_0^2}{\mu V_0} \sqrt{\frac{hcB_e m}{f(J,M_J)}} = \frac{1710.35 v r_0^2}{\mu V_0} \sqrt{\frac{MB_e}{f(J,M_J)}} \text{ cm}, \tag{4.106}$$

where r_0 is in cm, v in m/s, μ in debye, V_0 in V, B_e in cm^{-1} and M in atomic mass units. Using a field radius of 0.5 cm, a length of the quadrupole field of about 1 m is sufficient to focus CsF molecules in low-lying quantum states.

Using (4.101) or (4.105) for the wave numbers, the imaging properties of a given setup can be easily calculated using the lens equation (4.68) in complete analogy to the magnetic case.

We first consider the focusing of polar diatomic molecules (or linear polyatomic molecules) in a quadrupole field. Figure 4.13 shows a typical focusing curve for a TlF beam. Plotted is the measured intensity curve as a function of the voltage applied to the quadrupole field. The points have been measured by Bennewitz et al. (1964), the full line is a calculated curve assuming a total resolution of $\Delta V/V = 5\%$. At the upper margin of this figure, the voltages at which focusing is obtained for given states are indicated by short lines. As can be seen, the low-lying

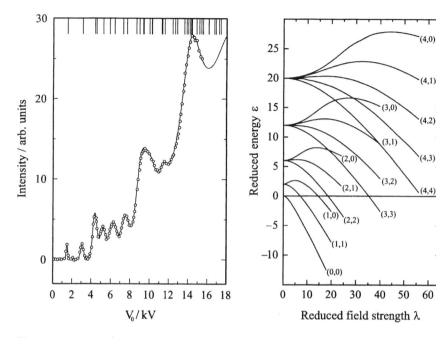

Fig. 4.13. Measured intensity of a TlF beam as a function of the applied quadrupole field voltage V_0 [Bennewitz et al. (1964)]. The full curve is calculated (beam velocity 148 m/s, halfwidth of the resolution $\Delta V_0/V_0 = 5\%$). The voltages at which the various quantum states (J, M_J) are expected to be focused are marked at the upper margin of the figures

Fig. 4.14. Reduced Stark effect energy ε of a rigid rotor as a function of the reduced electric field strength λ for J = 0 to 4. The reduced quantities are defined in the text. The quantum numbers (J,M_J) are given in brackets. The states with (J,M_J) and (J,–M_J) have the same energy [Kusch and Hughes (1959)]

states J,M = 1,0; 2,0 can be well resolved. For states with higher quantum numbers, the resolution becomes worse.

Apart from the experimental conditions (finite velocity resolution of the beam, finite width of the apertures used), the reasons for this decreasing resolution is of principle nature. The density of states increases with increasing J (as can be seen at the upper margin of Fig. 4.13), and the quadratic dependence of the Stark energy W upon the field is an approximation for low fields. With increasing field strength, depending on the state to be focused, the Stark energy deviates from the quadratic behavior, giving rise to imaging errors (spherical aberration). This is clearly demonstrated by Fig. 4.14, which shows the reduced Stark energy $\varepsilon = W/hcB_e$ for the states J = 1 – 4 as a function of the reduced field strength $\lambda = \mu E/hcB_e$ (B_e is the rotational constant) [Peter and Strandberg (1957), Kusch and Hughes (1959)].

The spherical aberration of a quadrupole field caused by higher order Stark terms in the expansion of the energy W can be compensated by an additional hexapole field in the middle of the quadrupole. Thus, the angle of acceptance and the transmission of the field can substantially increased [Everdij et al. (1973)].

Due to the large values of the partition function at high oven temperatures, only a small fraction of the beam molecules are in the low rotational states which can be focused with high resolution. For rotational energies $\varepsilon_r \ll kT$, this fraction of molecules in a specific J, M_J state follows from (I.2.57)

$$\frac{N_{J,M_J}}{N} = \frac{1}{Z_r}\exp\left(-\frac{hcB_e J(J+1)}{kT}\right) \approx \frac{1}{Z_r} = \frac{hcB_e}{kT} \ll 1. \qquad (4.107)$$

For CsF at a temperature of 1200 K, this fraction is 2×10^{-4}. Consequently, the intensity of particles in a well defined rotational state is very low, if an effusive beam is used, despite the fact that the large aperture of the quadrupole field allows for a considerable enhancement in flux (about three orders of magnitude). However, the intensity of state-selected beams can be increased by several orders of magnitude, if nozzle beams with carrier gas expansions are used [Borkenhagen et al. (1975)]. Apart from the increase in the total flux due to the gas dynamic expansion, and due to the fact that no velocity selector is required, the cooling of the beam particles during the expansion down to rotational temperatures of a few kelvin, increases the fraction of molecules in low-lying rotational states substantially (in the above example by two orders of magnitude).

By mounting two focusing fields in series, one-cycle sine wave trajectories are achieved. This yields a considerable increase in resolving power. With such a device, the states (3,0) and (3,1) also have been separated from each other [Borkenhagen et al. (1979)].

Two quadrupole fields in series with a microwave cavity placed between the fields have been employed by Bromberg et al. (1975) to generate CsF molecules in both a well defined rotational $(J,M_J) = (1,0)$ and vibrational state ($v = 0$ to 4). The first field focuses the $(J,M_J) = (2,0)$ state into the cavity, in which transitions $J = 2 \rightarrow J = 1$ in a specific vibrational state are stimulated by microwave radiation. The second quadrupole field focuses the molecules of the selected (v, J, M_J) state onto the image point.

As a further example, we consider the focusing of symmetric top molecules in a hexapole field. Since the focus length (4.103) depends on three quantum numbers and the velocity, and because the density of states is very high even at rather low rotational temperatures, it has been impossible for many years to resolve a state defined in all quantum numbers [Brooks and Jones (1966), Beuhler et al. (1966), Chakravorti et al. (1982), Parker et al. (1982), Jalink et al. (1986a), (1986b)]. In this situation, the hexapole field can be considered to be a selector for given values of the expectation value of $\langle\cos\vartheta\rangle$, defined by

$$\langle\cos\vartheta\rangle = KM/J(J+1). \qquad (4.108)$$

Thus, the hexapole field selects an ensemble of rotational states with an average orientation $\langle\cos\vartheta\rangle$. This follows from (4.103), which can be rewritten in the form

$$V_0 = \frac{\pi^2 r_0^2 v^2 m}{3\mu z_0^2} \frac{1}{\langle \cos\vartheta \rangle} \quad \text{with} \quad -1 \leq \langle \cos\vartheta \rangle < 0. \tag{4.109}$$

A given voltage V_0 supplies beam molecules with a given $\langle \cos\vartheta \rangle$; as V_0 is reduced in magnitude from a high voltage for which the precision angle is approximately 90° (i.e. $\langle \cos\vartheta \rangle \approx 0$) towards a lower value, the precision angle becomes smaller until $\langle \cos\vartheta \rangle = -1$ when V_0 is reduced to a threshold voltage V_{th} below which no molecules are focused. Figure 4.15 shows a typical "focusing curve" (beam intensity as a function of the voltage applied to the field electrodes) for a symmetric top molecule (CH_3I) [Kramer and Bernstein (1965)].

The focusing curve starts at the threshold voltage V_{th}, which follows from (4.105) for $\langle \cos\vartheta \rangle = 1$. In the vicinity of the threshold, the states with large values of J, K, and M_J are focused. The focused intensity increases with increasing voltage up to a maximum (not yet reached in Fig. 4.15), whose position depends on the rotational temperature. For even higher voltages, the intensity decreases since the population of rotational states with high rotational quantum numbers decreases. The states with large values of J and small values of K and M_J are focused at high voltages. The temperature dependence of the occupation numbers for J and K depends upon the type of the symmetric top (oblate or prelate) molecule [Herzberg (1968)]. The example used here (CH_3I) is a prelate symmetric top, for which only K values which are small compared to J are significantly populated. This determines the focusing curve in the vicinity of the threshold.

The molecules obtained with a given voltage V_0 having a known value of $\langle \cos\vartheta \rangle$, may be oriented in the laboratory system by the use of a weak dc electric field (about 10 to 100 V/cm), about whose axis they precess with an angle

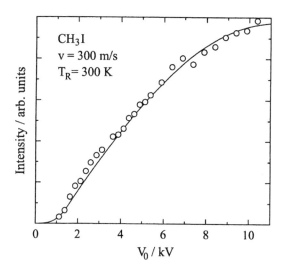

Fig. 4.15. Measured focusing curve of a velocity selected CH_3I beam in a hexapole field [Kramer and Bernstein (1965)]. The full line is calculated for a rotational temperature of 300 K and a total resolution of 10%

$$\bar{\theta} = \arccos\langle\cos\vartheta\rangle. \tag{4.110}$$

By reversing the polarity of the guiding field the molecules can be turned through an angle of 180°.

With nozzle beams, the rotational temperature in the beam and thus the number of states can be substantially reduced, but first attempts with a total resolution of 9% proved to be insufficient to resolve states well defined by all quantum numbers [Chakravorti et al. (1982)]. However, if the beam is expanded in a carrier gas, yielding a further reduction of the rotational temperature, and if the total resolution of the device is of the order of 1%, single quantum states $|J,K,M_J\rangle$ can be resolved and beams in well defined quantum states can be produced [Gandhi et al. (1986), Gandhi and Bernstein (1988), Kasai et al. (1993)].

As an example, Fig. 4.16 shows a focusing curve for CH_3I (expanded in Kr) at a rotational temperature of 4.5 K. The full line has been calculated using the linear Stark effect. While at low and intermediate voltages the agreement between measured and calculated values is quite good, deviations occur at high voltages, which are due to the neglect of higher terms in the expansion of the Stark energy. These deviations have also been observed by Kasai et al. (1993), who repeated these measurements. They were able to show that the deviations vanish (as far as the line position is concerned) if the second and third order Stark terms are taken into account [Che et al. (1994), Kasai et al (1993)]. The intensity in the $|1,1,1\rangle$ maximum corresponds to a particle flux of 10^{11} molecules/s.

In complete analogy, linear molecules with electronic angular momentum (e.g. NO) and linear $^1\Sigma$ molecules with excited bending vibration (e.g. N_2O) are also focused and oriented. In the latter case J, K, M_J is replaced by J, ℓ, M_J, in the former case K is replaced by the electronic angular momentum Ω (see Table 4.1) [Jalink et al. (1986a), (1986b), Brandt et al (1998), van Leuken et al. (1996)].

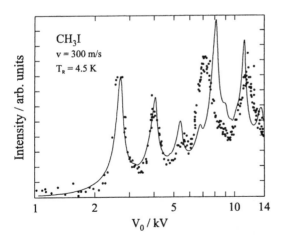

Fig. 4.16. Semilogarithmic plot of a focusing curve for a velocity selected CH_3I nozzle beam in a hexapole field. The full curve is calculated for a rotational temperature $T_R = 4.5$ K and a total resolution of 1% [Gandhi et al. (1986)]

4.3.4 Other Fields

Multipole fields with n >3 exert a force on the particles which increases nonlinearly with the distance from the field axis. As an example, we consider the motion of a molecule with linear Stark effect in a ten-pole field (n = 5). The trajectories describe radial oscillations with a wave number which depends upon the amplitude of the oscillation. Trajectories with large amplitudes of the radial oscillation have a smaller wave number than those with small amplitudes. Accordingly, these fields cannot be used to focus particles in a given quantum state onto a given point. But in some applications, this is not necessary. For example, in beam–maser spectroscopy a maximum concentration of selected molecules is desired to pass through the cylindrical cavity with a given angle of acceptance, independent of their focus length. In such cases, it may be advantageous to employ multipole fields with n > 3 such as octupole fields (n = 4) or 12-pole fields (n = 6) for molecules with quadratic Stark effect, while a 10-pole field is suited for molecules with linear Stark effect. Figure 4.17 (left half) shows trajectories of molecules with linear Stark effect in a ten-pole field and in the region behind the field. With this field a selected beam of given aperture angle can be kept within a radius of the order of the field radius over a length of the order of the field length. The transverse components of the velocity correspond to the angle of acceptance of the field. With a hexapole field of the same length and the same field radius (see right half of Fig. 4.17) the particles are focused onto the image point with the same maximum aperture angle, but the voltage required for a hexapole field is about a factor of 3/2 higher than for the ten-pole field.

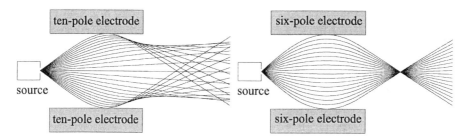

Fig. 4.17. Particle trajectories in a ten-pole field (*left half*) and in a hexapole field (*right half*)

4.4 Magnetostatic and Electrostatic Traps

The force exerted by an inhomogeneous magnetic or electric field on an atom or molecule due to its effective magnetic or electric dipole moment, which is discussed in Sects. 4.2.1 and 4.3.1, can also be used to store neutral particles, if their kinetic energy is smaller than their potential energy in the trap. Since typical trap

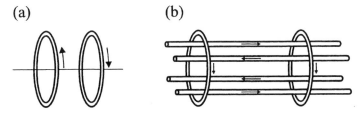

Fig. 4.18. Two magnetostatic trap configurations. (a) Magnetic quadrupole trap consisting of two coils with opposing currents. (b) Ioffe trap. The arrows indicate the direction of the currents [Bergeman et al. (1987)]

depths are of the order of 1 K, trapping depends essentially on substantial cooling and is, therefore, inextricably connected with laser cooling (see Chap. 5), which is restricted to atoms with an appropriate level diagram. Consequently, the static traps developed so far utilize magnetic fields. However, with new cooling mechanisms (thermalization of atoms in a cold buffer gas with subsequent evaporative cooling), it has been possible to store atoms, whose level diagrams do not allow laser cooling, in magnetostatic traps [Doyle et al. (1995), Kim et al. (1997), Weinstein et al. (1998)].

These cooling techniques can also be applied to molecules, and electrostatic field devices, therefore, may become important in the future.

In contrast to focusing of neutral particles in two-dimensional inhomogeneous fields, the exact dependence of the force as a function of the distance from the trap center plays no role. Instead, we distinguish simply between atoms in states that are drawn into a region of increasing field ("high field seekers") from those, which are drawn into regions of decreasing field strength ("low field seekers"). Since Maxwell's equations do not allow a maximum in the magnitude of a magnetostatic or electrostatic field in a source-free region, the high-field seekers cannot be held in a pure magnetostatic or electrostatic trap [Wing (1984)]. On the other hand, however, field configurations are possible which possess a local minimum in the magnitude of the field and these can be used to trap the low field seeking states (with negative effective dipole moment or positive energy in the field).

Two types of magnetostatic traps are frequently used [Bergeman et al. (1987)]: Three-dimensional quadrupole fields with rotational symmetry and vanishing field in its minimum, yielding a repelling force which is proportional to the distance from the trap center. In the magnetic case, they may be realized by two anti-Helmholtz coils (see Fig. 4.18). Two permanent magnets with opposite field orientation may also be used [Höpe et al. (1993), Vuletic et al. (1996)].

Another setup, which may be realized with a two-dimensional quadrupole field and two current-carrying coils (see Fig. 4.18), was originally proposed for the confinement of plasmas [Gott et al. (1962)]. Its application as a neutral particle trap was first proposed by Pritchard (1983). In the vicinity of the trap center this potential has an asymmetric well and the field at its minimum is not zero.

4.4.1 Three-Dimensional Quadrupole Fields

A potential of the form

$$U(x,y,z) = \alpha_x x^2 + \alpha_y y^2 + \alpha_z z^2 \quad \text{with} \quad \alpha_x + \alpha_y + \alpha_z = 0 \tag{4.111}$$

satisfies Poisson's equation (4.1). The two-dimensional quadrupole field follows from (4.111) for $\alpha_z = 0$, yielding necessarily $\alpha_x = -\alpha_y$. A three-dimensional quadrupole field follows for $\alpha_x = \alpha_y = \alpha$, yielding $\alpha_z = -2\alpha$. This potential has the form

$$U(x,y,z) = \alpha_x (x^2 + y^2 - 2z^2) \quad \text{or} \quad U(\rho, z) = \alpha(\rho^2 - 2z^2) \tag{4.112}$$

in cylindrical coordinates. It has rotational symmetry with respect to the z axis, the equipotentials are hyperboloids of revolution. The magnitude of the field (in the magnetic case) is given by

$$|\mathbf{B}| = |-\text{grad}U| = 2\alpha\sqrt{x^2 + y^2 + 4z^2} = 2\alpha\sqrt{\rho^2 + 4z^2}. \tag{4.113}$$

The surfaces of constant magnitude of the field are ellipsoids of revolution. Since this field vanishes at the origin of the coordinate system ($x = y = z = 0$) and is everywhere else different from zero, it can be used to trap neutral particles via their magnetic dipole moment [Wing (1980)].

In the electrostatic case this field is generated by three electrodes (see Fig. 4.19, right half): One electrode is a one-sheet hyperboloid of revolution, the other two electrodes are the two halves of a two-sheet hyperboloid of revolution. They may be obtained by rotating the hyperbolas shown in Fig. 4.19 (right half) around the z axis. Thus, the electrodes are identical with those of Paul's ion cage [Fischer (1959)]. In the magnetic case, as has already been mentioned, the field is generated by two coils of radius R at a distance 2A, carrying equal but opposing currents. This field can be expanded in the vicinity of the field minimum [Bergeman et al. (1987)], and is identical with the field described above as far as distances not too far from the field minimum are considered. This may also be seen from Fig. 4.19, which shows contours of $|\mathbf{B}|$ for both fields. α may be expressed by the current and the geometry of the device:

$$\alpha = \frac{3\mu_0 I A R^2}{(R^2 + A^2)^{5/2}} \quad (\mu_0 = \text{vacuum permeability}). \tag{4.114}$$

I is the current, 2A the distance of the two coils, and R their diameter. In SI units (A, m, tesla), $\mu_0 = 4\pi \times 10^{-7}$, in "mixed" units (A, cm, G) $\mu_0 = 4\pi/10$.

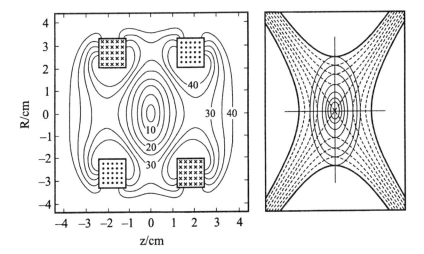

Fig. 4.19. Left half: Contours (*full lines*) of constant field B for the two coil magnetic quadrupole trap (numbers are the field strength in mTs) [Migdall et al. (1985)]. Right half: Cross-sectional view of an electrostatic quadrupole trap. The dashed lines are equipotentials, the full lines are contours of constant electric field. In both cases, the surfaces of constant field are obtained by rotation around the z axis. The anti-Helmholtz coils produce the same field in the vicinity of the trap center

The fact that $|\mathbf{B}|$ vanishes at the field minimum causes the problem of Majorana (spin-flip) transitions near the field zero due to the violation of the adiabaticity condition. Nonadiabatic transitions into high-field seeking states lead to particle losses, having the same effect as a "hole" in the trap. To close this hole, additional measures are required (see Sect. 4.5).

4.4.2 Ioffe Trap

A schematic diagram of the Ioffe trap is shown in Fig. 4.18 (b) [Bergeman et al. (1987)]. Transverse confinement is achieved from a four-wire quadrupole focusing field, while axial confinement is obtained from a two-coil "bottle field". Because of the nonzero minimum in $|\mathbf{B}|$, the Larmor frequency for even the coldest atoms can be made greater than the orbital frequency, so that the probability of Majorana transitions is drastically reduced.

The field in this trap near the origin can be expanded using cylindrical coordinates (ρ, z, φ) with the result

$$B_z = b_0 + b_2\left(z^2 - \rho^2/2\right) + \cdots; \quad B_\rho = -b_2 + c_1\rho\cos(2\varphi) + \cdots,$$

$$B_\varphi = -c_1\rho\sin(2\varphi) + \cdots. \tag{4.115}$$

The coefficients of this expansion can be expressed by the geometry of the device and the currents [Bergeman et al. (1987)]. For small values of z and ρ the field increases linearly with the distance ρ from the axis of symmetry, while it increases quadratically with z. The stored atoms, therefore, occupy a cigar-shaped volume in the trap.

If the four current-carrying wires are adjusted parallel to the gravitational field, the upper coil of the bottle field can be omitted, since the gravitational force can be used for axial confinement in this case [Monroe et al. (1990)].

The Ioffe trap configuration has been employed in the attempts to achieve Bose–Einstein condensation in a gas of spin-polarized hydrogen atoms, which began more than fifty years ago [Walraven and Hijmans (1994), Greytak (1995), Pinkse et al. (1998)]. The first observations of Bose–Einstein condensation have been reported recently [Fried et al. (1998)]. In this case, the magnetic fields are generated by superconducting coils. It has also been used to trap alkali atoms [Bagnato et al. (1987), Willems and Libbrecht (1995)].

Apart from the configurations discussed here, other designs such as the "spherical hexapole" trap with three wires on the surface of a sphere or the "baseball" trap in which coils follow the pattern of the seams on a baseball, are also possible [Bergeman et al. (1987)].

Electromagnets with iron cores [Desruelle et al. (1998)] or devices made from permanent magnets are an alternative to superconducting coils [Frerichs et al. (1992), Ricci et al. (1994), Tollet et al. (1995)]. The latter have the advantage that the field is free from noise and thus does not contribute to the heating of the stored particles. On the other hand, their field cannot be changed during operation. This disadvantage is avoided by electromagnets with iron cores, which are constructed similar to traps of permanent magnets (see Fig. 4.25). By appropriate control of the currents of the six electromagnets, it is possible to change from a three-dimensional quadrupole field to the Ioffe configuration during operation.

4.5 Nonadiabatic (Majorana) Transitions

An undesired effect when working with state selectors are nonadiabatic transitions from a selected state into other states. These transitions are caused by spatial field variations, which are experienced by the moving particles as time-dependent fields. If the Fourier analysis of such time-dependent fields contains an appropriate frequency in sufficient strength, the corresponding transition may be induced. To avoid these spontaneous transitions, the motion of the particles in the field has to be adiabatic: The direction of a magnetic field as seen by a moving molecule should not change at a rate comparable to or greater than the Larmor frequencies of the dipole moments involved (adiabaticity condition). These nonadiabatic transitions are often called "Majorana flops", since Majorana (1932) was the first to calculate the probabilities of transitions between the M states of a free spin in an external magnetic field. In early atomic beam experiments, these transitions were used to determine the signs of the nuclear moments and thus have been the first step towards the development of the atomic beam magnetic resonance method [Phipps and Stern (1931), Frisch and Segré (1933)].

No general or systematic method to avoid these undesired transitions is available, the experimentalist is entirely reliant on trial and error procedures. The general guiding principle requires that the particles should not experience large field variations over short distances, because the Fourier expansion of abrupt field changes contains a broad frequency spectrum. Locations which are particularly critical are the ends of a given field and the transition regions between two fields (for example the transition from the A field to the C field or the C field to the B field in an atomic beam resonance apparatus). Sharp edges should be avoided in any case. Usually, additional guiding fields are introduced in these regions in order to convert abrupt field changes into continuous field variations.

In the magnetic case, even the residual magnetism of steel apertures used for beam collimation, may lead to depolarization. Schmiedmayer et al. (1997), therefore, used silicon apertures instead of stainless steel, which reduces the number of nonadiabatic transitions considerably.

The particle losses in magnetostatic quadrupole traps have already been mentioned in Sect. 4.4. Since their field is zero at the trap center, the adiabatic condition is violated there on principle. If the region of very low field is small compared to the total volume of the trap, rather long storage times may still be achievable (of the order of a minute) [Davis et al. (1995a)]. However, with decreasing temperature, the attainable temperatures are limited by the "hole" in the trap.

Two methods have been applied to overcome the storage time limitation due to Majorana spin-flips. The first has been realized in the time-averaged orbiting potential trap, which adds to the quadrupole field a continuously changing bias field, that moves the location of the field zero around much faster than the atoms can respond. Its frequency is considerably lower than the Larmor frequency in the bias field, but much higher than the atom oscillation frequency in the trap potential [Petrich et al. (1995), Han et al. (1998)]. Thus, the particle motion is primarily governed by the time average of the bias field. The other method suppresses the trap loss by adding a repulsive potential around the zero of the magnetic field. This is accomplished by tightly focusing an intense blue-detuned laser that generates a repulsive optical dipole force (see Chap. 5) [Davis et al. (1995b)].

4.6 Technical Details

To realize a magnetic two-wire field with an iron magnet, a U-shaped yoke or a yoke in a double E shape, such as shown in Fig. 4.20, is usually used. The yoke is machined from two blocks of suitable iron, the two windings of copper wire are situated closest to the pole tips on the cross bars within the yoke in a casing of Araldite together with broad, water-cooled copper plates (to remove the ohmic heat). The material of the pole faces has a high saturation field strength in order to obtain high fields. The spacers shown in Fig. 4.20 between the two parts of the yoke are made from brass and prevent the pole tips partially drawing together when the field is applied.

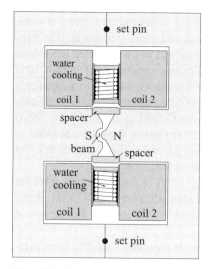

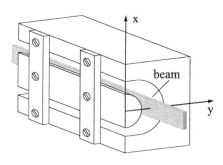

Fig. 4.21. Construction of an electrostatic deflecting field (two-wire field). The two cylindrical metal surfaces coincide with two of the equipotentials.

Fig. 4.20. Cross-sectional view of a magnetic two-wire field

The electrostatic two-wire field is realized with two cylindrical metal surfaces which coincide with two of the equipotentials. A typical deflecting field is shown in Fig. 4.21, for which the electrodes were constructed of duralumin and the insulating supports were of lavite. The surfaces of the electrodes should be highly polished in order to avoid electrical flashover.

In general, the pole faces of magnetic multipole fields are replaced by cylindrical surfaces of circular cross section in the vicinity of the field radius having a radius of curvature which coincides with the curvature of the exact hyperbolae at their vertices (see (4.12)).

As an example, we describe a hexapole field constructed from permanent magnets [Baum et al. (1988)]. Such a device has the advantage that it can be used in experiments which require ultrahigh vacuum (such as surface experiments), where problems may arise from the coils of an electromagnet. A disadvantage, of course, is that the pole tip field strength of such a field cannot easily be varied.

Therefore, their design and length have to be accurately adapted to the given application.

A cross section of the hexapole magnet is shown in Fig. 4.22. It consists of six shaped parts of magnetically soft material (Vacoflux 50, a cobalt–iron alloy) of approximately triangular cross section, having the desired curvature at their tip. These tips are positioned on each end with molybdenum-rod spacers which are retained in copper rings, defining the field radius ($r_0 = 2.5$ mm). The ring-shaped yoke is made from six fitting pieces from magnetically hard material (Vacomax 170, a rare earth–cobalt alloy), which guarantee the outer distance of the pole pieces and which are supported by appropriate copper rings (see Fig. 4.23). The total length of this assembly is 10 cm, the field strength at the pole tips (at the field radius r_0) is about 8000 Oe.

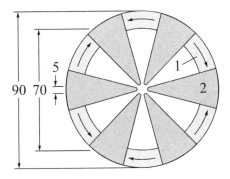

Fig. 4.22. Construction of a magnetic hexapole field using permanent magnets [Baum et al. (1988)] (dimensions in mm). (1) magnetically hard material (Vacomax 170), (2) magnetically soft material (Vacoflux 50). The arrows indicate the direction of magnetization

Fig. 4.23. Hexapole magnet assembly [Baum et al. (1988)]

This state selector has been used to produce a target beam of metastable ^{4}He(2^3S) atoms (Zeeman component $m_s = 1$) for electron scattering experiments and yields a density of state-selected atoms of the order of 10^7 atoms/cm^3 at a distance of 90 cm. Figure 4.23 shows the complete assembly.

Another way to realize multipole fields using permanent magnets is indicated in Fig. 4.24. Discrete, homogeneous pieces of suitably magnetized materials containing rare earths (e.g. SmCo) are joined together. Such a field does not vary continuously with the azimuthal angle, causing field contributions of higher multipoles [Halbach (1990)] and thus field errors, which may, however, be neglected in the vicinity of the field axis. The influence of the finite field length can also be neglected when working with thermal energy beams.

If, however, these multipole fields are used to focus slow atom beams produced by laser deceleration [Kaenders et al. (1996)], only very short fields are required, and the field length is no longer large compared to the field radius. In this case, the influence of the finite length of the field has to be carefully considered. Appropriate methods are described in the literature [Halbach (1990)].

In principle, an electromagnetic hexapole field is constructed similarly to the devices shown in Fig. 4.22 and Fig. 4.23. The shaped iron parts carrying the pole tips, however, have a smaller cross section in order to obtain more space for the copper coils wound around them.

The influence of field errors due to pole faces of circular (instead of hyperbolic cross section) has been studied in several investigations. Numerical techniques which solve Poisson's equation, analytical methods based on conformal mapping techniques, and semianalytic methods are available to calculate the field and the error harmonics for a given pole geometry to any desired accuracy [Grivet and Septier (1960), Sarma and Bhandari (1998), Sarma et al. (1999)]. As in the case of electrostatic multipole fields, which are discussed below, the usable field

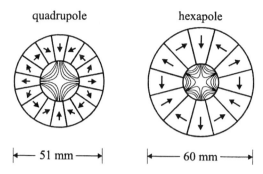

Fig. 4.24. Quadrupole and hexapole fields consisting of permanent magnets with field lines. The arrows indicate the direction of magnetization. For a detailed description see text [Kaenders et al. (1996)]

radius, within which the field errors are negligible, may be increased if the radius of the pole faces is somewhat larger than the radius of curvature of the exact hyperbolic surfaces at their vertex. These optimum radii, R_o, are identical with those of the corresponding electrostatic fields (see Table 4.2).

The construction of an electrostatic multipole field is relatively simple if the hyperbolic electrodes are replaced by rods of circular cross section with radius ρ_0 (see (4.12)) which corresponds to the radius of curvature at their vertex. The field errors due to this replacement reduce, as already mentioned in the magnetic case, the usable field radius and thus the angle of acceptance of the field.

This is similar to the situation in optics where lens aberrations diminish the usable lens diameter. The field radius capable of being used may be increased if the radius of the rods is made about 12% larger than ρ_0 (see Table 4.2) [Everdij et al. (1973), Anderson (1997)].

For quadrupole fields, the same result has been derived earlier [Dayton et al. (1954)], partly in connection with the quadrupole mass filter (see Chap. I.5), where the same problem is encountered [Denison (1971)].

The fact that no real molecule exhibits a pure linear or quadratic Stark effect causes similar "lens aberrations". The expansion of the Stark energy in powers of the field strength contains higher order terms, which — depending on the field strength — may contribute to the deflection and cause deviations from the harmonic motion [Anderson (1997)].

Table 4.2. Radii of the rod-shaped electrodes of a multipole field for optimum field approximation

Field	Optimum radius R_s	Radius of curvature ρ_0
Quadrupole	$R_o = 1.1458\, r_0$	$\rho_0 = r_0$
Hexapole	$R_o = 0.5602\, r_0$	$\rho_0 = r_0/2$
Octupole	$R_o = 0.3739\, r_0$	$\rho_0 = r_0/3$

If the radius of the electrodes is equal to the radius of curvature ρ_0 of the hyperbolae, the smallest distance Δ between the rods of a quadrupole field is given by $\Delta = 2r_0(2^{1/2} - 1) = 0.83r_0$, for a hexapole field it is $r_0/2$. Using the radii of Table 4.2, the minimum distance Δ is even smaller ($\Delta = 0.743r_0$ for a quadrupole field and $\Delta = 0.4397\ r_0$ for a hexapole field). This distance and the quality of the rod surfaces, which should be highly polished, limit the voltage difference between neighboring electrodes. Attainable field strengths are of the order of 10^7 V/m. Some devices have been operated close to the threshold of field emission.

In general, the rods are machined from stainless steel. At their ends, they are turned off to a smaller diameter in order to fix them to high precision spacers from insulating material. Usually, these parts are assembled on an optical bench, so that the state selector can be adjusted in itself outside of the vacuum system. Reported accuracies for this adjustment are of the order of ± 0.001 cm with respect to a total field length of 190 cm [Tsuo et al. (1979)].

The insulators between the rods should be shaped in such a way that the path for creeping currents between neighboring rods is as large as possible. The optical bench supporting the state selector is mounted into the vacuum system in such a manner that it can be accurately adjusted with respect to the bean path from outside the vacuum system. In the case of very long fields (> 1 m), often two (or more) equal devices are mounted in series in order to avoid field errors by transverse bending of the long rods. Field radius and field length are coupled. The larger the field radius, the larger the length required for the field in order to focus a given state with a given voltage. Field radii between 3 and 13 mm have been employed. For smaller field radii, the angle of acceptance becomes too small and the adjustment is complicated, larger radii require very long fields which may lead to intensity losses (e.g. by residual gas scattering).

In spite of the complicated machining, electrostatic fields with the correct hyperbolic electrodes and lengths up to 3 m (in parts of 1 m length) have been used [Gandhi et al. (1987), Gandhi and Bernstein (1988)].

The necessity of a central beam stop, which prevents particles moving along the field axis without experiencing a selecting force, has already been mentioned. This central obstacle is usually placed at a position where the particle trajectories have their largest distance from the axis. In order to avoid field disturbances, the beam stop (a small metal sphere) is fixed to a thin wire, which is inserted into the field along an equipotential line of zero potential. The radius of the obstacle is chosen such that the detector opening (or the defining aperture in front of it) is totally shadowed by the obstacle.

As an example of a practical design of a three-dimensional, inhomogeneous magnetic field for neutral particle trapping, an Ioffe configuration (see Sect. 4.4) constructed from permanent magnets is described in the following. Compared to corresponding devices utilizing superconducting coils, this trap is much simpler, but the attainable fields and field gradients are quite comparable [Tollett et al. (1995)]. Furthermore, this device offers more possibilities to admit laser beams for manipulation of the stored atoms. It consists of six axially magnetized cylindrical (Nd–Fe–B) permanent magnets (2.54 cm long, 2.22 cm ∅), which are arranged in

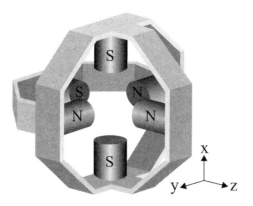

Fig. 4.25. Magnetostatic trap (Ioffe configuration) made from permanent magnets [Tollet et al. (1995)]. For a detailed description see text

pairs along three orthogonal directions of space (see Fig. 4.25). The four magnets in the (x,y) plane generate the quadrupole field, the two magnets in the z direction the dipole field of the Ioffe trap. An appropriate iron yoke carries the magnets and closes the magnetic circuit. The free space between opposite pole faces is 4.45 cm. The complete device is nickel-plated for use under ultrahigh vacuum conditions. This trap has been used to store Li atoms.

4.7. Applications of State Selection by Inhomogeneous Fields

Applications of inhomogeneous magnetic and electrostatic fields include many studies in atomic and molecular physics, but also in various other areas of physics and physical chemistry, ranging from surface science and chemical reaction dynamics to nuclear and high-energy physics. The purpose of this section is not to provide a comprehensive review, but rather to highlight the diversified capabilities of these fields in solving important physical problems.

4.7.1 Molecular Beam Magnetic and Electric Resonance Method

The atomic beam magnetic resonance technique, devised by Rabi et al. (1938), has been of great importance to the development of both atomic and nuclear physics. It dominated molecular beam research for more than twenty years, and an immense body of precise measurements of nuclear spins and moments as well as of the hyperfine interaction has been collected [Ramsey (1956), Nierenberg (1957)]. The principles of this technique are briefly elucidated in the following, because the basic concept — state selection — induced state transition — state analysis — is of fundamental importance and has been applied in many ways. Magnetic state selection and analysis may not only be replaced by appropriate electrostatic fields, but also by lasers [Penselin (1978)]. Transitions between two states cannot only be induced in radiofrequency or microwave fields, but also by atomic and molecular collisions [Toennies (1962)]. A schematic diagram of a magnetic resonance apparatus is shown in Fig. 4.26. An atomic beam, defined by the oven aperture and a collimating slit passes through two inhomogeneous magnetic fields produced by two magnets usually designated by the letters A and B (state selector and state analyzer). The effective magnetic moment μ_{eff} of the atoms and thus the deflecting force in the inhomogeneous fields depends on the magnitude and the sign of the quantum state of the atom (e.g. on J and M_J). If the gradient of the B field is opposite to that of the A field and if its magnitude is suitably selected, it can refocus the beam, which is deflected in the A field, back to the detector. A uniform magnetic field (C field) is placed between the two deflecting fields. If an oscillatory magnetic field is applied in the region of the C field, atomic transitions from the initial state 1 to the final state 2 will be induced when the oscillator frequency coincides with the corresponding atomic frequency in the field. As a result of a transition from state 1 to state 2 (connected with a change of the magnetic moment), refocusing is no longer possible and the intensity at the detector is reduced ("flop out" method). On the other hand, the magnitudes and directions of the A and B fields can be made such that refocusing only occurs for particles which have undergone a specific transition. In this case, a resonance increases the intensity at the detector ("flop-in" experiment).

In this fashion a radiofrequency spectrum of the atomic or molecular system is obtained. Energy differences at various values of the magnetic C field can be measured with high accuracy. As already mentioned above, these spectra yield important information on the atomic and molecular structure as well as on nuclear properties (spin, magnetic dipole moment, electric quadrupole moment, and magnetic octupole moment). All elements and their isotopes, including the highly refractory ones [Brenner et al. (1988)] and the artificial radioactive elements (with half-lives down to 30 min), have been investigated with this technique.

After having investigated the focusing properties of magnetic hexapole fields experimentally [Friedburg and Paul (1951), Friedburg (1952), Lemonick and Pipkin (1954)], these fields have been applied to atomic beam resonance experiments using beams of radioactive atoms. Because the atoms are detected by their radio-

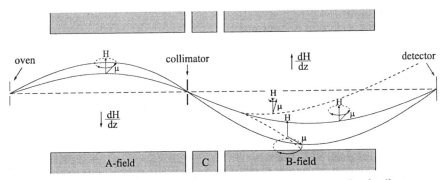

Fig. 4.26. Principle of atomic beam magnetic resonance measurements. For details see text

activity (see Chap. I.5), the intensity gain due to focusing causes a considerable reduction of the measuring time [Lemonick et al. (1955), Christensen et al. (1956)]. Later, these resonance experiments with magnetic hexapole fields were extended to paramagnetic molecules [Muenter (1992)].

An important field of application of electrostatic state selectors is the electric molecular beam resonance method, which was developed shortly after World War II in complete analogy to the magnetic resonance method [Hughes (1947)]. In the early experiments, the electrical analog to the two-wire magnetic deflecting field was frequently used [Braunstein and Trischka (1955)]. But already at the end of the 1940s, electrostatic quadrupole fields for state selection and analysis were developed [Trischka (1948), Gordon et al. (1954), (1955), Bennewitz et al. (1955), Schlier (1957), Trischka (1962), English and Zorn (1974), Muenter (1992)]. A very large number of molecules, also in excited electronic states, has since been spectroscopically investigated by this method, in a frequency range from some kHz up to several GHz. With these empirical data very detailed information about the electronic structure of molecules has been achieved [Muenter (1992)]. Numerous measurements have also been performed with weakly bound dimers and small clusters in order to obtain a deeper understanding of the cluster binding. More recent investigations concentrate on vibrationally excited molecules, using infrared laser radiation (e.g. of color center lasers) for excitation [Bass et al. (1987), (1990), Ebenstein et al. (1987), Shostak and Muenter (1991)].

A special resonance experiment testing the time-reversal symmetry by nuclear-spin resonance in a rotationally cold, state-selected supersonic beam of thallium fluoride molecules has been described by Hinds and co-workers. The resonance region is located between a state selector and a state analyzer, consisting of electrostatic quadrupole fields and a combination of static and radiofrequency electric fields [Cho et al. (1989), Schropp et al. (1987)].

4.7.2. Cesium Frequency and Time Standard

One of the oldest practical applications of the atomic beam magnetic resonance method is the cesium clock, an important frequency and time standard with an accuracy of 10^{-12}, which defines the second by means of a hyperfine structure transition in ^{133}Cs.

Until 1956 the second was defined as 1/86400 of the mean solar day and could be reproduced with an accuracy of 10^{-8}. Because of irregularities in the Earth's rotation one changed over to the ephemeris time which defines the second as a certain fraction of the time of the Earth's rotation around the Sun (tropical year) [Kartaschoff (1978)]. Even this definition did not result in an invariant unit of time measurement. Facing this problem of astronomical time definition, it is not surprising that the idea of a passive frequency standard by means of the atomic beam magnetic resonance method has been under discussion for some time [Rabi (1945)]. Several years later, the experimental foundations of the methods had been improved sufficiently [Ramsey (1949)] to begin the realization [Lyons (1952),

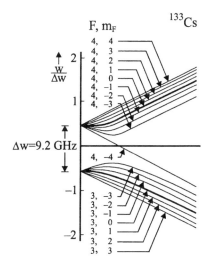

Fig. 4.27. Level diagram and Zeeman splitting as function of the magnetic field for ^{133}Cs atoms in the ground state

Zacharias et al. (1955)]. The first laboratory model of a cesium clock dates back to Essen and Parry (1955). After only one year the first commercial device entered the market. Finally, in 1967 the second was newly defined based on the hyperfine structure of the ground state of ^{133}Cs. Figure 4.27 shows the level diagram and the Zeeman splitting of the ^{133}Cs ground state.

Nuclear spin ($I = 7/2$) and electron spin ($J = 1/2$) result in a total spin F with the values $F = 3$ and $F = 4$. In a magnetic field the lower state ($F = 3$) splits up into seven, the upper state ($F = 4$) into nine substates, characterized by the quantum number m_F. The two states $F = 3$, $m_F = 0$ and $F = 4$, $m_F = 0$ show the least field dependency. For this reason this transition is used as the frequency standard.

Figure 4.28 shows the setup schematic of the frequency standard which is essentially an atomic beam resonance apparatus (see Fig. 4.10). The Cs beam is produced by an effusive oven held at a temperature of about 100 °C. The beam is ribbon-shaped (typically 0.5 mm wide and 12 mm high), and a multichannel aperture is used as a rule. An oven filling of 4 g lasts for about 5 years of service in commercial units. A much lower rate of consumption is reached in laboratory units [King and Zacharias (1956)]. The adjustment is such that only atoms in the $F = 4$, $m_F = 0$ state are guided from the A field into the C field. Those atoms which make a transition into the $F = 3$, $m_F = 0$ state proceed through the B field, reach the Langmuir–Taylor detector and are ionized. The ions are usually amplified by a multiplier and represent the output signal, after the background noise ions from the detector surface have been separated out by a crude mass spectrometer of low resolution. A suitable feedback circuit keeps the transition inducing frequency always at the frequency of maximum resonance.

As a rule, Rabi-fields are used as state selector and state analyzer, implemented usually by permanent magnets. However, six-pole fields have also been used because of the inherent intensity gain [Vanier and Audoin (1989)]. The homogeneous C field, usually normal to the beam direction, is typically 6×10^{-6} T and

244   4. State Selection

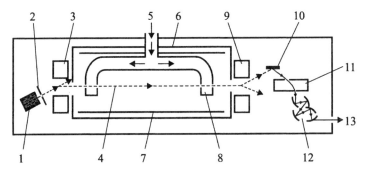

Fig. 4.28. Schematic view of a Cs frequency standard [Lainé (1992b)]. (1) Cs oven, (2) collimator, (3) A field, (4) Cs beam, (5) transition frequency feed (9.2 GHz), (6) magnetic shielding, (7) C field, (8) U-shaped microwave resonator, (9) B field, (10) surface ionization detector, (11) mass selection, (12) multiplier, (13) output signal

is often generated by suitable current-carrying conductors. Field inhomogeneities are lower than the field strength by a factor 100. The distance between the two locations where the high frequency is applied determines the linewidth. Laboratory units have used distances between 1 and 3.75 m and minimal transition linewidths are about 30 Hz. Commercial units use smaller distances (between 50 and 170 mm), resulting in correspondingly larger linewidths. The microwave system has to be carefully tuned. Both arms of the U-shaped resonator must have the same electrical length, and the phase of the microwave in both locations of interaction with the atomic beam must be the same within 10^{-3}.

In order to avoid loss of particles and a rise in noise through scattering of Cs atoms by the residual gas, the pressure inside the system must be below 10^{-8} torr, especially when the beam path is long. Therefore, closed commercial systems contain a getter pump so that a possible pressure increase over time can be corrected.

Besides the Cs clock described here, many other atomic and molecular transitions have been investigated concerning their suitability as a frequency standard. Numerous investigations refer to active frequency standards, with the ammonia maser and hydrogen maser being the best known. Rather than going into further details here, we refer readers to the literature [Lainé (1992b)]. As discussed in Sect. 5.3.1, the search for more accurate frequency standards still continues. Experiments with cold atoms open up new possibilities in this respect.

4.7.3 Atomic and Molecular Collisions

The first scattering experiment with magnetic state selection was already performed in 1962, measuring total scattering cross sections for collisions between gallium and rare gas atoms. The Ga atoms were polarized in a magnetic two-wire field and oriented in the beam direction by a homogeneous magnetic field superimposed to the target chamber [Berkling et al. (1962)]. By switching the current of the magnetic field between two values, the attenuation of the $M_J = 1/2$ and $M_J = 3/2$ beam component, respectively, has been alternately measured. From these measurements, the relative difference in the total cross section has been determined. The same collision pairs were re-investigated twenty years later over a wide range of velocities using state selection by a magnetic hexapole field. The total cross section of a given M_J state was measured as a function of the direction of the orientation field (parallel or perpendicular to the relative velocity). Detailed information on the anisotropy of the interaction potential has been determined from these and analogous investigations of indium atom–rare gas collisions [Hishinuma and Sueoka (1983), Hishinuma and Sueoka (1985)]. Similar experiments with partially selected beams (according to Zeeman states) have been carried out to measure the velocity dependence of integral cross sections for collisions of O, N, F, and Cl atoms and SO radicals with rare gas atoms [Aquilanti et al. (1988), (1997)].

Differential cross sections for spin exchange in atomic collisions (between alkali metal atoms) have been measured using the principle outlined at the beginning of this section [Beck et al. (1968), Pritchard et al. (1970)]. The alkali atoms are polarized in a magnetic two-wire field and cross a target beam; collisions leading to spin exchange are detected by analyzing the atoms scattered into a given angle by means of a second Rabi-field. The differential cross section for spin exchange depends on the difference of the potentials for the triplet and the singlet state of the collision pair, which has been determined from these measurements.

Zandee and Reuss (1977) used an atomic beam resonance apparatus of the type described above (Fig. 4.26) to select an ortho-hydrogen beam (a nozzle beam with 75% of the molecules in the $J = 1$ state) according to the magnetic quantum number m_J. Since the main contribution to the magnetic moment of the H_2 molecule is due to the nuclear spin I ($I = 1$), the molecular beam splits into three separate m_I components when passing through the A field. Each of these components consists of three nearly unsplit m_J components. To achieve a selection according to m_J, an rf transition with $\Delta m_J \neq 0$ is induced within the homogeneous C field. The subsequent B field focuses those particles to the detector, which have made a transition. Since the transition frequencies depend on m_J, one single m_J state can be modulated, if the rf field is modulated. Thus, a signal is observed at the detector with the modulation frequency which is proportional to the number of molecules in the state (I, m_I, J, m_J). In this manner, the anisotropy of various atom–molecule (H_2–Ne, Ar, Kr, and Xe) and molecule–molecule (H_2–H_2, N_2, and CO_2) interactions has been determined over a wide range of collision energies.

In reactive scattering experiments with alkali metal atoms, Stern–Gerlach magnets have also been used to separate the reaction products from the elastically and inelastically scattered particles [Herschbach (1966), Sholeen and Herm (1976), Sholeen et al. (1976), Behrens et al. (1976)].

A rather similar application is the separation of paramagnetic atoms from their dimers (and larger clusters), which are usually diamagnetic, in a beam in order to prepare a pure beam of dimers [Rosenberg (1939), (1940), Tarnovsky et al. (1993)]. Two-wire fields have also been used to determine the fraction of dimers in an atomic beam [Hundhausen and Pauly (1965)].

Shortly after the electric molecular beam resonance method was established, electrostatic quadrupole fields were used to measure the M_J dependence of integral collision cross sections in order to obtain information about the anisotropy of atom–molecule interaction potentials (CsF–, TlF–rare gases) [Bennewitz and Haerten (1969)]. Later, scattering experiments with state-selected NO molecules were carried out [Stolte et al. (1972), Thuis et al. (1979), van Leuken et al. (1996)]. Differential scattering cross sections (for small scattering angles) have also been measured for collisions between state-selected ($J = 1$, $M_J = 0$) LiF molecules with Ar atoms [Tsuo et al. (1977), (1979)].

Further measurements with electrostatic quadrupole fields have been concerned with the study of relaxation processes in gas dynamic expansions. As an example, we mention here investigation of rotational cooling of CsF molecules expanded with xenon as carrier gas [Malthan and Toennies (1974), Borkenhagen et al. (1975)]. Vibrational relaxation of CsF and LiF molecules, expanded in rare gases as well as in N_2O, NH_3, and SF_6 has also been studied in detail [Freund et al. (1971), Mariella et al. (1974), Bennewitz and Bues (1978)].

Using a setup analogous to the resonance apparatus shown in Fig. 4.26, in which the C field is replaced by a target beam and electrostatic quadrupole fields are used for state selection and analysis, numerous inelastic integral scattering cross sections have been measured for collisions of TlF and alkali halide molecules with various atoms and molecules [Toennies (1962), (1965), (1966), Beuhler and Bernstein (1969), Brooks (1976), Borkenhagen et al. (1975), (1979), Meyer and Toennies (1980), (1982)]. In other experiments, the state analysis after collision has been performed by laser-induced fluorescence [Dagdigian et al. (1979), Dagdigian and Wilcomb (1980), Bullman and Dagdigian (1984), Alexander and Dagdigian (1985), Schreel et al. (1993)].

Electrostatic state selectors have also been applied in many investigations of reactive collision processes, either to study the distribution of rotational states of the reaction products [Grice et al. (1970), Mosch et al. (1975), Hsu et al. (1975)] or their vibrational state distribution [Freund et al. (1971), Bennewitz et al. (1971), Mariella et al. (1973)]. In this case, an rf field is used in addition to the state selector, which induces $(J, M_J) \rightarrow (J, M_{J'})$ transitions.

Another application of electrostatic hexapole fields is reactive collisions with state-selected reactants. After the first experiments to produce spatially oriented molecules (CH_3I and $CHCl_3$) [Kramer and Bernstein (1965)], such beams have been used to study the orientation dependence of chemical reactions (e.g. K + CH_3I, Rb + CH_3I) [Brooks and Jones (1966), Beuhler et al. (1966), Beuhler and Bernstein (1969)]. The experimental results confirmed, at least qualitatively, the simple idea that the reaction is favored when the alkali atom approaches the molecule from the side where the iodine atom is located. In later experiments with more refined molecular orientation, these measurements were repeated as a function of the average angle of orientation (determined by the voltage applied to the electrodes of the hexapole field) of the molecules [Parker et al. (1981), (1982)]. Analogous experiments have been carried out to investigate reactions of oriented NO molecules with ozone [van den Ende et al. (1982)].

In further studies of reactive collision processes with state-selected reactants, the influence of the rotational energy on the reaction probability (for example in the reaction CsF + K → Cs + KF). The comparison with measurements using a beam of state-selected molecules in low-lying rotational states ($E_{rot} < 0.01$ eV) and thermal energy beams ($E_{rot} \approx 0.1$ eV) shows that the rotational energy has a similar influence on the chemical reactivity as the translational energy [Stolte et al. (1977), Zandee and Bernstein (1978)]. A strong dependence of the reactivity on the rotational energy has also been observed in the reaction NO + O_3 → NO_2 + O_2 [van den Ende and Stolte (1980)]. A survey on these experiments and their theoretical interpretation has been given by Sathyamurty (1983).

4.7.4 Cluster Investigations

Inhomogeneous magnetic fields find increasingly application in investigations of the magnetic properties of clusters and to determine their magnetic moments. The first experiments in this direction were performed with small iron and iron oxide clusters (Fe_N and $Fe_N O_M$) [Cox et al. (1986)], larger iron cluster (with 15 to 650 atoms) have been investigated by de Heer and co-workers [de Heer and Knight (1988), de Heer et al. (1989), (1990)]. These clusters, as well as cobalt clusters, show large magnetic moments, which, referred to the number of atoms N of the cluster, drastically exceed the values determined for the bulk material [Bucher et al. (1991a), Douglass et al. (1993), Bloomfield et al. (1993), Billas et al. (1994)]. Clusters of nickel show similar behavior [Louderback et al. (1993)], and even clusters of rhodium, a nonmagnetic transition metal, exhibit large magnetic moments, while palladium and ruthenium clusters are found to be nonmagnetic [Cox et al. (1994)]. Measurements of this kind not only allow the investigation of the magnetic behavior of matter between single atoms and the bulk material, but also establish detailed information on the cluster structure, since atoms on the cluster surface and atoms inside the cluster contribute in different ways to the magnetism.

Figure 4.29 shows an example of very precise measurements with nickel clusters, which have been carried out over a wide range of cluster masses (N = 5 to 740 atoms) [Apsel et al. (1996)]. The measured magnetic moment μ per cluster atom is plotted as a function of the number N of cluster constituents. The general behavior shows a decrease of the magnetic moment μ per cluster atom with increasing N from large values of μ to smaller values, tending towards the bulk value (0.61 μ_B), which, however, is not yet obtained at N = 100, as can be seen from Fig. 4.29. Even at N = 740 ($\mu = 0.68\ \mu_B$) there is still a difference between the cluster values and the asymptotic bulk value. The decrease of μ with increasing N is not monotonous, but shows pronounced maxima for clusters with a large surface and minima for rather compact clusters with a relatively small surface. This empirical result has been confirmed by cluster simulations, leading to a detailed understanding of the cluster structure and its changes when further atoms are added to the cluster. The determination of magnetic moments of clusters from the deflection in an inhomogeneous magnetic field requires a high resolving power

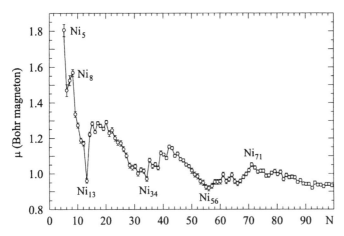

Fig. 4.29. Magnetic moment μ per cluster atom as a function of the number N of cluster constituents (for nickel clusters) [Apsel et al. (1996)]. For a description see text

of the setup, since all clusters practically have the same speed due to the expansion in a carrier gas, and the relative differences in the magnetic moment-to-mass ratio μ/m of neighboring masses are very low for large clusters. The high resolution is achieved by a relatively long magnetic field (25 cm) followed by a long drift path (118 cm) and a narrow focused laser beam (ArF excimer laser) for cluster ionization. Mass analysis is provided by a time-of-flight mass spectrometer. The laser beam is scanned through the deflected beam profile, and in each laser position the mass spectrum of the ionized clusters and thus their magnetic moment is measured [Bucher et al. (1991b)]. Analogous measurements have also been performed for Ni, Co, and Fe clusters with up to 700 constituents [Billas et al. (1996)].

Magnetic hexapole fields have also been used to separate small metal clusters (Li_N, N = 2–9) from each other, in order to perform resonance experiments with size-selected clusters [Hishinuma and Amano (1990)].

Deflection by inhomogeneous electrostatic two-pole fields has also been used to measure the electric polarizability of clusters as a function of their size [Knight et al. (1985), de Heer and Knight (1988), de Heer et al. (1989), Schäfer et al. (1995), Woenckhaus et al. (1996), Antoine et al. (1999)]. de Heer and co-workers, for example, measured the polarizabilities of Al_N clusters with N = 15 to 61 and compared the results with the predictions of a simple jellium model. While this model yields good agreement with the experimental data for alkali metal atoms, the measured polarizabilities of aluminum clusters are substantially smaller than predicted by the model [de Heer et al. (1989)].

4.7.5 Atomic Polarizabilities

Early in molecular beam studies, the deflection of atoms in inhomogeneous electric two-pole fields due to their induced dipole moment was used to investigate the polarizabilities of Cs, K, Li, H, and O atoms [Scheffers and Stark (1934), (1936)]. This early work has been continued afterwards in many attempts to increase the accuracy of the results [Drechsler and Müller (1952), Liepack and Drechsler (1956), Chamberlain and Zorn (1963)], but since the evaluation of the experimental data requires accurate knowledge of the beam geometry, its velocity distribution, and of the electric field and its gradient, the achievable accuracy is rather limited. A considerable increase in accuracy has been achieved by using mechanical velocity selectors [Hall and Zorn (1974), Crosby and Zorn (1977)]. Another method to increase the accuracy of the deflection measurements is the compensation method of Bederson and co-workers [Bederson et al. (1960), Salop et al. (1961), Pollack et al. (1964), Bederson and Robinson (1966), Tarnovsky et al. (1993)]. This method utilizes a combination of a magnetic and electric two-wire field, in which the pole faces of the magnet are simultaneously the equipotential surfaces of the electric field. Because the boundary conditions are the same for both the magnetic and electric field $(\nabla H)/H = (\nabla E)/E$ holds and field errors are eliminated. The magnetic and electric fields can be adjusted so that for at least one of the magnetic hyperfine states, a null deflection can be obtained. When this null condition is satisfied, the ratio of the magnetic moment to the induced electric dipole moment can be readily deduced from the values of H and E. A major advantage of this method is that the balance is independent of the particle speed.

4.7.6 State Analysis

Magnetic Rabi-fields offer a simple possibility to investigate the distribution of magnetic substates in a given molecular beam. This may be used either to check the state selection of another inhomogeneous field [Baum et al. (1988), Chan et al. (1988)], or to determine the efficiency of optical methods for state selection [Riddle et al. (1981), Lynn et al. (1986), Weiser and Siska (1987), Hammond et al. (1992), Oró et al. (1992)]. Another application of Rabi-fields is the state analysis of beams with no thermal state distribution, such as, for example, beams produced in discharge sources [Aquilanti et al. (1992), Alagia et al. (1997)]. The analysis of the state distribution in beams of paramagnetic molecules, which yields information about the distribution of rotational states in the beam, is yet another application of two-wire fields [Aquilanti et al. (1995)]. Although multipole fields are usually used in modern experiments on state selection, Rabi-fields are still of great importance in many cases, since much higher field gradients (up to a factor 6000) can be achieved. This is of particular significance in investigations of diamagnetic molecules, whose deflection is essentially due to their nuclear magnetic moment as has been discussed above for H_2 molecules [Muenter (1992)].

4.7.7 Gas–Surface Interaction

Oriented molecules have also been used for detailed studies of molecule–solid interactions and for the investigation of chemical reactions on surfaces. An example is measurements of the sticking probability of oriented NO molecules on Ni(100) and Pt surfaces [Fecher et al. (1990)]. The molecules are oriented in an electrostatic hexapole field, their orientation with respect to the surface is determined by an appropriate orientation field, as is described in Sect. 4.3.3. The molecules, therefore, are forced to approach the surface either with the O atom or the N atom ahead. Detecting the particles which are reflected by the surface and measuring their speed distribution, the sticking probability can be determined as a function of the molecular orientation. A considerable enhancement of the sticking probability is observed if the molecules approach the surface with the N atom in front [Müller et al. (1992), (1994a), Brandt et al. (1995)].

The same experimental technique has been used to study the reactions of oriented NO molecules with CO adsorbed at the surface. The reactivity as a function of the molecular orientation is measured by detecting the desorbing CO_2 molecules, which originate from the reaction [Müller et al. (1994b)]. To study the interaction of metastable $CO(a^3\Pi)$ molecules with LiF surfaces, hexapole fields have been used for state selection, while resonance-enhanced two-photon ionization has been used for state-specific analysis and detection [Jongma et al. (1997)].

4.7.8 Miscellaneous Applications

An important application of magnetic hexapole fields is the production of polarized electrons, which may be used, for example, in electron accelerators. The spin-polarized electrons are generated by photoionization of intense alkali beams (best suited is ^6Li), which are polarized in magnetic hexapole fields [Hughes et al. (1972), Alguard et al. (1979), Celotta and Pierce (1980), Kessler (1985)]. Such polarized atom beams, prepared in hexapole fields, have also been used to polarize electrons stored in traps utilizing spin-exchange collisions [Gräff et al. (1968), (1969)].

Furthermore, hexapole fields are frequently used to produce intense, spin-polarized target beams, either for experiments in atomic physics [Baum et al. (1988), Chan et al. (1988), Crowe et al. (1990)] or in high-energy and nuclear physics [Haeberli et al. (1982), Dunham et al. (1984), Slobodrian et al. (1986)]. Finally, spin-polarized atom beams produced by hexapole fields are an important part of ion sources for generating polarized ions for injection into particle accelerators [Schiemenz (1992), Schmelzbach (1994), Hertenberger et al. (1996), (1998), Mori (1996), Belov et al. (1996), Levy and Zelenski (1998)].

The deflection in magnetic hexapole fields may also be used for enrichment of isotopes, which can be polarized by optical pumping [Baum et al. (1980)], as has been demonstrated for ^6Li and ^7Li [Zhu (1984)]. The atoms of one isotope are pumped by circularly polarized laser light into the state $M_F = F_1$, while the atoms of the other isotope are pumped into the state $M_F = -F_2$ using a second laser. When passing through the magnetic hexapole field, the atoms of one isotope are focused, while those of the other isotope are defocused. The grade of enrichment depends only upon the degree of polarization achieved by optical pumping.

If the axis of a hexapole field is bent to a circular arc or to a full circle, the field in the vicinity of the axis remains unchanged, as long as the radius of curvature of the field axis R is large compared to the field radius r_0. Thus, particles with a given magnetic moment can be guided on a circular trajectory. Of course, the centrifugal force experienced by the atoms has to be added to the equations of motion, causing a coupling between the radial and azimuthal motion.

This principle, already recognized by Wolfgang Paul in the early 1950s [Friedburg and Paul (1951)], has finally led to the construction of a storage ring for neutrons [Kügler et al. (1985), Paul et al. (1989), Paul (1990)], after the high magnetic fields required to focus particles with very small magnetic moments became realizable by superconducting coils. Similarly, the axis of a electrostatic hexapole field can be bent to a circle in order to obtain a storage ring for polar molecules [Katz (1997)]. A curved magnetic octupole field has been used as an atomic low-pass filter to load a magnetooptical trap from a thermal energy effusive lithium beam [Ghaffari et al. (1999)].

A very important application of magnetic hexapole fields is molecular amplifiers, oscillators, and spectrometers based on stimulated emission at microwave frequencies (masers) [Goldenberg et al. (1960), Audoin et al. (1971), Wineland and Ramsey (1972), Lainé (1992a)].

Oriented molecules also offer a unique possibility to investigate the process of photodissociation in detail, since the averaging processes over the orientation of the molecules, which are usually required, become unnecessary. Spectroscopic studies with oriented molecules have also been carried out [Kaesdorf et al. (1985), Gandhi and Bernstein (1988), Bernstein et al. (1989), Ohoyama et al. (1996)]. Finally, chemical and energy transfer reactions in collisions between metastable rare gas atoms and oriented cyanides (CH_3CN, CD_3CN, and CF_3CN) have been investigated detecting the CN radicals formed in these reactions. A rather strong dependence upon the molecular orientation has been found [Kasai et al. (1989), Che et al. (1991), (1994), Kasai et al. (1995)].

Electrostatic hexapole fields have also been used in the production of radical beams from corona discharges (see Chap. I.4) to separate a desired radical from a mixture of other fragments emanating from the source [Hain et al. (1997), Weibel et al. (1997), (1998)].

Applications of inhomogeneous magnetic fields as velocity filters and in atom traps and mirrors are discussed in detail in Sect. 3.4 and Chap. 5, respectively.

4.8 Optical Methods for State Selection

Apart from state selection by inhomogeneous magnetic and electric fields, which has been described in the preceding sections, state selection based on pure optical methods is also possible. These methods have the advantage that their application to molecules is not restricted to low quantum numbers (J,M). Moreover, they are neither limited to atoms and molecules in their electronic ground state nor do they

252   4. State Selection

require a permanent electric or magnetic dipole moment of the particles to be selected. In the following, the basic principles of optical state selection are briefly described and illustrated by a number of examples. For more details, the reader is referred to the literature [Bergmann (1988), McClelland (1996)].

4.8.1 Optical Pumping

The basis of state selection by optical methods was established during the 1950s with the first investigations of optical pumping, a couple of years before lasers became available [Kastler (1950), (1954), (1957), Brossel et al. (1952), Bernheim (1965), Happer (1972)]. Optical pumping means the selective population or depopulation of atomic or molecular states by absorption of light. These states may be, for example, single rotational levels either in the electronic ground state or in excited states, but also selected fine structure, hyperfine structure, or Zeeman sublevels.

The change of the population density of the M levels due to polarized light causes a deviation from the thermal population at equilibrium and thus an orientation of the atoms or molecules. The highest degree of orientation (polarization) is achieved if only one M level is populated. If, in a nonthermal population, the two states ±M are equally populated, the orientation is called alignment.

The early experiments on optical pumping concentrated on attempts to align atomic angular momenta by a redistribution of their magnetic sublevel population due to the irradiation with polarized light. The resulting anisotropy of the population was used for spectroscopic investigations in the gas phase [Budick (1967), Demtröder (1976)]. To perform these experiments, the intense atomic resonance lines from spectral lamps (hollow cathode or microwave discharges) have been used. For molecules, one was left to fortuitous coincidences of intense atomic resonance lines with a single molecular transition. Therefore, only very few applications of optical pumping to molecules were performed before the advent of lasers.

4.8.2 Selective State Depopulation

The situation has greatly improved since tunable lasers became available which yield much more intensity and which can be tuned to the center of a molecular absorption line. New applications of optical pumping, especially in molecular physics and in molecular beam research, became possible. One of the first applications of optical state selection in molecular beams is a laser-spectroscopic alternative to Rabi's molecular beam resonance method (see Sect. 4.7.1). Figure 4.30 illustrates the basic principle. The two inhomogeneous magnetic fields A and B (state selector and state analyzer) of a Rabi-setup (see Fig. 4.26) are replaced by two regions A and B of optical pumping, realized by two laser beams (produced from the same laser using a beam divider) crossing the molecular beam at right angles. If the laser frequency is tuned to a given transition i → k, the state i is partly (or completely) depopulated in the optical pumping region A. This change in the population of state i is observed at the second optical pumping region B by a decrease of the fluorescence light. If an rf or microwave field C is placed between the two regions A and B of optical pumping, which induces electric or magnetic dipole transitions between the M states (indicated by k_1 and k_2 in the right half of Fig. 4.30) and the depopulated level i, the population of level i is increased and the intensity of the fluorescence light at region B increases, if the frequency of the C field is in resonance with the transition M → i.

This method allows the application of the molecular beam resonance method to metastable and short-living excited states, and it has been used, for example, to precisely investigate the hyperfine structure of various metastable atom states [Ertmer and Hofer (1976), Dembczynski et al. (1980), Neumann et al. (1987)].

The first application of this method to molecules was reported by Rosner et al. (1975). They measured the small fine structure splitting (< 100 kHz) of a rotational-vibrational state (v" = 0, J = 28) in the $X\,^1\Sigma_g^+$ ground state of the Na_2 molecule.

The transition $X\,^1\Sigma_g(v",J") \to B\,^1\Pi_u(v',J')$ accidentally coincides with a frequency of the argon ion laser, so that the absorbing level (0,28) is depopulated in the A region and the induced fluorescence at B is reduced. Transitions in the rf field at C lead to a partial repopulation of this state, and the resonance is observed by the increase of the fluorescence light at B. From the resonance frequency, the electric quadrupole moment of the selected state (0,28) can be determined.

An example of a microwave version of this resonance method is measurements of the hyperfine structure of the BaCl molecule in its electronic ground state [Ernst and Kindt (1983)]. If an electric field is superimposed on the microwave region C, highly accurate measurements of the Stark effect become possible, which have been used to determine the electric dipole moment of the CaCl molecule in its $^2\Sigma$ ground state [Ernst et al. (1983)].

Labeling of a given level by selective depopulation can be used for state-specific experiments in atomic or molecular states which are thermally populated. On the other hand, states with low or no population have to be selectively populated in order to perform state-specific investigations.

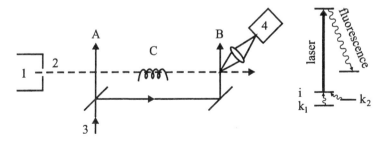

Fig. 4.30. Schematic of a molecular beam resonance setup with state selection and state analysis by optical pumping. (1) source, (2) molecular beam, (3) laser beam, (4) fluorescence detector [Demtröder (1976)]

4.8.3 Selective Population of an Atomic State

In the following, selective population of an M_F state by optical pumping is explained using a sodium atomic beam irradiated by laser light tuned to the $^2S_{1/2}(F = 2) \rightarrow {}^2P_{3/2}(F' = 3)$ transition. The level diagram and excitation scheme is shown in Fig. 4.31 [Dreves et al. (1981)]. The repeated excitation of the atoms in the F = 2 ground state with right-handed circularly polarized light ($\Delta M_F = +1$) (thick arrows) leads — together with the spontaneous emission (thin arrows) — to a preferred population of the higher M_F states of the F = 2 ground state until finally all atoms are in the $|F = 2, M_F = 2\rangle$ state after sufficient excitation cycles. The beam is completely polarized. Correspondingly, the use of counterclockwise circularly polarized light ($\Delta M_F = -1$) leads to the selection of the $|F = 2, M_F = -2\rangle$.

Complications are due to the hyperfine structure of the ground state, which has the effect that this system is not a real two-level system. Power broadening and a remaining Doppler broadening can lead to a loss of atoms to the F = 1 ground state (dashed arrows in Fig. 4.31). This problem is solved by additionally irradiating the atoms with a frequency corresponding to the transition from the F = 1 ground state to the excited F' = 2 state. This frequency can be generated from the same laser using an electro-optical modulator (see Sect. 5.1.3 [Cusma and Anderson (1983)]. Using an adiabatic rf transition in a magnetic field of medium strength the selected $|F = 2, M_F = 2\rangle$ state can be transferred into any other M_F state of the F = 2 multiplet [Dreves et al. (1983)].

Compared to state selection in an inhomogeneous magnetic field, the optical state selection described above has the advantage of yielding a higher intensity of selected atoms, since, in principle, all atoms can be brought into the desired state.

Apart from a direct application of the polarized alkali atom beams for the production of polarized electrons by photoionization, optically pumped alkali metal atoms play an important role in the generation of polarized H and D atom beams, which are subsequently ionized and injected into particle accelerators for experiments with polarized protons and deuterons in nuclear and high-energy physics [Schmelzbach (1994), Poelker et al. (1995), Stenger et al. (1997)]. Polarization of

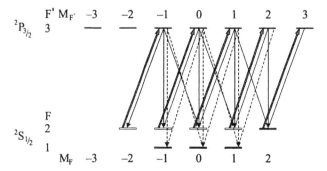

Fig. 4.31. Level diagram and excitation scheme for optical pumping of ^{23}Na with right-handed circularly polarized light. For a further description see text

the hydrogen atoms is achieved by spin-exchange collisions when the H atom beam traverses a spin-exchange chamber filled with alkali metal vapor, which is optically pumped with circularly polarized light. This principle has also been used to polarize fast atom beams [Schmelzbach (1994)]. Since spin-exchange collisions lead to a considerable fraction of polarized particles in excited states, a strong magnetic field is superimposed on the spin-exchange chamber in order to preserve the polarization due to the decoupling of spin and orbital angular momentum. Moreover, the attainable degree of polarization also depends upon the efficiency of the optical pumping process and on depolarization effects, which may be caused by wall collisions and the self-absorption of the radiation in the high alkali metal vapor density [Walker and Happer (1997)].

4.8.4 Selective Population of a Molecular State

As an example we consider the vibrational excitation of a molecule in its electronic ground state. In principle, three possibilities exist to excite a molecular beam by laser radiation: (1) The direct excitation with infrared photons from a laser of medium power, (2) the multiphoton infrared excitation with high-power lasers and (3) the indirect excitation, using, for example, the fluorescence of an electronically excited state (see Fig 4.33 A) or two-photon processes (see Sect. 4.8.5).

Due to the small absorption coefficients for infrared radiation, the direct excitation (1) of a molecular beam by an infrared laser is not very efficient. Despite this disadvantage, this method has already been employed in early experiments with vibrationally excited molecules [Karny and Zare (1978), Boughton et al. (1982), Hoffmeister et al. (1983)], and it is still used in modern experiments [Simpson et al. (1995), Orr-Ewing et al. (1997)]. One of the main applications of vibrationally excited molecules is the investigation of the effect of the vibrational energy on elementary chemical reactions. In the first experiments, scattering chambers were used for the target molecules in order to take advantage of the higher density in the bulk for infrared excitation. In this manner, the reactive scattering of Ba and Sr atoms with HF and DF molecules has been investigated [Karny and Zare (1978), Gupta et al. (1980)]. Using crossed beams, the reaction K + HF with vibrationally excited ($v'' = 1$) HF molecules in various rotational states j" has also been studied [Hoffmeister et al. (1983)]. In order to obtain a larger region of interaction for the infrared excitation, Rubahn et al. (1985) employed a coaxial, antiparallel arrangement of molecule and laser beam to study the influence of vibrational excitation on the fragmentation behavior of HF molecules due to electron impact ionization. Using a CW HF laser, about 10% of the HF molecules in the beam could be excited to the $v = 1$ state. Multipass devices have also been used to increase the excitation efficiency. In other experiments, the effect of vibrational excitation of SF_6 or HF molecules on the relaxation processes in gas dynamic expansions (in the region close to the nozzle) have been studied [Coulter et al. (1980), Ellenbroek et al. (1981)]. The energy is redistributed in collisions during the expansion and leads to the dissociation of dimers. Finally, the behavior of vibrationally excited molecules when impinging on solid surfaces has also been investigated [Misewich et al. (1985), Misewich and Loy (1986)].

An important application of direct infrared excitation is the optothermal spectroscopy of molecules in a molecular beam (see Sect. I.5.7.4), which takes advantage of the collision-free conditions in a beam and of the long radiative lifetimes of vibrational-rotational levels in the electronic ground state. The beam molecules are irradiated by infrared light from an appropriate tunable source (for example, a diode laser, a color-center laser, or an optical parametric oscillator). If the frequency of the infrared radiation corresponds to a molecular transition, the vibrational-rotational energy of the molecule is increased by the energy $h\nu$ of the absorbed photon. This extra energy causes an additional signal at the bolometer detector, yielding an indication for the transition and the excitation frequency.

Molecular beam experiments for the study of multiphoton excitation (2) started at the beginning of the 1980s [Schulz et al. (1979), Coulter et al. (1980), Luijks et al. (1981), Borsella et al. (1982)]. Beams of polyatomic molecules are best suited for excitation, since the large number of vibrational and rotational degrees of freedom as well as combination vibrations and anharmonic couplings compensate the decrease of the distance between the vibrational levels (due to the anharmonicity of the interaction) with increasing energy. A typical example is the SF_6 molecule, which can absorb several photons of the stretching vibration ν_3, if a high-power, CW CO_2 laser is used for excitation. This is due both to the coincidence between the emission frequency of the CO_2 laser and the excitation frequency, and the high oscillator strength of this vibration [Liedenbaum et al. (1988), (1989)]. Therefore, this molecule has been used to investigate reactive collisions between atoms and vibrationally excited molecules ($Na + SF_6^*$). An excitation efficiency of 30% can be achieved [Düren et al. (1994)].

The indirect vibrational excitation (3) is based on the following principle: Starting from a given rotational-vibrational level of the electronic ground state ($v'',J = 0,J$), a well defined rotational-vibrational state of a higher lying electronic level is excited. The subsequent spontaneous decay of the excited electronic state populates excited vibrational states of the electronic ground state. This method is not very selective, the population of the excited vibrational states depends on the variation of the Franck–Condon factors with v''. Accordingly, a distribution of vibrational states is excited [Bergmann et al. (1980), Fuchs and Toennies (1986), McGeoch and Schlier (1986)]. Since the laser-induced fluorescence depends also on the vibrational frequency v' of the electronically excited state, a variation of the population of the desired vibrational states v'' can be achieved by varying v'.

This method of vibrational excitation, which is often called "Franck–Condon pumping", has been used, for instance, to study the dependence of total integral cross sections on the initial vibrational state ($0 \leq v'' \leq 21$) for collisions of Li_2 molecules with rare gas and sodium atoms [Fuchs and Toennies (1986), Rubahn and Toennies (1988), Ziegler et al. (1991), Keller et al. (1992), Rubahn et al.

(1994)]. The total integral cross section σ(v",J") is the sum of the integral cross sections σ_α over all elastic, inelastic, and reactive channels α.

The experimental results show a large increase (e.g. 35% for the collision pair Na–Li$_2$ between v" = 0 and v" = 21) of the cross section with increasing vibrational quantum number v", which can be explained by the increase of the long-range, attractive part of the interaction potential with increasing internuclear distance in the Li$_2$ molecule [Rubahn and Bergmann (1990), (1993)].

Other applications of Franck–Condon pumping are measurements of the dependence of reactive collisions upon the vibrational energy of the molecules involved. Dittmann et al. (1992), for example, investigated the chemical reaction Na$_2$ + Cl → NaCl + Na* and found a linear increase of the reaction probability between v" = 0 and v" = 31. Electron impact ionization of diatomic molecules and its dependence upon the vibrational energy has also been studied using this excitation method [Külz et al (1995)].

The combination of the selective depopulation or population of a given level with the Doppler shift method for velocity-selection (see Sect. 3.3) makes a simultaneous state- and velocity-selection possible. If the exciting laser beam crosses the molecular beam at a given angle φ instead of 90°, only molecules having velocities in an interval around a velocity v given by

$$v = \frac{(\omega_L - \omega_0)c}{\omega_0 \cos\varphi} \tag{4.116}$$

will be state selected. ω_L is the angular frequency of the laser, ω_0 the resonance frequency, and c is the velocity of light.

4.8.5 Two-Photon Processes

The selective population or depopulation of a given state using a one-photon transition is the simplest method and is often used for state selection [Dispert et al. (1979), Hoffmeister et al. (1983), Vohralik and Miller (1985)], but processes involving two or more photons, which yield a higher selectivity in the vibrational level population, are also frequently employed. An example is the pumping of a vibrational level of the electronic ground state, which may be performed either in a resonant two-step process or a stimulated Raman process. In the former case, the molecules are first excited into an intermediate electronic state. Instead of working with the vibrational state distribution due to the spontaneous fluorescence from this intermediate state, a second laser with a frequency that stimulates a desired transition into a given vibrational level v" is employed, so that essentially only this level v" of the electronic ground state is populated (stimulated emission pumping) [Hamilton et al. (1986), Drabbels and Wodtke (1997)]. Compared to the population by spontaneous emission, the selectivity of the excitation by stimulated emission is substantially increased. Spontaneous decay of the intermediate, electronically excited state may populate other vibrational states of the electronic ground state and thus can cause a loss of particles from the pump cycle. To keep these losses as low as possible, the interaction time of the molecules with the lasers should be smaller than the spontaneous lifetime of the electronically excited state.

As an example of this "pump and dump" method, we describe in the following the preparation of NO molecules of a collimated pulsed beam in the $X\,^2\Pi(v=20)$ ground state, which has been used to measure state-specific inelastic scattering cross sections utilizing the crossed beam method [Drabbels et al. (1997)]. At the preparation zone 6 cm downstream of the nozzle the NO molecules are excited from the $X\,^2\Pi(v=0)$ ground state into the $B\,^2\Pi(v=5)$ state using pump light with a wavelength of about 198 nm. The dump laser, operating at a wavelength of 153 nm, is fired 5 ns after the pump laser pulse, stimulating transitions from the electronically excited state into the desired vibrational state $v=20$ of the electronic ground state.

Figure 4.32 shows a schematic drawing of the setup. The two beams are generated in pulsed sources, collimated by skimmers (at a distance of 2 cm from the corresponding nozzle), and cross each other at an acute angle. Since the amount of gas emanating from the pulsed beams is rather low, only one vacuum chamber (with a diffusion pump of 2000 l/s pumping speed) is required for the experiment. To generate the pump light the output of an argon ion (A) pumped single-mode ring dye laser (R) operating at Rhodamin 6G is pulse amplified in a three-stage dye amplifier by a frequency-doubled, Q-switched, injection-seeded Nd:YAG

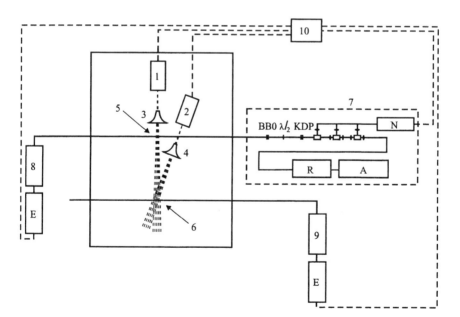

Fig. 4.32. Setup for measuring state-specific inelastic cross sections in collisions between vibrationally excited NO molecules and He atoms [Drabbels et al. (1997)]. (1) Pulsed NO source, (2) pulsed helium source, (3) and (4) skimmer, (5) preparation zone, (6) scattering region, (7) pump laser system (see text), (8) dump laser, (9) detection laser, (10) delay generator, (A) argon ion laser, (R) ring dye laser, (E) excimer laser, (N) Nd:YAG laser

laser (N) and is frequency-doubled in a KDP crystal to yield 15 mJ UV radiation around 300 mm. After passing a 266 nm λ/2 plate, the frequency-doubled and residual fundamental laser beams are mixed in a BBO crystal to yield radiation around 198 nm with an energy of 1.5–3.5 mJ/pulse and a bandwidth of about 200 MHz. This bandwidth is sufficient to prepare a given rotational and parity state.

At a time 5 ns after the pulse of the pump laser the pulse of the dump laser traverses the preparation zone in the opposite direction and stimulates the transition into the electronic ground state. An excimer-pumped dye laser operating on Coumarin-153 with a wavelength of 153 nm is used. Finally, the transitions of the state-prepared NO molecules into other states due to collisions with atoms from the target beam (helium) are detected and analyzed with a third laser (excimer-pumped dye laser) which passes through the crossing region of the two beams.

Using stimulated emission pumping it has also been possible for the first time to determine infrared radiative lifetimes of highly vibrationally excited molecules. The method is similar to the one described in Sect. 1.11.3, measuring the fraction of molecules that has undergone infrared radiation in a predetermined flight time by laser-induced fluorescence [Drabbels and Wodke (1997)].

In the case of stimulated Raman pumping none of the transitions is in resonance [DePristo et al. (1980), Shimizu et al. (1985), Valentini (1993)]. The population of the initial state is directly transferred to the final state (see Fig. 4.33). This process is not derogated from spontaneous emission, since the intermediate state is not populated. The transition probability, however, decreases with increasing distance from resonance, so that in many cases both processes have approximately the same efficiency.

Off-resonance stimulated Raman pumping depends of the variation of the molecular polarizability with the nuclear coordinates. Since the coordinate variation is generally slow, the selection rule $\Delta v = 1$ follows. Thus, a molecular beam can be excited into the first vibrational level $v = 1$. The frequency of the lasers used is not determined by the molecules to be excited, only the difference of the frequencies of the two lasers has to match the energy difference between initial and final state. When the laser frequencies come closer to resonance, a higher transition probability is achieved, and several vibrational quanta can be excited. Thus, the selectivity of the method decreases.

A completely different application of two-photon excitation is used to excite energetically high lying atomic or molecular states (e.g. Rydberg states). In this

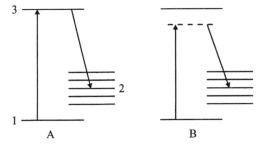

Fig. 4.33. Level diagram to illustrate two-stage pumping processes (A) and stimulated Raman pumping (B)

case, the sequential absorption of two photons is used [Feneuille and Jacquinot (1981), Eisel et al. (1983)]. Sequential absorption has also been employed to produce vibrationally excited molecular ions [Conaway et al. (1985)].

4.8.6 Photodissociation

Instead of using a transition between two bound states, a bound–free transition (e.g. photodissociation), can also be employed for state selection and orientation of molecules [van Brunt and Zare (1968), Bersohn ad Lin (1969), Zare (1972), de Vries et al. (1987), Miller et al. (1994), Suzuki et al. (1996), Mo et al. (1996)]. The basic principle is the following: Since the transition probability is proportional to the scalar product μE and thus proportional to $\cos\theta$, where θ is the angle between the direction of polarization E of the laser and the transition moment μ, molecules with $E \parallel \mu$ are preferentially dissociated and removed from the beam, while molecules with $E \perp \mu$ are not dissociated at all. Consequently, the molecules remaining in the beam are oriented. Since the excitation leads into a continuum, single vibrational-rotational states cannot be selected, in contrast to the optical pumping described above (see Fig. 4.34). Photodissociation, of course, may also be achieved by sequential absorption of two photons [Eppink et al. (1998)].

4.8.7 State Selection in Excited States

With the advent of tunable narrow-band lasers, especially CW dye lasers, it became possible to excite atoms into short-living states and to produce a stationary upper state population comparable to that of the ground state. In the simplest case of a two-level system, the ratio of the excited state number density n_e to the total number density n_0, is given (for not too high intensity fields) by

$$\frac{n_e}{n_0} = \left(2 + \frac{8\pi h}{3\lambda^3 u_\nu}\right)^{-1}, \qquad (4.117)$$

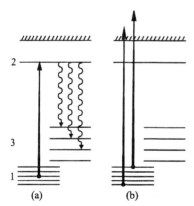

Fig. 4.34. State selection by selective population depletion utilizing optical pumping (a) and photodissociation (b) [Bergmann (1988)]

where λ is the exciting wavelength and u_ν the spectral radiation density of the laser. With laser intensities of less than 1 W/cm^2 and bandwidths of the order of the natural linewidth (≈ 10 MHz), a nearly equal population of ground and excited state particles may be achieved.

To explain this principle, we consider in the following the electronic excitation of sodium atoms (see Sect. 5.1.3) using linearly polarized laser light tuned to the transition between the hyperfine structure levels F = 2 of the $S_{1/2}$ ground state and F' = 3 of the $P_{3/2}$ state [Hertel and Stoll (1974), (1977)]. Figure 4.35 shows the level diagram of the two states involved, displaying their fine and hyperfine structure and the excitation and decay according to the selection rules.

To obtain the time-dependent population of the M_F states of the excited state and considering only times large compared to the spontaneous lifetime, a system of rate equations — taking into account the selection rules — can be established [Cohen-Tannoudji (1975), Hertel and Stoll (1977)]. The solution of this system of rate equations yields the population of the magnetic sublevels as a function of time, provided that the probabilities for spontaneous and induced transitions are known. After a number of excitation cycles, a stationary distribution w(M_F) is obtained, which has been calculated numerically [Hertel and Stoll (1976)] for the case discussed here. The resulting population numbers w(M_F) are compiled in Table 4.3. The numerical calculations show that this stationary distribution is already achieved after a time of about 30 spontaneous lifetimes, independent of the laser power, as long as the time for the spontaneous transition is small compared to the spontaneous lifetime. In the case of circularly polarized light, only one M_F sublevel of the excited state ($M_F = 3$ for σ^+ light and $M_F = -3$ for σ^- light) is populated (see Sect. 4.8.3 and Fig. 4.31) after a sufficient number of excitation cycles.

Table 4.3. M_F distribution for excitation with linearly polarized light

M_F	−3	−2	−1	0	1	2	3
w(M_F)	0	1/42	10/42	20/42	10/42	1/42	0

Since the beginning of the 1970s, numerous scattering experiments have been performed using atoms in short-living excited states, either to investigate the influence of the excitation energy on the collision process or to study the dependence upon the polarization of the atoms (or both) [Andersen et al. (1997)]. Elastic and inelastic collisions between atoms [Carter et al. (1975), Hüwel et al. (1982), Düren et al. (1984), Lackschewitz et al. (1986)], reactive scattering experiments [Rettner and Zare (1981), (1982), Mestdagh et al. (1987), Düren (1988), L'Hermite et al. (1991), Suits et al. (1992)], and investigations of energy transfer processes (transfer of electronic energy to molecular degrees of freedom) are a few examples [Hertel et al. (1976), Hertel (1982), Suits et al. (1992)]. Further examples are studies of associative ionization [Kircz et al. (1982), Brencher et al. (1988), Thorsheim et al. (1990), Klucharev and Vujnovic (1990), Meijer (1990),

262   4. State Selection

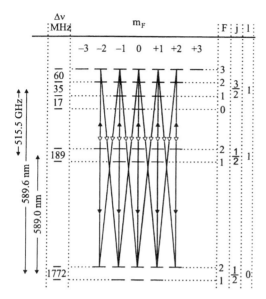

Fig. 4.35. Hyperfine structure of the Na(3 $^2S_{1/2}$), Na(3 $^2P_{1/2}$), and Na(3 $^2P_{3/2}$) states. The excitation scheme using linear polarized light follows from the selection rules (*full arrows*) $\Delta F = 0, \pm 1$, $\Delta M_F = 0$ for absorption, $\Delta M_F = 0$ for induced emission (*full arrows*) and for spontaneous emission (*open arrows*) $\Delta F = \pm 1$, $\Delta M_F = 0, \pm 1$

Weiner et al. (1990), Urbain and Pauly (1993)], of inelastic electron–atom scattering [Hertel and Stoll (1974), Hertel (1975), (1976), Hermann et al. (1977), Hanne et al. (1993)], of electron impact ionization [Trajmar et al. (1986), Tan et al. (1996)], and of charge exchange [Dowek et al. (1990)]. Finally, photoelectron spectroscopy of optically polarized atoms with VUV radiation shows a strong dependence upon the polarization of the atoms, which yields detailed information on the dynamics of photoionization [Dohrmann and Sonntag (1996), von dem Borne et al. (1997)].

The above-described method of producing a stationary number density of excited atoms in a beam can be extended to energetically high-lying states if sequential absorption of two or more photons is used. An example is the selective excitation of the F Rydberg states with $7 \le n \le 14$ of Rb atoms in a target beam, which has been used to study charge exchange processes [Deck et al. (1993), Salgado et al. (1997), Fisher et al. (1997)]. With three lasers (see the excitation scheme summarized in Table 4.4), selected Rydberg atoms with stationary number densities of the order of 10^{10} atoms/cm^2 are obtainable.

Table 4.4. Excitation of Rb atoms to the 10 F Rydberg state

Transition	Wavelength [nm]	Lifetime τ [ns]	Laser
$5\,^2S_{1/2}$ (F = 3) – $5\,^2P_{3/2}$ (F = 4)	780	27	Diode
$5\,^2P_{3/2}$ (F = 4) – $4\,^2D_{5/2}$ (F = 5)	1529	86	Color center
$4\,^2D_{5/2}$ (F = 5) – $10\,^2F_{7/2}$ (F = 6)	755	700	Ti-sapphire

In this four-level system the two first transitions are cyclic transitions which means that the spontaneous decay of the upper state can only lead to the initial state, so that no atoms are lost for the excitation cycle. The third excitation step leads, with a probability of about 50%, to the population of the 11 $^2D_{5/2}$ state (lying close below the 10 $F_{7/2}$ state) [Fehrenbach et al. (1995)], which decays rapidly into the 5 $P_{3/2}$ state and thus remains in the excitation cycle.

As soon as the lifetime of the excited state is larger than the time of flight of the atoms through the target region, i.e. for higher Rydberg states, a closed excitation cycle is no longer required, since a single excitation is sufficient. An example is state-resolved measurements of electron capture in collisions of multiply ionized Kr^{8+} ions with laser excited Rb(n,p) atoms (n = 15 – 25) of a thermal beam [Pesnelle et al. (1996)]. The spontaneous lifetime of the excited states range from 5 μs to 32 μs. For a beam velocity of 380 m/s, this corresponds to an average path between 2 and 12 mm, which is larger than the length of the target volume. In this experiment excitation is achieved by absorption of one photon, using a narrow-band (2 MHz), frequency-doubled, tunable, single-mode dye-laser.

4.8.8 Stimulated Raman Adiabatic Passage (STIRAP)

The techniques of selective excitation discussed so far have the common disadvantage that only a fraction of the atoms or molecules is transferred to the desired quantum state. This fraction is usually not higher than 50%. However, an excitation efficiency close to 100% may be achieved by utilizing the coherency of the laser light. This method, which was first proposed by Oreg et al. (1984) and which is termed stimulated Raman adiabatic passage ("STIRAP"), was experimentally realized several years later [Gaubatz et al. (1990), Rubahn et al. (1991), Schiemann et al. (1993), Bergmann and Shore (1995), Bergmann et al. (1998)].

This technique permits a complete transfer of the population of an initial state $|1\rangle$ into a final state $|3\rangle$ by a Raman-type coupling via an intermediate state $|2\rangle$ involving two radiation fields: The field of a pump laser which couples the initial state $|1\rangle$ with the intermediate state $|2\rangle$, and the field of a Stokes laser, which connects the latter with the final state $|3\rangle$. The simplest form of the excitation scheme is shown in Fig. 4.36.

The initial state, for example, can be a rotational level of the vibrational ground state of a molecule, and the final state can be a highly excited vibrational state. While the lifetimes of the initial and the final state have to be long, the intermediate state $|2\rangle$ may undergo spontaneous emission not only to the states $|1\rangle$ and $|3\rangle$, but also to other states which are summarized by level $|4\rangle$ in Fig. 4.36. To simplify the following considerations, the frequencies ω_P and ω_S of pump and Stokes laser, respectively, are assumed to be tuned to the corresponding transition

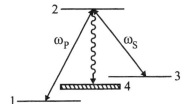

Fig. 4.36. Three-level system

frequencies. This is by no means a necessary assumption; it is only necessary that the combination of pump and Stokes frequencies are resonant with the two-photon Raman transition. A general treatment, for example, has been given by Fewell et al. (1997).

The temporal behavior of the two radiation fields is decisive, which is achieved in the case of pulsed lasers by a suitable time delay between pump and Stokes laser, while for CW lasers a spatial displacement of the two laser beams results in a corresponding temporal behavior for the moving beam molecules.

The importance of coherence and of the temporal behavior of the radiation fields becomes evident when considering a simple two-state system [Bergmann et al. (1992)]. If such a system interacts with a coherent, time-independent radiation field, the population oscillates between the lower and the upper state with the Rabi frequency $\Omega = \mu E/\hbar$ (μ is the transition dipole moment, E the field strength). While the time-averaged populations of both levels is 50%, the instantaneous population oscillates between 100% and 0%. Thus, at periodic time intervals, a 100% population of the upper state is achieved, if spontaneous decays can be neglected on the time scale determined by the Rabi frequency. In principle, therefore, the radiation field has only to be switched off at an appropriate time to obtain a complete population of the upper state. The condition for the time is given by

$$\int \Omega(t) dt = N\pi, \tag{4.118}$$

where N is an integer, odd number [Shore (1990)].

The basic principle sketched above cannot simply be transferred to a three-level system using the sequential excitation $|1\rangle \rightarrow |2\rangle$ and $|2\rangle \rightarrow |3\rangle$, since spontaneous decay is generally not negligible. Instead, a quantum-mechanical treatment of the complete system atom–radiation field is required. Due to the strong coupling of the three states by two coherent radiation fields, this quantum-mechanical treatment cannot utilize perturbation methods. Both the energy of the radiation fields and the interaction energies have to be considered in the Hamiltonian. Therefore, instead of the pure eigenfunctions of the three-level system without radiation field ("bare states"), the eigenfunctions of the complete Hamiltonian are obtained ("dressed states"). If the photon numbers of the pump and Stokes lasers are n_P and n_S, respectively, the three degenerate bare states (for vanishing radiation field $\Omega = 0$) of the total system are described by the following state vectors: $|1, n_P, n_S\rangle$, $|2, n_P-1, n_S\rangle$, and $|3, n_P-1, n_S+1\rangle$. The cancelation of the degeneracy by the radiation field leads to eigenstates $|a^{0,\pm}\rangle$, which are linear combinations of the bare states with coefficients which depend on the Rabi frequencies and the detuning of the laser frequencies with respect to the resonance frequencies (which we have assumed to be zero). In detail, the following eigenfunctions $|a^{0,\pm}\rangle$ and eigenvalues $\omega^{0,\pm}$ are obtained:

$$\left|a^{\pm}\right\rangle = \frac{1}{\sqrt{2}}\left(\sin\Theta|1\rangle \mp \cos\Theta|3\rangle\right), \quad \left|a^{0}\right\rangle = \cos\Theta|1\rangle - \sin\Theta|3\rangle, \tag{4.119}$$

with

$$\cos\Theta = \frac{\Omega_S}{\sqrt{\Omega_S^2 + \Omega_P^2}}, \quad \sin\Theta = \frac{\Omega_P}{\sqrt{\Omega_S^2 + \Omega_P^2}}, \quad \omega^0 = 0, \quad \omega^\pm = \pm\sqrt{\Omega_S^2 + \Omega_P^2}. \quad (4.120)$$

The state $|a^0\rangle$ connects the states $|1\rangle$ and $|2\rangle$, and its population can be achieved by a suitable temporal behavior of the radiation fields. At the beginning of the interaction with the radiation field $|\langle 1|a^0\rangle|^2 = 1$ is required, while at the end of the interaction $\langle a^0|3\rangle|^2 = 1$ has to be satisfied. Therefore, at the beginning we should have $\cos\Theta = 1$ and at the end of the excitation $\sin\Theta = 1$. According to (4.120), this requirement is fulfilled with the following temporal behavior of the laser fields

$$\begin{aligned}\Omega_S &\gg \Omega_P \quad \text{at the beginning of the interaction,} \\ \Omega_S &\approx \Omega_P \quad \text{in the middle of the interaction,} \\ \Omega_S &\ll \Omega_P \quad \text{at the end of the interaction.}\end{aligned} \quad (4.121)$$

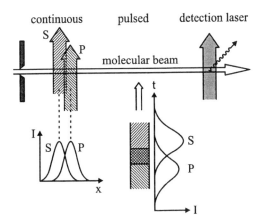

Fig. 4.37. Experimental realization of STIRAP techniques [Bergmann et al. (1992)]. For a detailed description see text

This means, experimentally, that the atoms have first to be exposed to a sufficiently strong field of the Stokes laser, before the interaction with the pump laser begins. After an appropriate time of overlap of both lasers, the interaction with the Stokes laser has to finish before the interaction with the pump laser terminates. With CW lasers, this is achieved by a spatial displacement between the two lasers, while for pulsed lasers a temporal delay between the two laser pulses is required, as is shown schematically by Fig. 4.37. It is important to note that the whole process should be an adiabatic one. The change of the mixing angle Θ determined by the temporal behavior of the Rabi frequencies should occur slowly.

For the case considered here, where the laser frequencies are tuned to the resonance frequencies, this can be expressed by the condition

$$\Omega\Delta\tau \gg 1, \tag{4.122}$$

where $\Delta\tau$ is the time of interaction. If (4.121) ad (4.122) are satisfied, the population of the initial state is adiabatically transferred via the state $|a^0\rangle$ into the final state, without populating the intermediate state and losses due to spontaneous emission are eliminated. The efficiency of the transfer process is independent of small fluctuations of the laser intensities and has been approved in numerous applications, both for atoms [Goldner et al. (1994), Weitz et al. (1996), Martin et al. (1996), Kulin et al. (1997), Süptitz et al. (1997)] and molecules [Sussmann et al. (1994), (1995), Külz et al. (1996), Kuhn et al. (1998)]. The same principle has also been used for coherent momentum transfer in order to deflect an atomic beam by stimulated light forces (see Chap. 5) [Theuer and Bergmann (1998)].

5. Slow Atom Beams, Traps, and Atom Optics

While it has been possible for more than 70 years to produce molecular beams with hyperthermal energies (see Chap. 1), the reverse process, the generation of slow atoms, has been out of reach for many decades. Attempts to use the gravitational field of the earth to decelerate the slowest particles of a beam with Maxwellian speed distribution (see Sect. 3.5) failed, as did efforts to slow down molecules in an arrangement of many inhomogeneous electrostatic, time-dependent two-pole fields in series (Stark-decelerator) [King (1959)]. Only recently. the latter approach has been successfully demonstrated [Bethlem (1999)]. The proposal of a rotating source [Hundhausen and Harrison (1966)], in which the circumferential speed is opposite to the speed of the effusing atoms and thus subtracted from it, has never been realized experimentally.

The situation changed during the 1970s, when intense, tunable lasers with narrow bandwidths became available. It was readily recognized that laser light can exert forces on atoms which are strong enough to decelerate atom beams to arbitrarily low velocities [Hänsch and Schalow (1975)]. Only a few years later, Ashkin (1978) proposed the first atom trap based on the interaction of atoms with laser radiation. Milestones of experimental development in the following years have been the first slowing down of an atomic sodium beam [Phillips and Metcalf (1982)], the first three-dimensional cooling of an ensemble of atoms [Chu et al. (1985)], and finally the storage of slow atoms in a trap [Chu et al. (1986)].

In the course of this development it has been found that "light forces" cannot only be used to slow down atom beams or to cool and store atoms in traps, but also to control and manipulate atoms and their motion completely [Pritchard (1991)]. Slow atom beams may be collimated, focused [Bjorkholm et al. (1978), Balykin et al. (1988), Sleator et al. (1992)], deflected, diffracted by gratings, reflected by mirrors [Balykin and Letokhov (1989), Berkhout et al. (1989), Roach et al. (1995), Hughes et al. (1997a), (1997b)], guided in hollow glass fibers (optical waveguides) [Dowling and Gea-Banacloche (1996), Renn et al. (1996), Ito et al. (1997), Balykin (1999)], and finally trapped in "optical lattices", produced by superposition of standing-wave fields [Hemmerich and Hänsch (1993), Grynberg et al. (1993), Jessen and Deutsch (1996)].

The forces used to manipulate atoms are the photon recoil connected with resonance absorption [Gordon and Ashkin (1980)], the electrical dipole force due to the dipole moment induced by the radiation field [Dalibard and Cohen-Tannoudji (1985)], the magnetic dipole force experienced by an atom with permanent magnetic dipole moment in an inhomogeneous magnetic field (see Chap. 4), and the force due to a spatial gradient of light polarization, based on optical pumping of atoms with complex hyperfine structure [Dalibard and Cohen-Tannoudji (1989), Ungar et al. (1989)]. The latter force also plays an important role in cooling the atoms below the Doppler limit. Finally, gravitation is also used for atom manipulation.

Numerous physical applications have already resulted from this development, ranging from quantum optics, atom optics and interferometry, atom lithography, atomic collision processes, and Bose–Einstein condensation to the physics of correlated quantum effects, precision measurements of fundamental constants, and the exploration of new frequency standards which will increase the accuracy of existing standards by several orders of magnitude.

A detailed description of this new field, which has undergone an impetuous development during recent years, is far beyond the scope of this book. Therefore we restrict this chapter to the description of the fundamental principles and basic techniques used in this new field, especially to those used in the generation of slow atom beams and in atom optics. For further details, the reader is referred to numerous reviews [Kasevich et al. (1991), Phillips (1992), (1995), Chu (1992), Metcalf and van der Straten (1994), (1999), Sengstock and Ertmer (1995), Müller et al. (1995), Demtröder (1996), Dowling and Bea-Banacloche (1996), Jessen and Deutsch (1996), Savage (1996), Bradley and Hulet (1996), Adams and Riis (1997), Bigelow (1998), Chu (1998), Phillips (1998), Cohen-Tannoudji (1998), Dalfovo et al. (1999)].

5.1 Radiation Pressure Forces

The forces exerted on an atom by a resonant or near-resonant radiation field can be uniformly described both by quantum mechanics and by semiclassical approximations [Cook (1979), Gordon and Ashkin (1980), Dalibard and Cohen-Tannoudji (1985)]. Here we use the result of the semiclassical approximation of Cook (1979) to give an intuitive interpretation on the basis of elementary considerations.

The forces experienced by a two-level atom from a monochromatic radiation field of amplitude $E(r,t)$ and phase $\theta(r)$ may be written in the form

$$\mathbf{F} = \frac{\hbar}{A^2 + 4(\Delta + \dot{\theta})^2 + 2\Omega^2} \left(A\Omega^2 \nabla \theta - (\Delta + \dot{\theta}) \nabla \Omega^2 \right). \tag{5.1}$$

A is the transition probability from state $|2\rangle$ to state $|1\rangle$, $\Delta = \omega - \omega_0$ is the detuning of the field frequency ω against the resonance frequency ω_0, and

$$\Omega = \mu E(\mathbf{r})/\hbar \tag{5.2}$$

is the Rabi frequency (μ is the transition dipole moment). The first term in (5.1) describes the recoil force of the photons, which is often termed the spontaneous force. Other designations are "scattering force" or "radiation pressure force". For a plane wave, the phase θ and its time derivative $d\theta/dt$ are given by $\theta = \mathbf{kr}$ and $d\theta/dt = \mathbf{k} d\mathbf{r}/dt = \mathbf{kv}$. Accordingly, the force has the direction of the wave vector $\mathbf{k}$. Furthermore, the force is a Lorentz function of the frequency ω, centered around the Doppler-shifted eigen frequency $\omega_0 + \mathbf{k\dot{r}}$ of the atom with a halfwidth

$$\Gamma = \sqrt{A^2 + 2\Omega^2}, \tag{5.3}$$

which is determined by the power broadening of the natural linewidth. The second term in (5.1) is called the dipole force. Since the field intensity I is proportional to Ω^2, this force has the direction of the field intensity gradient ∇I (direction of increasing field intensity), while the sign of the force is given by the sign of the detuning Δ. If the laser is red-detuned with respect to the resonance frequency ($\Delta < 0$), the atoms will be attracted from the region of highest intensity. In the case of blue-detuning ($\Delta > 0$), the atoms are repelled from the location of highest intensity.

Both parts of the force (5.1) can be derived from elementary considerations as is shown in the following sections.

5.1.1 Photon Recoil Force

A two-level atom, exposed to a resonant laser field for a given time t, can absorb and re-emit a large number of photons, provided that the time t is large compared to an excitation cycle. The population numbers N_1 and N_2 of the two states attained after a sufficiently long time can be calculated from the rate equations (I.5.107) and (I.5.108) yielding

$$N_2(t \to \infty) = N_1(0)\frac{S}{1+2S} \quad \text{and} \quad N_1(t \to \infty) = N_1(0)\frac{1+S}{1+2S}, \tag{5.4}$$

where the saturation ratio S is given by

$$S = \left(\frac{B}{A}\right)\rho(\nu). \tag{5.5}$$

A and B are the Einstein coefficients of spontaneous emission and induced absorption, respectively, $\rho(\nu)$ is the spectral energy density of the laser. Accordingly, the fraction of excited atoms may be written as

$$\frac{N_2}{N} = \frac{N_2}{N_1 + N_2} = \frac{S}{1+2S}. \tag{5.6}$$

Since the rate of spontaneous emissions is $AN_2 = N_2/\tau$, where τ is the lifetime of the excited state, the time Δt required by an excitation cycle is

$$\Delta t = \frac{1+2S}{S}\tau \quad \text{with} \quad \lim_{S \to \infty} \Delta t \to 2\tau. \tag{5.7}$$

The minimum time of an excitation cycle $\Delta t_m = 2\tau$ is achieved for large saturation ratios ($S \to \infty$). With each absorption and each emission of a photon, its momentum $\hbar k$ is transferred to the atom, where k is the wave vector of the photon. While for induced absorption and emission the vector k is conserved, so that the momenta of absorption and emission compensate each other, the spontaneously emitted photons are randomly distributed over the total solid angle 4π. Consequently, these momenta add up to zero in the time average. Therefore, only the momenta due to the excitation, which all have the same direction (determined by the laser beam) sum up to a total momentum $\mathbf{p}$:

$$\mathbf{p} = nm\Delta\mathbf{v} = n\hbar\mathbf{k}. \tag{5.8}$$

n is the number of excitation cycles during the interaction time t, given by

$$n = \frac{t}{\Delta t} = \frac{S}{1+2S}\frac{t}{\tau}. \tag{5.9}$$

Thus, the spontaneous recoil force $\mathbf{F}_R$ of one excitation cycle is

$$\mathbf{F}_R = \frac{\mathbf{p}}{n\Delta t} = \frac{\hbar \mathbf{k}}{\tau}\frac{S}{1+2S}. \tag{5.10}$$

Since the saturation ratio S can also be expressed by the Rabi frequency Ω and the spontaneous transition probability A

$$S = \Omega^2/A^2, \tag{5.11}$$

(see, for example, Demtröder (1996)), (5.10) is identical with the first term of (5.1), since we have assumed resonance ($\Delta + \mathbf{kv} = 0$).

Up to now, the spontaneous force has predominantly been used to decelerate, deflect, and focus atomic beams. Disadvantages of the spontaneous force are its statistical nature (giving rise to recoil heating) and its limitation due to saturation. These disadvantages are avoided by the stimulated recoil force, which has found an increasing number of applications for the past several years. [Nölle et al. (1996), Söding et al. (1997), Goepfert et al. (1997), Theuer and Bergmann (1998)]. Different techniques are employed. One method uses short, opposite, and time-delayed (with no overlap) laser pulses; the first pulse excites the beam atoms (a two-level atom system is assumed), the second pulse cancels the excitation by stimulated emission. Since the laser pulses have opposite directions, the photon momenta of absorption and stimulated emission have the same direction and with each pair of pulses the momentum $2\hbar k$ is transferred to the atom. Thus, the total force is $F = 2\hbar k/\Delta t$, where Δt is the time between two successive double pulses. This force is only limited by the experimentally achievable repetition frequency of the double pulses. The two pulses with opposite direction are usually produced by reflecting the first laser pulse from a mirror [Nölle et al. (1996)].

Another technique utilizes the coherent momentum transfer according to the STIRAP principle (see Sect. 4.8.8) [Theuer and Bergmann (1998)]. The pump and Stokes lasers also have opposite directions in order to avoid a compensation of the photon recoils due to absorption and emission.

5.1.2 Optical Dipole Force

The electrical field of a light wave induces a dipole moment **M** in an atom with the polarizability α which is proportional to the field strength **E**. In the case of an inhomogeneous field, the induced dipole moment experiences the force

$$\mathbf{F}_d = -(\mathbf{M} \cdot \text{grad})\mathbf{E} = -\alpha(\mathbf{E} \cdot \nabla)\mathbf{E}, \tag{5.12}$$

which can be rewritten in the form

$$\mathbf{F}_d = -(\alpha/2)\text{grad}E^2 - \alpha(\mathbf{E} \times (\nabla \times \mathbf{E})). \tag{5.13}$$

The second term of this equation vanishes when averaging over a cycle of the optical oscillation. Thus, the mean dipole force reduces to

$$\langle \mathbf{F}_d \rangle = -(\alpha/2)\text{grad}E^2. \tag{5.14}$$

The polarizability α depends on the frequency ω of the radiation field. Using the Kramers–Kronig relationship for the index of refraction, this dependence may be expressed in the following form

$$\alpha(\omega) \propto \frac{\Delta}{(\Delta + \mathbf{k}\mathbf{v})^2 + (\Gamma/2)^2}, \tag{5.15}$$

where Γ is the saturation-broadened linewidth according to (5.3). Accordingly, (5.14) corresponds to the second term of (5.1), if we take into account that the intensity I is proportional to the square of the field strength E^2

$$\mathbf{F}_d \propto -\frac{(\Delta + \mathbf{k}\mathbf{v})}{4(\Delta + \mathbf{k}\mathbf{v})^2 + A^2 + 2\Omega^2}\text{grad}I. \tag{5.16}$$

As already mentioned in Sect. 5.1.2, the sign of the dipole force depends on the sign of the laser detuning Δ. Accordingly, if the dipole force is used to store atoms at the location of highest (lowest) radiation intensity, the laser has to be red- (blue-) detuned with respect to the resonance frequency.

Contrary to the photon recoil, which requires a two-level system and thus is not applicable to molecules (see Sect. 5.1.3), the dipole force can be used to exert considerable forces on molecules of thermal energy beams [Seideman (1997), Stapelfeldt et al. (1997), Sakai et al. (1998)]. Using far-off resonance radiation, the force is independent of the particular level structure of the molecule and is, therefore, applicable to all molecules or atoms [Miller et al. (1993)]. Moreover, by employing the intense field from a pulsed laser to induce the molecular dipole moments, light-induced forces are obtained that are many orders of magnitude larger than the forces exerted upon atoms by standard CW laser beams. For example, a focused beam of a Nd:YAG laser ($\lambda = 1.06$ μm) acts on a CS_2 nozzle beam like a one-dimensional collective lens and thus may be employed for beam focusing.

The strong electric field in the center of a focused laser beam can also polarize macroscopic particles, which are drawn toward the focus (location of highest intensity), provided that the frequency of the laser is lower than the absorption frequency of the particles. Since the laser focus can be easily moved, large organic molecules and even living cells and bacteria can be accurately moved within aqueous solutions without damaging the organisms [Ashkin et al. (1987)]. Therefore, this technique has found widespread application in cellular biology ("optical tweezers") [Kasevich et al. (1991), Chu (1991), (1992), Liu et al. (1996)]. Optical tweezers may also be used to measure the forces associated with transport and adhesion of the particles to be moved [Finer et al. (1994)].

The dipole force can also be used to generate both transmission and reflection gratings for atoms, utilizing intense standing laser waves, as is discussed in detail in Sect. 5.4.5. Moreover, by superposition of two or three standing laser waves two- and three-dimensional lattices can be produced. With appropriate polarization of the laser waves, atoms can be cooled to temperatures low enough to allow their trapping and storage in the intensity antinodes of the lattice field. By simultaneous optical pumping nearly all atoms reach the energetically lowest Zeeman level and thus the energetic lowest state of the potential. These atoms, held in place by the radiation field, form an atom lattice with the regular structure of a solid body at a density corresponding to high vacuum. These optical lattices offer new possibilities for fundamental research. For instance, the quantized vibrations of the localized atoms can be studied by stimulated Raman spectroscopy. This is achieved by passing a weak probe laser beam through the optical lattice and measuring its transmitted intensity as a function of the probe laser frequency [Jessen et al. (1992), Grynberg et al. (1993), Hemmerich et al. (1994)]. Information about the lattice structure is obtained from Bragg scattering measurements [Weidemüller et al. (1995), Birkl et al. (1995), Dahan et al. (1996)].

As an example, Fig. 5.1 shows the spatial distribution of the potential of a two-dimensional optical lattice. It is produced by superposition of three coplanar,

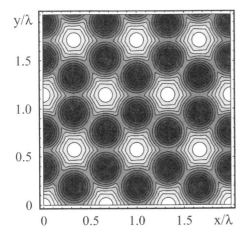

Fig. 5.1. Spatial distribution of the potential of a two-dimensional optical lattice [Grynberg et al. (1993)]. More details are described in the text

linear polarized laser beams of equal intensity with an angle of 120° between each pair of beams. The minima of the potential (dark regions) lie on a hexagonal lattice at locations where the resulting light polarization is purely circular [Grynberg et al. (1993)].

5.1.3 Optical Beam Slowing by Photon Recoil

A collimated atom beam counterpropagating to a resonant CW laser beam absorbs radiation and, according to (5.8), its average velocity decrease Δv (per excitation cycle) due to the photon recoil is given by

$$\Delta v = -\frac{\hbar \omega}{mc}. \tag{5.17}$$

Accordingly, the atoms of a collimated beam can be decelerated [Hänsch and Schalow (1975)]. However, it has to be considered that the Doppler-shifted absorption frequency

$$\omega = \omega_0 (1 - v/c), \tag{5.18}$$

changes with the atom velocity v during the deceleration process. Thus, either the laser frequency

$$\omega(t) = \omega_0 (1 - v(t)/c) \pm \Gamma \tag{5.19}$$

must be synchronously tuned with the changing velocity in order to remain within the linewidth Γ of the transition ("chirped slowing"), or the absorption frequency has to be suitably varied along the deceleration path ("Zeeman slowing"). The usefulness of both techniques has been experimentally demonstrated simultaneously

Table 5.1. Deceleration of a thermal energy sodium beam (T = 800 K)

Beam property	Numerical value
Most probable speed	$v_w = 760$ m/s
Resonance wavelength	$\lambda = 589$ nm
Photon energy	$h\nu = 2.1$ eV
Lifetime of the excited state	$\tau = 16$ ns
Force per excitation cycle	5.3×10^{-5} dyn
Recoil speed due to one photon	$v = 3$ cm/s
Total number of photons required	$n = 25\,000$
Minimum deceleration time (S→∞)	$\Delta t = 800$ μs
Deceleration path	$s = 30$ cm
Maximum acceleration	$a = -10^6$ m/s$^2 = -10^5$ g

[Ertmer et al. (1985), Prodan et al. (1985)]. Typical parameters for the deceleration of a thermal energy atom beam (Na) are compiled in Table 5.1.

An alternative technique makes use of a broadband, modeless laser whose light contains the whole frequency spectrum required by the deceleration process ("white light slowing") [Moi (1984), Liang and Fabre (1986), Hoffnagle (1988), Zhu et al. (1991)]. Another method uses isotropic laser radiation, produced by coupling laser light into a cavity with high diffuse reflectivity. As the atoms slow down while passing through the cavity, they compensate for their changing Doppler shift by preferentially absorbing photons at a varying angle to their direction of motion ("isotropic light slowing") [Ketterle et al. (1992), Aardema et al. (1996)]. Compared to the former techniques, the latter methods require more laser power in order to achieve the same degree of saturation on the whole spectrum.

Zeeman Slowing. The change of the absorption frequency ω_0 is achieved by using an appropriate magnetic field which depends on the path z of the atomic beam (see Fig. 5.2) [Metcalf (1986), Barett et al. (1991), Ketterle et al. (1993)]. Since the

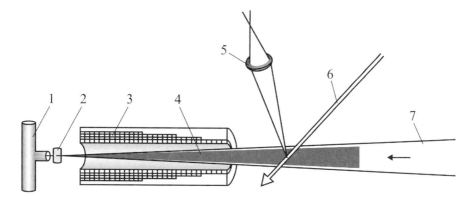

Fig. 5.2. Laser cooling of atoms using a fixed laser frequency and Zeeman tuning of the absorption frequency [Chu (1992)]. (1) oven, (2) collimator, (3) tapered solenoid, (4) atom beam, (5) collecting optics for fluorescence measurements, (6) probe laser, (7) cooling laser

Doppler shift along the deceleration path is given by

$$\Delta\omega(z) = \omega \frac{v_0}{c} \sqrt{1 - \frac{2a}{v_0^2} z}, \tag{5.20}$$

the longitudinal magnetic field must have the following z dependence:

$$B(z) = B_0 \sqrt{1 - 2az/v_0^2}. \tag{5.21}$$

Atoms with the initial speed v_0 enter the field at $z = 0$, where the value of the magnetic field is B_0. a is the negative acceleration of the atoms by photon recoil in m/s². Experimentally, the z-dependent field is realized by a tapered solenoid, whose number of turns depends on z in such a manner that (5.21) is satisfied. All atoms reach their final velocity at the same location at the field exit. Hence, a continuous beam of slow atoms is obtained.

The speed distribution of the slow atoms can be measured with a second laser using the Doppler method described in Sect. 3.3. The intensity of this probe laser has to be sufficiently low in order not to affect the speed distribution. Higher velocity resolution can be achieved by time-of-flight measurements. As an example, Fig. 5.3 shows measured speed distributions of a sodium beam without (A) and with (B) laser slowing [Molenaar et al. (1998)]. The time-of-flight method used in these measurements yields a resolution of $\Delta v < 1$ m/s, while the Doppler method is limited by the lifetime of the excited state to about $\Delta v \approx 6$ m/s. As can be seen from Fig. 5.3 (insert), the decelerated atoms have a narrow, approximately Gaussian-shaped speed distribution.

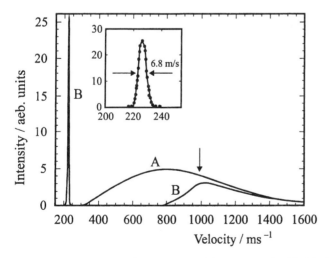

Fig. 5.3. Measured speed distribution without (A) and with (B) deceleration laser. The arrow indicates the maximum speed at which atoms are decelerated. The insert shows the speed distribution (B) of the slow atoms in an enlarged scale [Molenaar et al. (1998)]

Chirped Slowing. To keep the laser frequency ω in resonance with the constant frequency ω_0 of the atomic transition, a temporal behavior according to

$$\frac{d\omega}{dt} = -\frac{\omega_0}{c}\frac{dv}{dt} \qquad (5.22)$$

is required; this follows immediately from (5.19). The velocity change per unit time for optimum deceleration follows from (5.17) and (5.7)

$$\frac{d\omega}{dt} \cong -\frac{\hbar\omega(t)}{2mc\tau_k}. \qquad (5.23)$$

Substituting (5.23) into (5.22) yields a differential equation for the frequency ω(t) which has the solution

$$\omega(t) = \omega(0)\exp(\alpha t) \cong \omega(0)(1+\alpha t) \quad \text{with} \quad \alpha = \frac{\hbar\omega_0}{2mc^2\tau} \ll 1. \qquad (5.24)$$

In practice, the frequency tuning of the cooling laser is achieved by sidebands, which are generated from the laser by electro-optical modulation [Kelly and Gallagher (1987), Baum et al. (1980)]. An alternating electric field applied to a birefringent crystal ($LiTaO_3$, for instance) modulates its index of refraction and generates a phase modulation of the laser beam and thus sidebands symmetrically to the carrier frequency at a distance which is determined by the modulation frequency. One of these sidebands is used for cooling, its frequency chirp is matched to the time-dependent Doppler shift by changing the modulation frequency appropriately in time.

Up to now, we have tacitly assumed two-level atoms requiring only one laser frequency for excitation. An example of such an atom which offers a closed two-level system is Mg^{24}, which has zero nuclear spin and thus no hyperfine structure. The transition from the $^1S_0$ ground state to the $^1P_1$ excited state corresponds to a

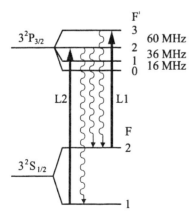

Fig. 5.4. Na $3\,^3S_{1/2} \rightarrow 3\,^2P_{3/2}$ transition. L1 cooling laser, L2 additional pump laser transition

wavelength of 285 nm. Due to the short lifetime of 2 ns of the excited state and the high transition frequency (recoil velocity 5.6 cm/s) Mg^{24} atoms with an initial speed of 1200 m/s can be stopped within a path of 11 cm [Sterr et al. (1992)].

For atoms with a degenerate ground state, laser deceleration is more complicated. Figure 5.4 shows as an example the level diagram of sodium atoms which are frequently used in cooling experiments. The transition between the hyperfine structure states F = 2 of the $^2S_{1/2}$ ground state and F' = 3 of the $2P_{3/2}$ excited state represents a true two-level system, as long as no spectral overlap between the two levels involved exists.

The frequency distance between these two levels is 60 MHz. Power broadening of the laser line and frequency fluctuations may lead to a partial overlap, so that particles can be optically pumped into the F = 1 state. These particles are lost from the excitation process and cannot be further decelerated. With an additional laser frequency, obtained by modulation with a second frequency so that a sideband matches the F = 1 → F' = 2 transition, these atoms are pumped back to the initial state [Ertmer et al. (1985), Sengstock and Ertmer (1995)].

Simultaneously, the particles which initially populate the F = 1 state (3/8 according to the statistical weight of this level) are also made available to the cooling process.

Figure 5.5 shows schematically the longitudinal speed distribution at different times of one sweep of the cooling frequency. At the beginning, the Maxwellian velocity distribution is observed, at the end of the frequency sweep all atoms below a given limiting speed are decelerated to a narrow velocity distribution. In contrast to Zeeman slowing by a magnetic field, chirped slowing yields a pulsed beam of slow atoms. To separate the slow atoms from the nondecelerated particles, another laser can selectively deflect the slow beam component out of the main beam and the cooling laser light [Nellessen et al. (1989c)].

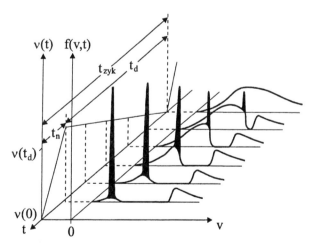

Fig. 5.5. Longitudinal speed distributions during the time t_d of one sweep of the slowing frequency v(t). After having finished the sweep, the laser frequency is set back to its initial value during the time t_n [Sengstock and Ertmer (1995)]

While the deceleration of Na atoms requires a single-mode CW dye laser pumped by an argon-ion laser, rubidium or cesium atoms can be slowed down by a GaAs diode laser. This reduces the experimental expenditure substantially, and simplifies the frequency modulation, which can be realized by a current modulation of the laser diode [Sheeby et al. (1989), Barett et al. (1991), Boiron et al. (1996)]. Metastable rare gas atoms such as He* and Ar* also have suitable transitions within the tuning range of commercial GaAs diode lasers and are often used for experiments with cold atoms [Metcalf (1989), Ertmer et al. (1996)].

The restriction of this technique to two-level atoms, or systems which can be made equivalent to two-level atoms, excludes its application to molecules. After optical excitation from a well defined rotational-vibrational state (v, J) of the electronic ground state into a corresponding state (v', J') of the electronically excited state, only a small fraction of the molecules returns by spontaneous emission to the initial state (v, J). Most molecules go back to other rotational-vibrational levels of the electronic ground state, making a further excitation impossible. To overcome this restriction, a cooling scheme for Cs_2 molecules has been proposed recently which requires a very complex laser system [Bahns et al. (1996)].

Up to now, we have only considered the influence of the mean force in the longitudinal direction. Perpendicularly to the beam direction a transverse heating process is observed. Although the photon emission recoil momenta add up to the mean value zero, the mean square value does not. Instead, the mean square value of the total momentum increases with increasing number n of spontaneously emitted photons according to

$$\langle p^2 \rangle = \frac{h\nu}{c} \sqrt{n}. \tag{5.25}$$

In typical experiments with Na atoms the transverse velocity spread is of the order of 5 m/s. This velocity spread reduces the beam density and thus the brightness of laser-decelerated beams. However, by applying suitable laser fields, the slow beams can be compressed or focused again [Tanner et al. (1988)]. A possibility is offered by standing laser waves perpendicular to the beam direction. If the frequency of the standing wave is somewhat lower than the absorption frequency of the atoms, atoms with a transverse speed component experience a repelling recoil force, since they absorb the wave propagating in the opposite direction with a higher probability than the wave propagating in the same direction. A more efficient transverse cooling is achieved by using a transverse two-dimensional optical molasses [Balykin and Letokhov (1987)] (see Sect. 5.2.1) or magnetooptical compression zones ("atomic funnel") [Riis et al. (1990)] (see Sect. (5.2.4).

5.1.4 Atomic Beam Deflection by Photon Recoil

The recoil of an atom when absorbing a photon was first measured by Frisch (1933) using the deflection of a thermal energy sodium beam crossed at right angles by a light beam from a resonance lamp. According to (5.8), the angle of deflection θ due to one photon is given by

$$\mathrm{tg}\theta = \frac{\Delta v}{v} = \frac{h\nu}{mcv} = \frac{h}{\lambda\, mv}. \tag{5.26}$$

m and v are the mass and velocity of the atom, respectively, ν the resonance frequency, λ the wavelength, h Planck's action quantum, and c the velocity of light. If this experiment is repeated with intense laser radiation, which illuminates a volume of length L of the beam, multiple excitation occurs and the angle of deflection increases correspondingly. In a two-level system the number n of excitation cycles is given by (5.9), resulting in an angle of deflection

$$\mathrm{tg}\theta = n\frac{h\nu}{mcv} = \frac{h\nu}{mcv}\frac{S}{1+2S}\frac{t}{\tau} = \frac{Lh\nu}{mv^2 c\tau}\frac{S}{1+2S}, \tag{5.27}$$

where we have replaced n by (5.9) and (in the last part of (5.27)) the interaction time t by L/v. Hence, the deflection is inversely proportional to the square of the beam velocity.

A number of deflection experiments with thermal energy atom beams have been performed using lasers [Schieder et al. (1972), Jacquinot et al. (1973), Nebenzahl and Szöke (1974), Abate (1974), Bernhardt et al. (1976), Citron et al. (1977)]. As an example, Fig. 5.6 shows the deflection of a velocity-selected thermal energy sodium beam irradiated with laser light corresponding to the transition

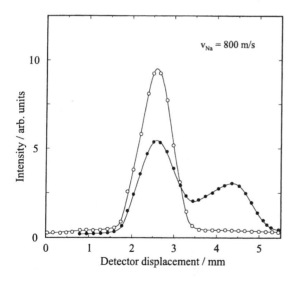

Fig. 5.6. Intensity distribution of a Na beam as a function of the detector displacement without (*open circles*) and with (*full circles*) transverse laser irradiation [Düren et al. (1975)]. For a detailed description see text

$$3^2 S_{1/2}(F=2) \rightarrow 3^2 P_{3/2}(F'=3) \tag{5.28}$$

(see Fig. 5.4), which can be considered as a two-level system if a laser of high stability with sufficiently small bandwidth is used [Düren et al. (1975)]. Since the initial state, due to its statistical weight, is populated by 5/8 of all atoms and since no additional pump laser is used, the maximum fraction of sodium atoms that can be deflected is 62.5%. Thus, the beam profile shown in Fig. 5.6 splits into two portions.

Measuring the deflection at different beam velocities, (5.27) has been experimentally confirmed [Düren et al. (1975)].

Using suitable mirror arrangements, so that the laser beam crosses the atomic beam several times, the angle of deflection can be increased. The deflection is limited to the transverse velocity $v_\perp$ at which the Doppler effect shifts the atoms out of resonance. As has already been discussed in the preceding section, all sodium atoms can be deflected if the beam is irradiated by additional laser light corresponding to the transition

$$3^2 S_{1/2}(F=1) \rightarrow 3^2 P_{3/2}(F'=2) \tag{5.29}$$

(see Fig. 5.4), so that the atoms of the $F = 1$ ground state are also excited and losses in other states are avoided. The two laser lines can be generated from one laser using an electro-optical modulator. Another method of two-line excitation with one laser can be achieved by changing the single-mode laser into a two-mode laser, as has been described by Strohmeyer (1990). Using the first method, the photodeflection has been used to separate Na atoms from Na_2 dimers from a beam containing both particles [Chapman et al. (1995)].

Since the deflection is velocity dependent, a slit may be introduced into the deflected beam which allows only atoms of a given velocity interval to pass, similar to the case of magnetic deflection in a two-wire field (see Sect. 3.4). Thus, a velocity selection can be achieved. Much more elegant is a method in which only atoms of a given speed interval are deflected from the main beam [Weiquan et al. (1995)]. In this case, the deflection laser crosses the atomic beam at an angle somewhat different from $\pi/2$ so that — due to the Doppler effect — only particles of a given velocity range are deflected from the initial beam direction forming a velocity selected beam.

Moreover, the absorption of photons is isotope specific, so that the deflection can also be used for isotope separation [Bernhardt et al (1974)]. Using a laser-decelerated atom beam, the slow atoms can be selectively deflected from the rest of the beam resulting in a nearly monochromatic slow beam [Nellessen et al. (1989a,b,c), (1990)]. As already mentioned, the deflection is limited by the Doppler shift of the deflected atoms. To overcome this limitation, it is necessary to compensate for this effect. An elegant solution of this problem is a pure geometrical one. Its principle is illustrated in Fig. 5.7. A convergent laser beam, focused by a cylindrical lens, crosses the atomic beam. For an appropriate choice of the

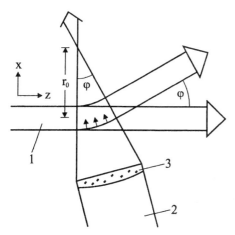

Fig. 5.7. Laser deflection with a focused laser beam. (1) atomic beam, (2) laser beam, (3) cylinder lens [Sengstock and Ertmer (1995)]. For details see text

experimental parameters, the beam is deflected in such a manner that the actual **k** vector of the laser is orthogonal to the local velocity of the atoms, so that the excitation is Doppler free at each location [Nellessen et al. (1989a)].

The atoms follow a circular trajectory determined by the compensation of the local spontaneous force and the centrifugal force. Equating the spontaneous force (5.10) with the centrifugal force yields the relationship

$$\frac{mv_0^2}{r_0} = \frac{h\nu}{c\tau}\frac{S}{1+2S}, \qquad (5.30)$$

from which the radius r_0 of the orbit follows as

$$r_0 = \frac{mcv_0^2\tau}{h\nu}\frac{(1+2S)}{S}. \qquad (5.31)$$

The angle of deflection is equal to the angle of convergence φ of the laser beam (see Fig. 5.7).

The field of force produced by the laser light provides radial focusing [Ashkin (1970)], similar to a cylindrical condenser for charged particles. Atoms of a given velocity v_0, having a direction which is not tangential to the nominal circle, are focused to it after an angle of $\pi/3^{1/2}$ (104°). This is due to the fact that atoms, which have a larger orbital radius r, get closer to resonance and thus experience a stronger force which redirects them to r_0. On the other hand, atoms following a trajectory with a radius $r < r_0$, get more out of resonance so that the deflecting force decreases. Since the radius of the trajectory depends on the atom velocity, the velocity distribution can be measured in the focusing plane.

Apart from the spontaneous force, the stimulated recoil force has also been used to deflect atomic beams [Nölle et al. (1996), Theuer and Bergmann (1998)].

5.2 Trapping and Cooling of Atoms

An important application of slow atoms is their confinement and further cooling in traps, which require a position-dependent force directed to a center. Accordingly, the potential of these forces has a minimum at the trap center. Already during the 1970s, the first suggestions for trap devices were discussed [Letokhov et al. (1976), Ashkin and Gordon (1979)]. A wealth of proposals for atom traps in many variants followed [Phillips (1992), Chu (1992),], and still new variants are added [Vuletic et al. (1996), Kim et al. (1997a), Wasik and Grimm (1997), Hughes et al. (1997a), (1997b), Ovchinnikov et al. (1997)].

The forces which can be exerted on neutral atoms and may, therefore, be utilized for trapping, have already been mentioned in the introduction: The photon recoil force, the electric dipole force, the magnetic dipole force experienced by a permanent magnetic dipole in an inhomogeneous magnetic field, and the gravitational force. Compared to forces which can be used to confine charged particles, these forces are very weak. The strongest force is the recoil force, since it can be enhanced by the absorption of many photons. Consequently, the potential minima achievable in neutral atom traps are very shallow. Typical values range from several millikelvins for magnetostatic traps and optical dipole traps to a few kelvins for magnetooptical traps. Therefore, atoms have to be precooled below this temperature before they can be stored in a trap.

According to the forces mentioned above, atom traps can be divided roughly into the following groups:

- Magnetostatic traps, which require a permanent magnetic dipole moment. These traps are described in Sect. 4.4.
- Magnetooptical traps, which utilize a combination of magnetostatic fields and radiation pressure.
- Optical dipole traps based on the induced electric dipole moment of the atoms.
- Traps in which one of the above principles is combined with the gravitational force (gravito-optical traps, for example).

The first trapping of neutral atoms succeeded in a magnetic quadrupole trap [Prodan et al. (1985), Migdall et al. (1985)], generated by two anti-Helmholtz coils (see Sect. 4.4) at the end of a Zeeman slower. The maximum speed of the atoms, which could be trapped, was 3.5 m/s, corresponding to a temperature of 17 mK.

5.2.1 Optical Molasses and Cooling Mechanisms

Shortly after, three-dimensional atom clouds were cooled to much lower temperatures, illuminating the atoms from all directions with light tuned slightly below the resonance frequency. This was achieved experimentally by intersecting three orthogonal standing waves [Chu et al. (1985)]. Other devices use the overlap region of six laser beams (generated from one laser) propagating in all directions of space $\pm x$, $\pm y$, and $\pm z$ Since the laser frequency ν is smaller than the resonance frequency

v_0 of the atoms, an atom which leaves the intersection region of the laser waves, always experiences a retarding force, since the counterpropagating laser wave is Doppler-shifted closer to resonance, whereas the laser wave propagating in the atom direction is shifted away from resonance. Therefore, the atoms experience a retarding force which is proportional to their velocity, providing a viscous damping of their motion. Due to this frictional force the atoms move similar to in a viscous medium. Accordingly, such devices are called optical molasses. This cooling mechanism (Doppler cooling) is counteracted by a diffuse heating process due to the spontaneously emitted photons (see (5.25)). The equilibrium of these processes determines the achievable lower limit of the temperature T_D (Doppler limit), which is given by

$$T_D = \frac{h\Gamma}{2k}. \tag{5.32}$$

Γ is the natural linewidth and k is Boltzmann's constant. Typical values are of the order of a few hundred microkelvins (240 µK for sodium atoms).

As a result of further experimental investigations of optical molasses, it soon became apparent that with certain polarizations of the light much lower temperatures could be achieved [Lett et al. (1988)]. Theoretical studies initiated by these experiments showed that for sufficiently cold atoms further, very efficient cooling mechanisms become effective [Dalibard and Cohen-Tannoudji (1989), Ungar et al. (1989), Weiss et al. (1989)]. These mechanisms are based on the interaction of the magnetic sublevels with the different polarization components of the light (sub-Doppler cooling). Thus, temperatures more than an order of magnitude lower than the Doppler temperature T_D can be achieved. The temperature limit T_r follows from a quantum-mechanical treatment which predicts a minimum temperature corresponding to an r.m.s. momentum of the order of the atomic recoil associated with the absorption or emission of a single photon hv/c. This yields

$$T_r = \frac{h^2 v^2}{2kmc^2}, \tag{5.33}$$

where k is Boltzmann's constant and c the velocity of light. Typical temperatures are of the order of 10 µK. These sub-Doppler cooling mechanisms are also termed polarization gradient cooling. Depending on the light polarization, two mechanisms become effective, which are distinguished by their field configuration (lin $\perp$ lin and $\sigma^+\sigma^-$). In the one-dimensional case, lin $\perp$ lin is a geometry with two counterpropagating laser beams, which are linearly polarized in orthogonal directions. On the other hand $\sigma^+\sigma^-$ is a geometry with two counterpropagating laser beams circularly polarized in orthogonal directions. The first mechanism is generally referred to as Sisyphus cooling.

The recoil temperature T_r can also be expressed as an r.m.s. velocity $v_r = hv/mc$. To obtain temperatures lower than T_r, the atoms must be excluded

284   5. Slow Atom Beams, Traps, and Atom Optics

from further excitation as soon as their velocity is smaller than v_r (subrecoil cooling) [Salomon (1991), Kasevich and Chu (1992), Cohen-Tannoudji (1992), Werner et al. (1993), Lawall et al. (1995), Lee et al. (1996), Ketterle and van Druten (1996), Adams and Riis (1997)]. This principle is often called "velocity space optical pumping" and the excitation-free region near $v = 0$ is referred to as a dark state. In practice, this can be achieved either by decoupling atoms close to $v = 0$ from the radiation field using destructive interference between two terms which contribute to the excitation rate (velocity-selective coherent population trapping), or by using a two-photon transition of extremely narrow linewidth to further cool atoms with velocities other than those close to $v = 0$ (stimulated Raman cooling). Both subrecoil schemes, which will not be discussed here in detail, can be used to obtain temperatures of the order of 100 nK.

Another cooling mechanism which does not rely on lasers is evaporative cooling, which was first proposed to cool spin-polarized H atoms in magnetostatic traps [Hess (1986), Ketterle and van Druten (1996)]. The basic principle is simple: In a collision between two stored atoms, which have roughly the same initial energy, the final energies may have quite different values. If the hotter atom is allowed to escape from the trap, the net effect, after rethermalization, is to leave the remaining ensemble colder.

5.2.2 Atom Traps

The optical molasses described above is not a trap, since there is no position-dependent central force which holds the atom cloud together. According to the optical Earnshaw theorem, it is fundamentally impossible to build a trap from a static configuration of laser beams based solely on absorption and spontaneous emission, in which the spontaneous force is proportional to the photon flux [Ashkin and Gordon (1979), (1983), Ashkin (1984)]. This is in complete analogy to the corresponding theorem in electrostatics, which states that it is impossible to trap charged particles with static fields.

Magnetooptical Traps. If the proportionality between the spontaneous force and the photon flux is altered in a position-dependent way, a stable trap can be

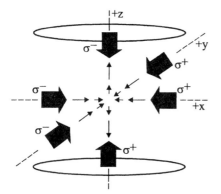

Fig. 5.8. Principle of a magnetooptical trap [Sengstock and Ertmer (1995)]

obtained [Pritchard et al. (1986)]. This concept is the basis of the magnetooptical traps [Raab et al. (1987), Wallis et al. (1993)], which have found widespread application. A schematic view of such a trap is shown in Fig. 5.8. A spherical magnetic quadrupole field produced by two anti-Helmholtz coils (see Sect. 4.4) is added to the configuration of six laser beams which generate the optical molasses. Each pair of counterpropagating laser beams (tuned below resonance) is circularly polarized in opposite directions σ^+ and σ^-, driving $\Delta m = +1$ and $\Delta m = -1$ transitions, respectively. An atom moving away from the center is Zeeman-shifted closer to resonance with a beam which pushes it back. This can be most easily understood by considering only one dimension, such as the x direction, assuming an atom with $J = 0$ to $J' = 1$ transition. Since the magnetic field has the form $B(x) = cx$, the $m = +1$ sublevels of the excited state experience Zeeman shifts which are linear with the distance from the origin.

An atom with a positive x coordinate sees the σ^- beam closer to resonance, absorbs more σ^- than σ^+ photons and is pushed back towards $x = 0$. Similarly, an atom with $x < 0$ absorbs more σ^+ than σ^- photons and is also pushed towards $x = 0$. Consequently, the atoms experience a restoring force in addition to the frictional force of the optical molasses. Therefore, at sufficiently low velocities and for small deviations from the equilibrium position, the atoms perform damped harmonic oscillations about the trap center. The first magnetooptical traps were loaded from slowed atomic beams.

A few years later it was shown that magnetooptical traps can collect atoms directly from a room temperature vapor [Monroe et al. (1990)] or from a thermal atomic beam [Anderson and Kasevich (1994)]. The tail of the Maxwellian speed

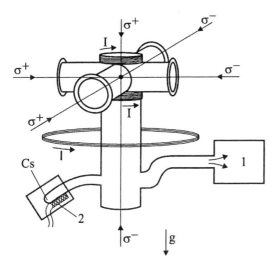

Fig. 5.9. Vapor cell with magnetooptical trap (upper part) for collecting Cs atoms from the vapor phase. (1) ion pump (1l/s), (2) Cs reservoir with cooling element. The large coil in the middle is part of a magnetic Ioffe trap, the four vertical current bars providing horizontal confinement are not shown. Vertical confinement at the top is achieved by the Earth's gravitational field [Monroe et al. (1990)]

286   5. Slow Atom Beams, Traps, and Atom Optics

distribution contains enough atoms slower than the capture velocity v_c of the trap to provide a sufficient loading rate. Figure 5.9 shows an example of such a device. Atoms from room temperature cesium vapor (6×10^{-9} torr) are captured directly by a magnetooptical trap, cooled, and then transferred to a magnetic trap (Ioffe configuration, see Sect. 4.4) for further experiments. In this way it has been possible to trap 1.8×10^7 spin-selected cesium atoms in the Ioffe trap at a temperature of 1.1 K. The small closed volume of this device makes it particularly attractive for experiments with rare isotopes.

Optical Dipole Traps. In the first realization of an optical dipole trap, a red-detuned dye laser with a power of 220 mW was focused to a spot size of 10 µm at the center of a cloud of sodium atoms cooled by optical molasses [Chu et al. (1986)]. The radiative heating of the atoms by the trapping light was compensated by alternating switching between the trapping laser and the cooling molasses beams with a frequency between 100 kHz and 2 MHz. In later devices much larger detunings have been used in order to reduce the photon absorption rate [Miller et al. (1993), Lee et al. (1995), Adams et al. (1995)]. Radiation heating, however, is a fundamental problem when working with red-detuned laser light, since the atoms always stay at the location of highest intensity. Even if large detunings are used, radiative heating is unavoidable and usually requires additional measures for compensation [Boiron et al. (1998)].

The capacity of optical dipole traps using red-detuned laser light is rather small (between several hundred to about 10^5 atoms have been reported) and the anisotropic potential well is very shallow (a few mK or less).

A somewhat larger volume and a nearly isotropic trap potential can be achieved by using two intersecting, focused laser beams [Adams et al. (1995)].

In blue-detuned dipole traps, atoms are repelled from the region of high intensity and therefore very low photon absorption rates (and thus low heating by the trapping light) can be achieved, since the atoms spend most of their time outside the laser field, especially if hard-wall potentials (similar to the case of light-based atom mirrors) are used. Examples of some devices are shown in Fig. 5.10.

Lee et al. (1995) used two intersecting sets of V-shaped sheets of far blue-detuned laser light (see Fig. 5.10a). The light sheets are 600 µm long and 10 µm thick. The atoms are confined within the four-sided, inverted pyramid formed by the crossed laser sheets by the repulsive dipole force of its "walls" and by the Earth's gravitational field (gravito-optical trap). Due to its relatively large volume,

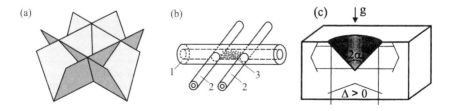

Fig. 5.10. Optical dipole traps

more than 5×10^5 atoms could be trapped. In another device (see Fig. 5.10b), the atoms (3) are radially confined within the hollow space of a blue-detuned doughnut mode laser beam (1) produced by an appropriate mode converter. Axial confinement is achieved by two additional, blue-detuned laser beams (2) [Kuga et al. (1997), Torii et al. (1998), Kuppens et al. (1998)].

Finally, Fig 5.10c shows another example of a gravito-optical trap. It consists of a circular cone (opening angle 2α) ground into a block of glass. A blue-detuned beam incident from the bottom of the glass block is totally reflected at the cone, since its angle satisfies the relation $\cos\alpha \geq 1/n$, where n is the index of refraction of the glass. The evanescent wave at the cone surface acts as an atom mirror and forms, together with the gravitational force, an atom trap [Dowling and Gea-Banacloche (1996)].

5.2.3 Atom Traps as Sources for Slow and Cold Atoms

In many experiments with cold atoms, atom traps are often used as a source for generating cold atom beams with diameters of a few µm and a tunable translational velocity. A simple way to realize such a beam utilizes the free fall of atoms from a trap over a well defined height. This method has been used, for example, to determine the temperature in the trap by absorption measurements at different distances from the trap [Marty and Suter (1995)]. The free fall is also often used in mirror experiments with cold atoms (e.g. "atomic trampoline", gravito-optical traps) [Roach et al. (1995), Landragin et al. (1996), Hughes et al. (1997a), (1997b)]. Another possibility is the acceleration of the stored atoms out of the trap by a suitable laser pulse. If the direction of the laser pulse is opposite to the Earth's gravitational field, the atoms are launched on ballistic trajectories, forming an "atomic fountain" (see Fig. 5.11). Since the atoms spend a long time in the vicinity

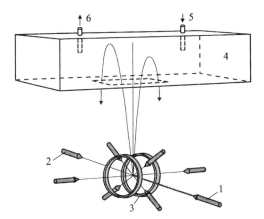

Fig. 5.11. Schematic of an "atomic fountain" for precision spectroscopy. A laser pulse accelerates atoms of a magnetooptical trap against the gravitational field. A microwave cavity in the region of the turning point induces transitions, which are detected with high precision. (1) atom beam for trap loading, (2) deceleration laser, (3) magnetooptical trap, (4) cavity, (5) microwave emitter, (6) microwave detector [Kasevich et al. (1989)]

of their turning point, such devices are discussed as new frequency standards [Clairon et al. (1991), Nakagiri (1996), Santarelli et al. (1999)]. The accuracy of such devices achieved so far is of the order of 2×10^{-15}.

Magnetooptical traps which are loaded directly from the vapor phase (see Fig. 5.9), are well suited as sources of intense and highly collimated slow atom beams [Lu et al. (1996)]. Figure 5.12 shows a schematic diagram of such a device, which consists of a conventional magnetooptical trap. However, in the center of one of the six cooling laser beams, a tubular volume of 1 mm diameter is free of light, so that the stored atoms are forced by the counterpropagating beam to leave the trap. Their velocity depends on the number of absorbed photons. Atoms having a divergence angle large enough to leave the light-free region get into the laser beam and are pushed back into the trap. The achievable atom flux of the beam is essentially determined by the capture rate of the trap. The latter is optimized by using laser beams of large cross section (4 cm $\varnothing$). An additional laser beam (3) (plug laser) can be used to interrupt the atomic beam. In this manner, a Rb beam with a continuous flux of 5×10^9 atoms/s has been produced, having a diameter of 0.6 mm and a velocity of 14 m/s ($\Delta v/v = 20\%$). Pulsing the plug laser, a pulsed atom beam can also be produced.

After having achieved Bose–Einstein condensation in ensembles of stored atoms, atom traps can also be used as sources of coherent matter waves, as has been impressively shown by the interference between atoms from two separate traps releasing two Bose–Einstein condensates by free fall [Andrews et al. (1997)]. Such a source for atoms in a coherent state is the simplest form of an "atom laser" [Kleppner (1997)].

Atoms having an energy level scheme which does not allow for optical cooling, may also be stored in magnetic traps if they have a magnetic moment [Bergeman et al. (1987), Ketterle and van Druten (1996)]. The loading technique utilizes a cold buffer gas (^{3}He or ^{4}He) to thermalize the species to be trapped to energies below the trap potential depth (about 1 K) by elastic collisions [Doyle et al. (1995), Kim et al. (1997b), Weinstein et al. (1998)]. On the other side, the temperature must be high enough to result in a vapor pressure of the buffer gas (in the mtorr range) which allows sufficient collisions for thermalization. Afterwards, the helium is cryo-pumped away (to pressures below 10^{-15} torr) by decreasing the he-

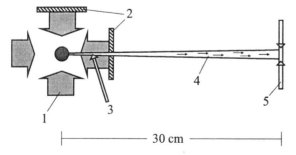

Fig. 5.12. (1) Magnetooptical trap as source for cold, slow atom beams [Lu et al. (1996)]. (2) mirror, (3) plug laser, (4) atom beam, (5) standing light wave for beam detection

lium temperature, leaving a thermally isolated, trapped sample. A further decrease of the temperature can then be achieved by evaporative cooling.

If the atom species to be trapped is vaporized by laser ablation, the loading process is simplified, since the cell containing the trap can be kept at constant temperature. This temperature corresponds to the final temperature of the loading cycle described above, so that the buffer gas vapor pressure is below 10^{-15} torr. A pulse of the ablation laser not only vaporizes the species to be trapped, but also helium atoms condensed at the walls of the cell, yielding a collision rate sufficiently high to obtain a fast thermalization of the atom species. Simultaneously, the low temperature of the cell provides for a sufficiently fast remove of the buffer gas from the gas phase. Using this technique, 4×10^{10} chromium atoms, for example, have been stored in a magnetic trap at a density of 4×10^{11} atoms/cm^3 [Weinstein et al. (1998)].

This combination of laser ablation, thermalization in a cold buffer gas, storage in a magnetostatic or, in the case of molecules, an electrostatic trap, and subsequent evaporative cooling offers, among others, the possibility to build sources for cold beams of a large variety of atoms and molecules.

5.2.4 Methods of Beam Compression

For many applications (e.g. in atom interferometry, atomic collision processes, or atom optics), continuous, cold beams of high spectral brightness with a well defined thermal or subthermal energy are required. The laser techniques described in the foregoing sections can also be used to produce such beams.

An often-employed technique is transverse laser cooling [Balykin and Letokhov (1987)]. The atomic beam leaving the source is directly captured by two counterpropagating pairs of laser beams tuned below resonance, having directions perpendicular to the atomic beam. Thus, a two-dimensional optical molasses is formed which reduces the beam divergence and increases the atom flux in the forward direction.

A strong compression of the beam is achieved by the "atomic funnel" first employed by Riis et al. (1990), which is essentially a two-dimensional magneto-optical trap (in a transverse direction with respect to the atom beam) and a moving optical molasses in the direction of the atom beam [Nellessen et al. (1990), Swanson et al. (1996), Weyers et al. (1997), Schiffer et al. (1997), McIntyre et al. (1997)]. Figure 5.13 illustrates the basic idea. Four current-carrying wires generate a longitudinal quadrupole field, while two pairs of counterpropagating, circularly polarized laser beams of elliptical cross section capture the atoms (from a slowed thermal beam or from the vapor phase), which have a velocity below 20 m/s, and accelerate them towards the axis of symmetry. Along the axis, a pair of counterpropagating laser beams with appropriate frequency offsets cool the atoms in a moving frame (moving optical molasses). In this way, a sodium beam has been produced with a flux of 10^9 atoms/s, a diameter of 1.5 mm, and a temperature of

290   5. Slow Atom Beams, Traps, and Atom Optics

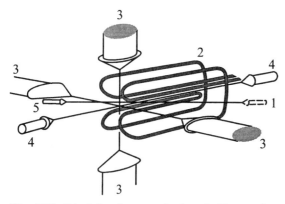

Fig. 5.13. Principle of an atomic funnel. (1) atom beam, (2) quadrupole field, (3) laser beams of the two-dimensional magnetooptical trap, (4) laser beams of the moving molasses, (5) deceleration laser

200 μK with a tunable velocity between 1 and 5 m/s. Its spectral brightness surmounts that of the original beam by a factor of 3000 [Riis et al. (1990)].

At constant magnetic field strength, the compressing radial force exerted on the atoms by the funnel arrangement decreases with increasing compression (decreasing distance of the atoms from the beam axis). To obtain a constant radial force independent of the beam diameter, Nellessen et al. (1990) used a compression stage with a quadrupole field that increases linearly along the beam axis. The field is generated by permanent magnets. The atoms are collected at low field (50 G/cm), allowing a larger capture range, and leave the compression stage at a high field (500 G/cm). With this device, Nellessen et al. were able to compress the diameter of a slow sodium atom beam (v = 50 m/s) from 3 mm to 43 μm, increasing its brightness by more than three orders of magnitude.

A related technique has been employed by Berthoud et al. (1998) to produce a continuous cesium atom beam. A two-dimensional magnetooptical trap with sub-Doppler cooling in three dimensions collects atoms from Cs vapor. The extraction of the beam utilizes the drift velocity of the atoms due to a constant magnetic field. The measured beam flux is 1.3×10^8 atoms/s and thus approximately equal to the capture rate of the trap, so that practically all captured atoms are transferred to the beam. Tunable atom beams of rather low velocities (0.5–4 m/s) can be produced, having temperatures T ≤ 200 μK, as has been measured by Ramsey fringes in a microwave cavity. A similar device for Rb atoms, yielding beam fluxes about two orders of magnitude higher, has been described by Dieckmann et al. (1998).

Thermal energy beams can also be substantially cooled in the axial direction to narrow their velocity spread, using the method of moving optical molasses. This has been demonstrated by Faulstich et al. (1992) using a metastable Ar beam. Starting with a nozzle beam (speed ratio S = 10, v/Δv = 5), the velocity resolution could be increased up to v/Δv = 890 at a beam velocity of v = 580 m/s.

Using the Zeeman slowing technique together with suitable compression and cooling methods, the best performance achieved to date is about 2×10^{10} slow cesium atoms/s (in the range of 35–120 m/s), with a record spectral brightness of up to 7×10^{19} atoms $s^{-1} cm^{-2} sr^{-1}$ [Lison et al. (2000)]. This is comparable to the brightness of conventional supersonic sources for alkali atom beams. The brightness of supersonic He sources, however, exceeds 10^{22} atoms $s^{-1} cm^{-2} sr^{-1}$, and multi-watt lasers, for comparison, have a spectral brightness of the order of 10^{27} photons $s^{-1} cm^{-2} sr^{-1}$.

5.3 Examples of Applications

Slow atoms, their optical cooling, deflection, compression, and trapping at temperatures below µK offer a multitude of new applications in atomic physics. These, in turn, have been an important motivation behind the development of atom manipulation techniques. Examples are discussed in the following sections.

5.3.1 Precision Spectroscopy and Frequency Standards

High-resolution spectroscopy of neutral atoms is an important application of optical cooling techniques, since its accuracy can be substantially increased by using laser-cooled atoms [Sengstock et al. (1993), (1994), Oates et al. (1996), Kokkelmans et al. (1997)]. This improvement in accuracy arises from the increase of the maximum possible interaction time of the atoms with the radiation field (about 1 s compared to 10 ms in conventional beam experiments), since the maximum achievable accuracy is determined by the uncertainty principle $\Delta E \Delta t = \hbar$. Furthermore, the quadratic Doppler shift (5.34) can be drastically reduced:

$$\frac{\delta v}{v} = \frac{1}{2}\left(\frac{v}{c}\right)^2. \tag{5.34}$$

An example of the use of cold atoms in precision spectroscopy is the measurement of the natural linewidth of the $3\,^2S_{1/2} \rightarrow 3\,^2P_{3/2}$ transition of sodium [Oates et al. (1996)], with which a long existing discrepancy of 1% between calculated and measured values of the lifetime of the $3\,^2P_{3/2}$ state has been clarified.

The extreme accuracy of high-resolution spectroscopy is necessary to investigate fundamental problems in many areas of modern physics by using methods of atomic physics. A few examples of applications are the test of weak interaction in parity violation experiments [Bouchiat (1991), Sapirstein (1995)], tests of quantum electrodynamics [Basini et al. (1989), Lepage (1995)], or statements of general relativity. Of practical importance are the already mentioned efforts to replace the cesium frequency standard by a new standard with an accuracy which is several orders of magnitude higher than the cesium clock. The goal is an accuracy of 10^{-16}. The accuracy obtained so far with atomic fountains is 1.4×10^{-15} [Ghezali et al. (1996), Simon et al. (1998), Santarelli et al. (1999)]. The main factors limiting the achievable accuracy of these devices are the following:

- Atomic collisions: The energy levels of colliding atoms shift, and they shift differently for the two hyperfine levels of the clock transition. This gives rise to a density-dependent frequency shift [Gibble and Chu (1992), (1993), Kokkelmans et al. (1997)]. Frequency shifts of several mHz have been measured for densities of the order of 10^9 cm^{-3}.
- The Stark shift in the average electric field of the thermal radiation from the walls of the vacuum chamber, which is essentially black-body radiation [Itano et al. (1982)]. Its relative value $(-1.69\times10^{-4}\times(T/300)^4)$ is known with an accuracy of only 10% [Bauch and Schröder (1997)]. This is insufficient for the desired accuracy [Simon et al. (1998)].

To establish these limiting factors with sufficient accuracy and thus derive frequency corrections, a great deal of the present work is concerned with investigations of both the collision-induced level shifts and the Stark shift of the hyperfine structure splitting in atomic fountain experiments [Simon et al. (1998)].

5.3.2 Atomic Collision Processes

The study of collisions between cold atoms is not only important to understand the collisional energy shift of atomic levels, but also from the standpoint of fundamental research as well as for the elucidation of the dynamical processes in atom traps and their interaction with radiation fields [Walker and Feng (1994), Heinzen (1995), Weiner (1995), Lett et al. (1995), Gensemer et al. (1997), Weiner et al. (1999)]. Collisions between cold atoms are characterized by the relationship

$$\lambda \geq r_0. \tag{5.35}$$

λ is the de Broglie wavelength and r_0 characterizes the range of the interatomic potential. In this energy region the scattering is dominated by pronounced quantum effects, for example "orbiting" resonances (quasibound states within the centrifugal barrier of the interaction potential) in the total cross section and interference structures in the differential cross section [Metcalf and van der Straten (1994), Arndt et al. (1997), Boesten et al. (1997), Westphal et al. (1996), (1997)].

In atom traps, these subthermal collisions are of great practical importance, since they are responsible for particle losses from the trap and for heating processes [Suominen (1996), Suominen et al. (1998)]. Up to now, trap experiments provide the only possibility for an indirect investigation of atomic collisions at extremely low energies [Marcassa et al. (1993), (1996), Hoffmann et al. (1996), Mastwijk et al. (1998)], but a number of direct collision experiments, utilizing either a slow atom beam and the atoms confined in a trap as a target, or crossed beam arrangements of two slow beams, is under progress in several laboratories. A first collision experiment, which is briefly discussed in Sect. 3.5, has been reported recently [Legere and Gibble (1998)].

5.3.3 Bose–Einstein Condensation

Another temperature and density region can be defined by the relationship

$$\lambda^3 n > 1 \tag{5.36}$$

(n is the atom density in the trap), where the de Broglie wavelength is not only large compared to the range of the potential, but also compared to the average distance between the stored atoms.

Under these conditions, Bose–Einstein condensation occurs [Burnett (1996), Lewenstein and You (1996)] (see Sect. I.2.2). Thus, with decreasing temperature, more and more atoms turn to the lowest quantum state. All atoms occupy the same minimum volume h^3 in phase space determined by the uncertainty principle. Attempts to experimentally realize this quantum effect in a weakly interacting gaseous system, which is characterized by the condition $|na^3| \ll 1$ (a is the effective range of the interaction), began more than 15 years ago [Lewenstein and You (1996)]. The first experiments concentrated on spin-polarized hydrogen atoms in a magnetic trap [Doyle et al. (1991)] using evaporative cooling. Recently, the first observations of Bose–Einstein condensation in this system have been reported (Fried et al (1998). By combining laser and evaporative cooling for alkali atoms stored in magnetooptical traps, experiments on Bose–Einstein condensation were already successful three years earlier [Anderson et al. (1995), Davis et al. (1995), Bradley et al. (1995)]. Many other experiments and theoretical studies followed [Han et al. (1998), Hau et al. (1998), Dalfovo et al. (1999)].

To measure the momentum distribution of the atoms in the trap after the cooling process, the trap is switched off so that the atom cloud can expand freely according to the speed distribution of the atoms (absorption time-of-flight imaging). Figure 5.14 presents an example [Ernst et al. (1998)].

Measuring the density of the cloud by spatially resolved light absorption along the axial coordinate, the spatial distribution of the atom ensemble at a well defined time after switching off the trap can be determined. Neglecting the atom–atom interaction, the expansion is ballistic and the imaged distribution of the expanding cloud can be directly related to the initial velocity distribution. Figure 5.14 shows the light absorption, which is proportional to the particle density, along a diameter of a ^{87}Rb cloud for different temperatures in the trap 26 ms after switching off the trap [Ernst et al. (1998)]. (a) shows the Gaussian distribution of an uncondensed atom cloud at a temperature of 1.2 µK. With decreasing temperature, a sharp peak in the center of a broader distribution occurs, and the distribution of the uncondensed fraction becomes narrower as can be seen from (b) which corresponds to a temperature of 310 nK and (c) corresponding to 170 nK. Finally, (d) shows the distribution of a pure condensate of 10^5 atoms without a visible contribution of uncondensed atoms. The sharp peak observed in the velocity distribution below a certain critical temperature is clear evidence for Bose–Einstein condensation.

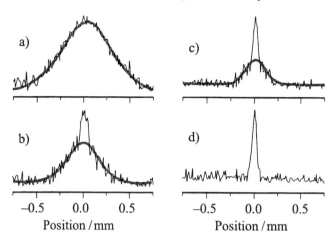

Fig. 5.14. Density distributions of an atom cloud as a function of the axial coordinate at different temperatures, 26 ms after switching off the trap [Ernst et al. (1998)]. For a detailed description see text

Another possibility to detect Bose–Einstein condensation experimentally is a direct measurement of the density of the atoms in the trap by means of dispersive light scattering [Andrews et al. (1996)].

Two-dimensional absorption images of the density distribution of such an atom cloud show that the uncondensed component expands to an isotropic cloud due to its isotropic speed distribution, independent of the initially elliptical shape of the cloud caused by the anisotropy of the trap potential. In contrast, the central peak of condensed atoms retains the anisotropy of the trap. Its ellipticity, however, reverses in time. The reason for this is the uncertainty principle: Along the more strongly confined direction of the trap, the atoms have a larger momentum distribution and thus expand faster, whereas in the direction of less spatial confinement the momentum distribution is narrower, resulting in a slower expansion of the atom cloud [Han et al. (1998)].

The observation of Bose–Einstein condensation in dilute atomic gases is undoubtedly one of the most exciting developments in atomic physics since the invention of the laser. Its study holds the promise of revealing new macroscopic quantum phenomena of coherent matter that can be understood from first principles, because the weak particle interaction renders the theory tractable and refined experimental techniques for detailed investigations are available. Such fundamental investigations may also advance our understanding of superconductivity and superfluidity in more complex systems.

However, the most fascinating future prospect is that Bose–Einstein condensation (BEC) is the first step in the production of coherent matter wave sources or "atom lasers". Various proposed principles for atom lasers include coherent coupling of atoms out of the condensate with simultaneous "pumping" of the condensate, and optical cycling of cold atoms into "lasing" atomic modes.

Recent developments are very promising. While the first atom lasers produced an output that was more like a burst of atoms than a beam, propelled out of the trap by gravity [Andrews et al. (1997)], Mewes et al. (1997) demonstrated the first output coupler for Bose-condensed atoms using short pulses of radiofrequency radiation to create Bose condensates in a superposition of trapped and untrapped hyperfine states. This configuration produced a pulsed beam of coherent atoms, but the beam could still only point downward, similar to a continuous atom beam demonstrated by Bloch et al. (1999) based on the same principle. Hagley et al. (1999), however, extracted sodium atoms from a trapped Bose–Einstein condensate by a coherent, stimulated Raman process, in which optical Raman pulses drive transitions between trapped and untrapped magnetic sublevels, giving the output-coupled BEC fraction a well defined momentum, so that arbitrary beam directions are possible. Although this atom laser is not fully continuous — pulsed Raman lasers are used — the emitted atom pulses overlap enough to form a nearly continuous beam.

Atomic beams with laser-like precision open up many applications, ranging from atomic clocks of unprecedented accuracy to atomic holography, a technique that could be used to grow three-dimensional nanostructures by interference of atom beams.

Fermi atoms (^{6}Li (I = 1) and ^{40}K (I = 4)), which are less abundant among the atoms which can be optically manipulated, have also been stored and cooled in magnetooptical traps [Abraham et al. (1997), Cataliotti et al. (1998)]. They offer the possibility to investigate the properties of an ultracold gas of fermionic atoms. At phase-space densities similar to the one achieved in BEC experiments, the behavior of a degenerate Fermi gas can be studied and, at still lower temperatures, the superfluid transition may be observed [Butts and Rokhsar (1997)].

5.3.4 Photoassociative Spectroscopy

Photoassociative spectroscopy of laser-cooled trapped atoms is another application of cold atoms which has been successfully used to investigate high-lying molecular vibrational states. The basic principle of this technique is the following: A photoassociation laser passes through an ensemble of ultracold atoms confined in a trap. If its frequency ω_p coincides with the frequency of a transition between an unbound state of a colliding pair of ground-state atoms and a bound state of an excited molecular state, an excited molecule may be formed. These newly formed diatomic molecules decay back into the dissociative continuum of the unbound collision pair (ω_F) or into a highly vibrational state of the bound electronic ground state of the molecule (ω_B) (see Fig. 5.15). Both processes result in a reduction of the number of atoms in the trap, observable by a decrease in trap-laser-induced fluorescence, which is monitored by a photodiode. By tuning the laser frequency ω_p, the photoassociation spectrum can be measured. Since the overlap between unbound states of the collision pair and the excited state has its maximum at large internuclear distances, photoassociative spectroscopy is well suited to investigate

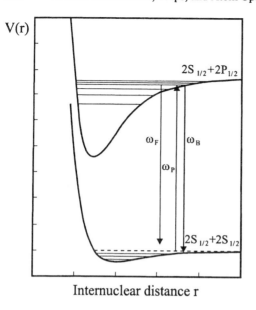

Fig. 5.15. Schematic potential diagram and transitions for photo-associative spectroscopy of alkali atoms. For a further description see text

highly excited vibrational states, which are difficult to access with conventional spectroscopic methods. Therefore, photoassociative spectroscopy can be used to obtain detailed information on the long-range interatomic potential and is thus an important complement to molecular spectroscopy. Furthermore, it can provide most accurate knowledge about many other molecular and atomic properties, such as ground-state dissociation energies and atomic radiative lifetimes [Jones et al. (1996), Wang et al. (1997a), (1997b)]. Feshbach resonances in cold atom scattering have also been observed recently using photoassociative spectroscopy [Courteille et al. (1998)]

Finally, photoassociative spectroscopy allows an accurate determination of the s-wave scattering length [Abraham et al. (1995), (1997)], an important parameter in the theoretical description of Bose–Einstein condensation [Boesten et al. (1997)]. This is due to the fact that the densities are sufficiently low in these experiments that only two-body interactions are important, while the temperature is low enough to describe the elastic interaction by a single parameter, the s-wave scattering length.

A future aspect of photoassociation may be the production of state-selected, translationally extremely cold molecules [Vardi et al. (1997), Fioretti et al. (1998), Julienne et al. (1998)].

Other recent activities are laser trapping and cooling of short-lived radioactive atoms. These atoms require improved trap-collecting techniques, because short-lived radioactive atoms are available only in limited quantities. On the other hand, trapping of short-lived radioactive isotopes is well suited for carrying out fundamental investigations of weak interaction symmetries in β decay and to measure the isotopic dependence of parity nonconservation in heavy atoms [Simsarian et al. (1996), Lu et al. (1997), Guckert et al. (1998)].

5.3.5 Atom Lithography

Present-day production of microelectronic circuits is dominated by optical lithography, since this parallel process allows large throughput in large-scale production applications [Taniguchi (1996)]. However, below 100 nm a limit is reached in principle, because there are neither suitable light sources nor suitable optical imaging components. Electron and ion lithography represent alternatives which have allowed the production of structures in the nanometer range for some time already. However, these serial processes are much less conducive to mass production. Thus, they are typically used for special applications, e.g. masks for optical lithography mass production. The search for further alternatives is therefore of utmost importance.

Progress achieved in the manipulation of neutral atoms has led to successful attempts to create nanostructures on surfaces with a resolution well below optical wavelengths [Timp et al. (1992), McClelland et al. (1993), Gupta et al. (1995), Berggren et al. (1995) (1997), Celotta et al. (1996), Thywissen et al. (1997)]. Two different procedures for "atom lithography" are under investigation: An atomic beam's intensity is modulated spatially by optical means and is used (1) for the direct deposition and growth of a structured layer or (2) for a modification of the wetting properties of the surface covered with a suitable protective layer in accordance with the beam's modulation pattern. This pattern is then transferred to the surface by chemical methods.

Direct Deposition from the Atomic Beam. Figure 5.16 explains the first-mentioned method. A laser-collimated atomic beam passes through a standing laser wave pattern which is slightly detuned out of resonance towards the blue. This causes the atoms to be focused at the nodes of the standing wave and then deposited on the substrate immediately behind, resulting in a line grating. Of the atoms which can be manipulated chromium is best suited, because it has a low mobility on the surface and is stable in air. For example, McClelland et al. (1993) made a chromium line grating on silicon with bars of 65 nm wide and 34 nm high with a

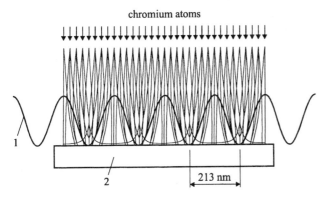

Fig. 5.16. Schematics of laser focusing of chromium atoms for deposition of a line grating. (1) standing laser wave, (2) silicon substrate

298   5. Slow Atom Beams, Traps, and Atom Optics

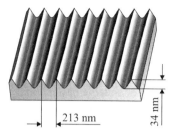

Fig. 5.17. Chromium grating from laser focused atom deposition The substrate area shown is 2×2 µm [McClelland et al. (1993)]

distance between the bars of 213 nm (see Fig. 5.17). In the meantime linewidths down to 38 nm have been reached [Celotta et al. (1996)]. Employing two superimposed standing laser waves perpendicular to each other one can also deposit regularly spaced point patterns (80 nm diameter, 13 nm high) [Gupta et al. (1995), Celotta et al. (1996), Schulze et al. (1997)].

Use of Monomolecular Protective Layers. The second-mentioned method in atom lithography starts with a 30 nm gold layer on a silicon substrate. It is covered with a 1.5 nm layer of dodekanthiol ($CH_3(CH_2)_{11}SH$) molecules, which form a hydrophobic, self-aggregated, extremely inert monolayer on the gold surface. If a molecule of the monolayer is hit by a metastable helium atom, the emitted UV photon destroys the molecule, and the gold layer is exposed. A structure thus "written" can be transferred to the gold layer below by chemical means. The structure is determined by the spatial distribution of the metastable helium atoms parallel to the surface. This can be realized by an optically induced depopulation of the metastable helium beam in accordance to a pre-established pattern [Berggren et al. (1995)]. An edge uncertainty of 30 nm has been achieved, corresponding exactly to the thickness of the gold layer [Nowak et al. (1996), (1997)]. This suggests that the resolution is limited by the etching process. The dose necessary for "writing" of the structure proved very low with 0.3 He^* atoms per thiol molecule.

Cs atoms have also been used for an analogous process. They modify the netting capability of the monomolecular protective layer (nonanthiol $CH_3(CH_2)_8SH$) on top of the gold layer, so that a wet-etching agent can dissolve the gold layer underneath at those places hit by Cs atoms [Kreis et al. (1996), Berggren et al. (1997)]. The edge uncertainty is larger than that of the metastable helium experiment and lies between 40 and 50 nm, but clearly below 100 nm. With 5 to 9 cesium atoms per thiol molecule the dose for "writing" a structure is significantly higher than for the metastable helium. However, cesium beams of high intensity are easily produced.

5.4 Atom Optics

Atom optics is a rapidly developing field of atom physics, in which atoms in atomic beams, in complete analogy to optics, are manipulated like photons in light beams [Pritchard (1991), Adams (1994), Müller et al. (1995), Baldwin (1996), Aspect (1997)]. Correspondingly, one attempts to transfer the possibilities which are provided by optical techniques to influence light beams to atomic beams. The optical techniques are based on a series of optical elements such as lenses, prisms, mirrors, gratings, and others, which can be combined to more complex atom-optical systems.

5.4.1 Atom-Optical Elements

Of this analogy, the conventional atomic beam technique was able only to utilize the straight-lined propagation of the beams, using slits or holes to form a beam with a given angle of divergence from the atoms emanating from a source. Thus, in the language of optics, the atomic beam technique remained in the state of a pinhole camera for many years. At the beginning of the 1950s, the first atom-optical elements were introduced into molecular beam techniques with magnetic and electric multipole fields, which act as lenses for atoms and molecules. However, their index of refraction is very close to unity for thermal energy molecular beams (see Sect 4.2.5), and they have been the only optical elements in molecular beam techniques for many years.

This situation changed since laser methods for atom manipulation were developed [Balykin and Letokhov (1989), Pritchard (1991)]. Progress in microstructure techniques, which allow the fabrication of slits, wires, transmission gratings, and zone plates with dimensions in the 100 nm range [Yen et al. (1992), Pritchard (1993), Savas et al. (1995), Rooks et al. (1995)] have contributed to this development. Using these atom-optical elements, the basic experiments of optics, diffraction from a slit, from a double slit, and from a wire could be performed during the last decade with atomic beams of thermal energy [Carnal and Mlynek (1991b), Miret-Artés et al. (1998)].

In the following, we will first review the elements of atom optics available at present together with some examples of their application. Then, as an example for a more complex atom-optical system combined from several elements, we shall discuss an atom interferometer and some of its applications.

Table 5.2 presents a comparison of optical elements for photons and atoms, which are described in the following sections. Only a few of these elements can be applied to arbitrary atoms and molecules. These are, in particular, the components

Table 5.2. Comparison of optical elements for photons and atoms

Photons	Atoms
Lens	Multipole field, zone plate
Mirror	Crystal surface, liquid He, evanescent light wave
Phase grating	Standing light wave
Amplitude grating	Nanofabricated grating
Prism	Light, magnetic two-pole
Beamsplitter	Grating

which are based on nanofabricated structures, such as nanofabricated gratings and zone plates. Other components can only be used for atoms which can be manipulated by laser light.

5.4.2 Lenses

Magnetic and Electrostatic Multipole Fields. Magnetic and electric multipole fields can be used as atom lenses, provided that the particles have a magnetic or electric dipole moment. This is described in detail in Chap. 4. The radial velocity v_r, which can be focused for a given axial velocity v_0, follows immediately for a magnetic hexapole field from Sect. 4.2.5

$$v_r/v = \sqrt{\mu H_0/E}, \tag{5.37}$$

where H_0 is the magnetic field at the pole tips, μ is the magnetic moment, and E the kinetic energy of the atoms. For field strengths which may be generated easily in the laboratory and a magnetic moment of the order of 1 Bohr magneton, the ratio (5.37) is of the order of 10^{-2}, if thermal energy beams are used. The focusable solid angle $\Delta\Omega$ is given by

$$\Delta\Omega = \pi \frac{\mu H_0}{E}. \tag{5.38}$$

Table 5.3. Focal length f and focusable solid angle $\Delta\Omega$ for two magnetic hexapole fields at different translational energies

Beam and field data	Thermal beam	Slow beam
Magnetic moment μ [Bohr magneton]	1	1
Field strength H_0 [Oe]	7000	7000
Field radius r_0 [cm]	0.5	0.3
Field length L [cm]	20	0.5
Translational temperature [K]	1000	1.5
Focal length [cm]	30.2	0.67
Solid angle $\Delta\Omega$ [steradian]	1.5×10^{-3}	0.99

For thermal energy beams, this solid angle is about 10^{-3}. Focal lengths are of the order of 20 to 50 cm. For slow atom beams, the situation is quite different. This is illustrated by Table 5.3, which shows characteristic properties of hexapole fields for beams of different translational temperatures.

Fresnel Zone Plates: Using nanofabrication technology, transmission zone plates of nanometer dimensions can be produced [Ito et al. (1998)], which are also commercially available. Their main field of application is x-ray microscopy. They can also be used for molecular beams of thermal energy, as was first demonstrated with beams of Na and metastable He atoms [Tennant et al. (1990), Carnal et al. (1991a), Keith and Rooks (1991), Ekstrom et al. (1992)]. The best performance so far, however, has been achieved recently with ground-state ^{4}He beams using tiny nozzles (5 μm ∅) and miniature skimmers with hole diameters down to 1 μm (see Chap. I.4) [Doak et al. (1999)].

Typically, these zone plates consist of a thin membrane of silicon or silicon nitride with appropriately spaced rings etched away (see Fig. 5.18).

Simply speaking, a zone plate is a circular grating with a grating constant which decreases with increasing distance from the center in a suitable way, so that for each order +s a focus at a distance f_s from the zone plate is obtained [Hecht (1987)]. Additionally, each order has a conjugated focus at the same distance f_s behind the zone plate (–s), from which the divergently diffracted rays seem to emanate. In practice, the alternating series of transparent and opaque concentric rings of a zone plate are supported by small radiating bars (see Fig. 5.18), which have negligible influence on the imaging properties except that they reduce the particle flux by about 10%. The radius of the rings increases with the square root of their order number n, if the numbering starts at the disk center with n = 0. Consequently, the transmission is given by

$$T = 0 \quad \text{for} \quad r_{2n} \leq \rho \leq r_{2n+1}, \quad T = 1 \quad \text{for} \quad r_{2n+1} \leq \rho \leq r_{2n+2}, \tag{5.39}$$

with $r_n = n^{1/2} r_1$ (n = 0, 1, 2 ⋯ n_{max}–1). r_1 is the radius of the innermost opaque zone, $2n_{max}$ is the total number of zones, and ρ the distance from the zone plate center. The distance a of an object in front of the zone plate is related to the distance b of its image behind the zone plate by the lens equation

$$\frac{1}{a} + \frac{1}{b} = \frac{1}{f_s}, \tag{5.40}$$

where the focal length f_s is determined by

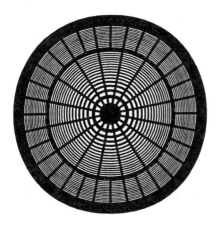

Fig. 5.18. Schematic view of a microzone plate made from gold (0.21 mm ∅, 0.5 μm thick). The radial struts support the grating structure [Carnal et al. (1991a)]

$$f_s = \frac{r_1^2}{s\lambda} = \frac{r_1^2 mv}{sh}. \tag{5.41}$$

λ is the de Broglie wavelength, m the particle mass, v the particle speed and h is Planck's constant. Accordingly, the focal length depends both on the velocity of the particles and on the order s of diffraction. The fraction ϕ_s of the particle flux ϕ diffracted into the s-th order may be written as

$$\frac{\phi_s}{\phi} = \frac{1}{4}\left[\frac{\sin(s\pi/2)}{s\pi/2}\right]^2. \tag{5.42}$$

The intensity of the even orders vanishes, while the intensity of the odd orders decreases proportional to $1/s^2$ with increasing s. Therefore, first order diffraction dominates, and the zone plate thus separates the transmitted beam into three main channels, the focused (s = +1) channel, the undeflected channel (s = 0), and the "defocused" channel (s = −1).

Due to the finite aperture of the zone plate, the image of a point-shaped source is limited to a diameter D_s, given by the outermost open zone width Δr_{min}

$$D_s = r_1/s\sqrt{8n_{max}} \approx \Delta r_{min}/s. \tag{5.43}$$

Apart from the spherical lenses discussed here, cylindrical lenses have also been used in the first molecular beam experiments with zone plates [Keith and Rooks (1991), Ekstrom et al. (1992)].

The microzone plates used by Doak et al. (1999) have the following data: Total diameter 0.27 mm, r_1 = 5.2 µm, Δr_{min} = 100 nm (zone plates, in which the finest rings are 50 nm wide, are also possible [Rehbein et al. (1999)]). With these zone plates, focusing of thermal energy ^{4}He beams was impressively demonstrated: At only moderate demagnification (factor 0.4), a focused spot diameter of ≤ 2 µm has been achieved, while without a zone plate the atoms would illuminate an area of about 400 µm diameter. This follows immediately from the background due to the unfocused zeroth order contribution of the beam. Thus, an improvement over previous work by a factor of 10 in resolution, 10^3 in intensity, and 10^8 in focused beam density has been achieved. The background due to both the zeroth order and higher orders can be reduced by a central obstacle behind the zone plate.

A zone plate may also be used as a velocity filter for molecular beams. If a narrow aperture is located at an appropriate distance behind the zone plate, only particles of a given velocity interval Δv around a given nominal velocity v_0 can pass. This is completely analogous to a magnetic hexapole field. The zeroth order has to be eliminated by a central beam stop. The higher diffraction orders cause sidebands with transmitted intensities which decrease strongly with increasing order. In contrast to a magnetic hexapole field, where the velocity passing through the sidebands is smaller than the nominal velocity, the velocities transmitted by the sidebands of a zone plate are higher than the nominal velocity. They are given by sv_0 (with s = 3, 5, 7, ...), as follows immediately from the focal length (5.41).

Focusing by the Dipole Force. As has already been mentioned in Sect. 5.1.2, the dipole force can also be used to exert a focusing force on atoms and molecules [Miller (1993), Stapelfeldt et al. (1997), Sakai et al. (1998)]. A far-off resonance radiation is typically used, so that the force is independent of the special level structure of the molecules. This method has recently been demonstrated successfully with CS_2 and I_2 nozzle beams, using intense pulsed Nd:YAG lasers (λ = 1.06 µm) or CO_2 lasers (10.6 µm).

5.4.3 Mirrors

Special Surfaces: Atom mirrors of several sorts have been demonstrated during recent years: Single gold crystals, for example, can specularly reflect rare gas atoms, [Doak (1989)], while for cold, spin-polarized hydrogen atoms suitably curved surfaces covered with a film of superfluid ^{4}He are used as mirrors [Berkhout et al. (1989), Berkhout and Walraven (1993), Luppov et al. (1993)]. The curvature of the surface allows an additional focusing of the particles. The reason for specular reflection from these surfaces is the very weak atom–surface interaction, which causes a low sticking probability for atoms having a small momentum component normal to the surface.

Another possibility for specular reflection and focusing of molecular beams is offered by appropriately prepared silicon surfaces, as has been shown recently. Holst and Allison (1997) used a thin slice (50 µm thick, 18 mm ∅) cut from crystalline silicon, Si(111), to produce a concave mirror in order to focus a thermal energy He beam. This method is based on a treatment of the silicon surface with a solution of NH_4F at room temperature, resulting in an extremely smooth surface of silicon monohydride (Si(111)-(1×1)H) with an average roughness smaller than 0.07 Å [Higashi et al. (1991)]. These surfaces reflect helium atoms specularly and their reflectivity is maintained for months if the surfaces are kept under high vacuum conditions. The thin crystal wafer can be electrostatically deformed into an approximately parabolic mirror, whose focal length can be changed by changing the deforming electric field. Detailed calculations of the focal properties and aberration coefficients can be found in the literature [Wilson et al. (1999)].

Magnetic Mirrors. Magnetic mirrors for neutral particles were already discussed many years ago by Vladimirskiĭ (1961) in connection with cold neutrons. He pointed out that a periodic array of elements of alternating magnetic polarity (such as a planar array of parallel wires alternately carrying electric current in opposite directions (see Sect. 4.1.3)) generates a magnetic field, whose absolute value B depends on the distance z from the surface as

$$B \propto e^{-kz} \quad \text{with} \quad k = 2\pi/\lambda. \tag{5.44}$$

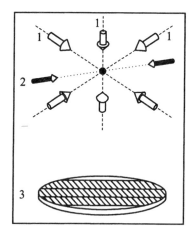 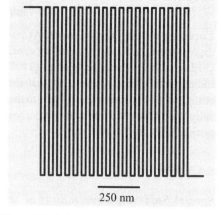

Fig. 5.19. Magnetic mirror experiment. (1) trap lasers, (2) pump laser, (3) magnetic mirror

Fig. 5.20. Scheme of a serpentine array of current-carrying wires. For a description see text [Johnson et al. (1998)]

λ is the spatial period. A neutron, with its magnetic dipole moment antiparallel to the field and approaching the field adiabatically, is repelled from such an array. This proposal was rediscovered in the 1990s [Opat et al. (1992)] and realized by writing sinusoidal signals on magnetic storage material such as strips of audio-tape or a floppy disk [Roach et al. (1995), Hughes et al. (1997a), (1997b)], by constructing an assembly of permanent rare-earth magnets of alternating polarity [Sidorov et al. (1996)], and by microfabricated arrays of current-carrying wires (see Fig. 5.20) [Johnson et al. (1998), Lau et al. (1999)].

Sinusoidal signals with a period smaller than 1 µm and field strengths larger than 0.1 T can be written on magnetic storage material, yielding a strong, short-range surface potential. These mirrors may also be deformed to concave mirrors.

The first realization of a magnetic mirror is schematically shown in Fig. 5.19 [Roach et al. (1995)]. Strips of an audio tape with a sinusoidal magnetization (λ = 9.5 µm) are glued to a glass substrate. A magnetooptical trap for Rb atoms which is loaded from the vapor phase is located 24.5 mm above the mirror. When the trap is switched off, the Rb atoms fall freely out of the trap and are optically pumped by an additional laser (2) to increase the population of the positive M_F sublevels of the F = 3 ground state. The orientation of the atoms is maintained by a weak magnetic field in the region between the trap and the mirror. At the mirror surface, the atoms are reflected and return to the trap, where they are collected again and detected by their fluorescence. From these measurements, the reflectivity of the mirror has been determined to be 94%.

Microfabricated arrays of current-carrying wires (electromagnetic mirrors) (see Fig. 5.20) have advantages over mirrors based on permanent magnets [Drndic et al. (1998), Johnson et al. (1998), Lau et al. (1999)]. The magnetic field may be readily varied, switched or modulated by altering the current in the wires. The counterwound geometry has an inherently small inductance which results in a fast time constant (≤ 100 ns). In practice, such devices are fabricated by photo-lithography, followed by chemical electrodeposition of the conducting metal (Cu, Au, Ag, or Nb), and subsequent annealing. The sapphire substrate is cooled to cryogenic temperatures to reduce electrical resistivity and increase thermal conductivity and to maximize the current capacity of the device.

Spatial periods down to 6 µm and field strength of about 0.1 T are possible, comparable to the values achievable with magnetic storage material.

Light Mirrors. A plane wave passing from an optical medium with an index of refraction n into vacuum with an angle of incidence γ with respect to the normal of the dielectric–vacuum interface, is refracted such that the outgoing angle γ" is connected to the angle of incidence γ by Snell's law of refraction

$$\sin \gamma'' = n \sin \gamma. \tag{5.45}$$

This relationship can only be satisfied by a real angle γ" as long as the condition

$$\sin \gamma \leq 1/n = \sin \gamma_0 \tag{5.46}$$

is fulfilled. If the angle of incidence is larger than a given limiting angle γ_0, the light is totally reflected. The expression for the refracted wave can formally still be used if complex angles γ" are permitted. In this case we obtain from (5.45)

$$\cos \gamma'' = \sqrt{1 - \sin^2 \gamma''} = \sqrt{1 - n^2 \sin^2 \gamma} = i\sqrt{n^2 \sin^2 \gamma - 1}, \tag{5.47}$$

and the electric field **E** at the boundary of the refracted wave

$$\mathbf{E} = \mathbf{E}_0 \exp(i(\mathbf{kr} - \omega t)) = \mathbf{E}_0 \exp(i(xk_0 \sin \gamma'' + zk_0 \cos \gamma'' - \omega t)) \tag{5.48}$$

can be written in the form

$$\mathbf{E} = \mathbf{E}_0 \exp\left(-zk_0 \sqrt{n^2 \sin^2 \gamma - 1}\right) \exp(i(xnk_0 \sin \gamma - \omega t)), \tag{5.49}$$

if the coordinate system is chosen in such a manner that the plane of incidence is the x,z plane and the boundary plane is the x,y plane. $\mathbf{k}_0$ is the wave vector (in vacuum) and k_0 its absolute value. A corresponding expression is obtained for the magnetic field strength of the wave. Thus, at total reflection, a wave is observed in the optically thinner medium, which propagates along the boundary plane in the x direction, having an amplitude which decreases exponentially with the distance z from the boundary plane ("evanescent wave"). After the distance

$$z_0 = \left[k_0 \sqrt{n^2 \sin^2 \gamma - 1}\right]^{-1} \tag{5.50}$$

the amplitude of the evanescent wave is reduced by the factor 1/e. With (5.50) and the abbreviation $q = nk_0 \sin\gamma$ for the x component of the wave vector, the electric field strength of the evanescent wave can be written as

$$\mathbf{E} = \mathbf{E_0} \exp(-z/z_0) \exp(i(qx - \omega t)). \tag{5.51}$$

Since this evanescent wave has a strong gradient in the z direction, its field can be used as an atom mirror [Cook and Hill (1982), Balykin et al. (1988), Esslinger et al. (1993)]. As is pointed out in Sect. 5.1.2, the force exerted on an atom by a light wave is proportional to the intensity gradient and the frequency detuning Δ.

The latter also determines the sign of the force. To reflect the atoms off the surface, the light of the totally reflected wave must be blue-detuned with respect to the resonance frequency of the atom. To avoid deviations from specular reflection, the spontaneous force has to be small compared to the dipole force. This requires a large frequency detuning Δ (compared with the natural width of the levels). To be reflected from the mirror surface, the normal velocity component of the atoms has to be lower than the height of the potential barrier, which is given by the frequency detuning Δ and the interface. For sodium atoms, this limiting speed component is of the order of 500 cm/s. Accordingly, thermal energy atomic beams can only be reflected at very small glancing angles $\alpha = \pi/2 - \gamma$ ($\alpha < 7 \times 10^{-3}$ rad for Na).

Although the reflection takes place at some distance from the surface, the demands upon the planeness of the surface are rather high. If the surface roughness becomes comparable to the wavelength of the incoming atoms, diffuse reflection dominates [Henkel et al. (1995), (1997)]. Figure 5.21 shows a possibility to realize an evanescent wave mirror.

When calculating the repulsive potential of an evanescent wave mirror, the attractive van der Waals potential of the surface has to be taken into account, as has been shown by Landragin et al. (1996). They measured the reflecting power of an evanescent wave mirror using a "trampoline configuration" in order to determine the van der Waals force experimentally.

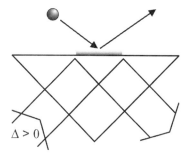

Fig. 5.21. Atom mirror. The atom reflects off the strong field gradient on the vacuum side of the interface, due to the evanescent field of the totally reflected laser beam

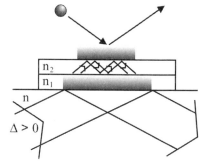

Fig. 5.22. Enhancement of the evanescent wave by dielectric layers [Dowling and Gea-Banacloche (1996)]

The intensity of the evanescent-wave field can be substantially enhanced by coating the vacuum–dielectric interface with two additional thin-film dielectric layers [Kaiser et al. (1994), Seifert et al. (1994a), Ke et al. (1997)] (see Fig. 5.22). The first dielectric layer has an index of reflection $n_1 < n$, the second one the index $n_2 > n_1$. Since the top layer (index n_2) is bounded above by vacuum (index 1) and below by a medium of index $n_1 < n_2$, it acts as an infinite planar dielectric slab waveguide. In the original dielectric of index n total reflection still occurs, since $n > n_1$. Radiation from the evanescent wave tunnels through the first layer of index n_1 and is trapped in the second waveguide layer of index n_2, where it will undergo many total internal reflections between its boundaries. Thus, a field much larger than the original field can build up in this layer, yielding a considerable enhancement of the intensity and its gradient at the vacuum–dielectric interface.

Another possibility to enhance the field strength at the dielectric–vacuum interface utilizes surface plasmons. These are electromagnetic charge-density waves propagating along a metallic surface [e.g. Raether (1988)]. In this case, the boundary of the dielectric is coated with a thin metallic layer. With appropriate polarization, wavelength, and angle of incidence, the evanescent wave produced by total reflection can excite surface plasmons, so that a large fraction of the incoming light is transferred to the plasmon wave, resulting in a correspondingly large evanescent wave on the vacuum side of the metal layer [Esslinger et al. (1993), Feron et al. (1993), Seifert et al. (1994b)].

Moving Mirrors. While a mirror which is fixed in space can only reverse the sign of the momentum of a reflected particle, a moving mirror may also transfer momentum to the incident particles and thus influence their trajectories in a desired way. This is what a tennis player usually does with his racket. If in the simplest case, for example, a mirror moves with the constant speed v_S, a particle arriving with the speed v_a has the speed $v_r = v_a + 2v_S$ after reflection. If more complex movements of a mirror can be realized, a moving mirror is a universal atom-optical instrument to manipulate atoms [Arndt et al. (1996)].

To move mirrors mechanically with the necessary velocity is practically impossible, if more than a constant velocity is required. The latter can be done by a rotating device, as has recently been proposed to produce slow helium atoms [Doak et al. (1997)]. However, in the case of a light mirror, a movement of the mirror surface is equivalent to a temporal modulation of the amplitude of the

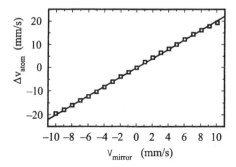

Fig. 5.23. Velocity change Δv caused by a "moving" mirror as a function of the equivalent mirror speed [Arndt et al. (1996)]

308   5. Slow Atom Beams, Traps, and Atom Optics

evanescent wave and thus easily realizable. In the first experiments with slow cesium atoms from a magnetooptical trap the usefulness of this method has been demonstrated.

An example is presented in Fig. 5.23, which shows the velocity change Δv of Cs atoms due to the reflection from a modulated light mirror as a function of the equivalent mirror velocity.

5.4.4 Atom Waveguides

The specular reflection of atoms in the field of an evanescent wave can also be used to produce waveguides for atoms. Such a waveguide consists essentially of a dielectric hollow fiber, which may be cladded with a metallic layer at its outside, as is shown schematically in Fig. 5.24.

The light of a blue-detuned laser is coupled into the ring-shaped annular core region of the hollow fiber and propagates practically without any attenuation by total internal reflection at the boundaries between the hollow inner space and the cladding (or vacuum, if no cladding is used). Thus an evanescent wave at the wall of the inner hollow cylinder is generated. Atoms which move through the hollow space, are specularly reflected from the wall and thus guided through the hollow fiber. This principle, first proposed by Savage et al. (1993), has been elaborated in detail in further investigations [Marksteiner et al. (1994), Ito et al. (1995)], and finally experimentally realized [Renn et al. (1996), Ito et al. (1996), (1997)].

This technique has the advantage that the atoms are guided through a region free of light except for the short times of reflection. Thus, using a sufficient detuning, the influence of the spontaneous force is minimized. Furthermore, the thin repulsive potential of the evanescent field at the wall of the hollow fiber can be approximated by an infinite potential step, which simplifies the quantum mechanical treatment of the motion of very cold atoms through the waveguide. If the de Broglie wavelength becomes comparable to the diameter of the waveguide, the description is completely analogous to that of a (scalar) microwave propagating through a cylindrical metal waveguide [Dowling and Gea-Banacloche (1996)].

Another possibility to build a waveguide for atoms was first proposed by Ol'Shanii et al. (1993) and experimentally realized a few years later [Renn et al. (1995)]. It is based on the fact that a hollow glass fiber (without cladding) can support electromagnetic modes of a light wave whose field intensity has its maximum

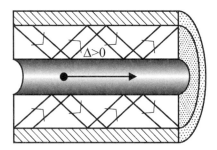

Fig. 5.24. Hollow fiber with metal cladding as waveguide for atoms [Dowling and Gea-Banacloche (1996)]

along the axis of the fiber hollow. If the light is red-detuned with respect to the atomic resonance, the atoms traveling through the hollow fiber experience a force which is directed to the fiber axis and keeps them away from the fiber wall.

This scheme has the same disadvantages as any device for optical manipulation of atoms based on red-detuned light: For the whole time of their passage the atoms are localized in a region of high field strength, and thus undergo many spontaneous excitation and emission processes. These lead to heating and diffusion which may result both in a loss of atoms to the wall and a loss of coherence.

Instead of guiding the atoms in the evanescent waves developed inside hollow optical fibers, the optical dipole potential of a blue-detuned hollow laser beam may also be used [Yin et al. (1998), Xu et al. (1999)].

Another guiding of cold neutral atoms based on magnetic trapping by a thin current-carrying wire, has been demonstrated recently [Denschlag et al. (1999)]. Atoms in high-field-seeking states can be trapped by the field of the wire and move in Kepler-like orbits around the wire ("Kepler guide"). This follows immediately from the wire field resulting in a force $F \propto 1/r^2$ (see Sect. 4.1.1). If the atoms also have momentum parallel to the wire, they will spiral along it, guided by the magnetic field. Combining the field of the wire with a homogeneous magnetic bias field B_b perpendicular to the wire, the circular magnetic field of the wire is exactly canceled along a line parallel to the wire at a distance $r = (\mu_0/2\pi)(I/B_b)$. Around this line the magnetic field increases in all directions and forms a tube with a magnetic field minimum on its axis. Atoms in the low-field-seeking state can be trapped in this tube and guided along the wire ("side guide"). The same principle has been used to build a miniaturized magnetic trap [Fortagh et al. (1998)]. In this case, the field of the wire is combined with a spherical magnetic quadrupole field, and atoms which are first trapped in the quadrupole field can be gradually transferred to the wire trap by changing the field currents in an appropriate way. Instead of a current-carrying wire, a ferromagnetic wire can also be used for guiding and trapping of neutral atoms [Tkachuk (1999)].

5.4.5 Diffraction Gratings

Free Standing Nanofabricated Transmission Gratings. The angular distribution of the intensity I of a monochromatic plane wave (wavelength λ) diffracted from a transmission grating with a grating constant d is, according to Kirchhoff's theory of diffraction, given by

$$I = \frac{\sigma^2 d^2 L}{\lambda} \left[\frac{\sin(\pi a \sigma d/\lambda)}{\pi a \sigma d/\lambda} \right]^2 \left[\frac{\sin(Z \pi d a/\lambda)}{Z \sin(\pi d a/\lambda)} \right]^2 = \frac{\sigma^2 d^2 L}{\lambda} F_1 F_2, \quad \text{with} \tag{5.52}$$

$$a = \sin\theta - \sin\theta_0. \tag{5.53}$$

θ_0 and θ are the angles of incidence and emergence (with respect to the normal of the grating), respectively, L is the length of the grating slits, Z is the number of illuminated grating slits, and σ is the fraction of the grating constant d which is transparent (the width σd is transparent, $(1-\sigma)d$ is opaque). The factor F_1 describes the diffraction pattern of a single grating slit. This factor determines the envelope of the intricate diffraction pattern, which is determined by F_2. If the number Z of illuminated grating slits is sufficiently high, we have $F_2 \approx 1/Z^2 \ll 1$ for all values of a, except for those values satisfying the relationship

$$da/\lambda = \pm n, \qquad (5.54)$$

where n is an integer. For these values of a we have $F_2 = 1$. n is called the order of the diffraction image. Inserting (5.54) into (5.52) yields

$$F_1 = \frac{\sin(n\pi\sigma)}{n\pi\sigma}. \qquad (5.55)$$

It is immediately obvious that, with a given choice of σ, certain orders in the diffraction spectrum can be suppressed or enhanced. In the case, for example, of $\sigma = 0.5$, the even orders do not occur in the spectrum, since they coincide with the zeros of the enveloping slit function F_1. For $\sigma = 1/k$ (where k is an integer), the orders $n = k, 2k, 3k \cdots$ do not occur in the spectrum.

For normal incidence ($\theta_0 = 0$), the angle θ, at which the main maximum of the n-th order is observed, is, according to (5.54), given by

$$\sin\theta = n\lambda/d. \qquad (5.56)$$

Located between two main maxima are $Z - 2$ secondary maxima of low intensity (because $I \propto 1/Z^2$) and $Z - 1$ zeros (minima); the corresponding angles obey the relationship

$$\sin\theta = n\lambda/Zd. \qquad (5.57)$$

The resolving power of the grating $\lambda/\delta\lambda$ is determined by the number Z of illuminated grating slits:

$$\lambda/\delta\lambda = nZ. \qquad (5.58)$$

Introducing the de Broglie wavelength

$$\lambda = h/mv \qquad (5.59)$$

into (5.56), the relationship between diffraction angle and particle velocity follows as

$$\theta = \frac{nh}{mvd} = n\frac{3.9614}{dMv} \text{ mrad}, \qquad (5.60)$$

where we have replaced $\sin\theta$ by its argument θ, since the angles are very small for thermal energy beams. In the numerical part of (5.60) the grating constant d is measured in cm and the velocity v in cm/s. M is the molecular weight. Hence, the angle of diffraction is inversely proportional to the particle velocity.

The smallest grating constants which have been used so far for molecular beam diffraction experiments are 1000 Å [Chapman et al. (1995), Savas et al. (1995)]; hence, the diffraction angles (5.58) for thermal energy beams are correspondingly small, especially for heavy particles. Thus, diffraction experiments require a high angular resolution of the apparatus. After the first diffraction experiments [Keith et al. (1988), Carnal et al. (1991b)], the experimental technique was well developed and many new results have been obtained from investigations with free-standing transmission gratings, as is shown by the following examples.

In various experiments, highly resolved diffraction patterns (using free-standing transmission gratings with a 200 nm grating constant) have been observed with He, D_2, and H_2 nozzle beams, partly up to the 12-th order. These experiments demonstrated impressively the narrow velocity distribution of highly expanded nozzle beams and the high temporal coherence. Figure 5.25 shows an example. Moreover, it has been possible to resolve the diffraction peaks of small clusters, which are observed at angles

$$\theta = n\frac{h}{Ndmv} = n\frac{\lambda}{Nd}, \qquad (5.61)$$

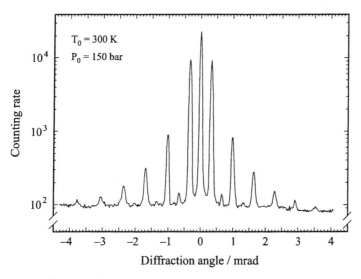

Fig. 5.25. Diffraction pattern of a He beam (grating constant 200 nm and $\sigma = 1/2$). Thus, the even orders are suppressed

where N is the number of cluster constituents. In this way it has been possible for the first time to experimentally demonstrate the existence of the extremely weakly bound He_2 van der Waals molecule [Schöllkopf and Toennies (1994), (1996)].

An example is shown in Fig. 5.26. A helium beam, expanded from a stagnation temperature of 5.2 K, with a de Broglie wavelength of about 6.4 Å for the atoms, has been diffracted from a 200 nm grating. Helium dimers and trimers are clearly resolved. Using 100 nm gratings, larger clusters (for example $(H_2)_N$ with N up to 29) have also been resolved [Schöllkopf and Toennies (1997)]. Compared to other techniques (such as collisional methods for cluster size selection, see Chap. 2) diffraction has the advantage of being nondestructive, which is very important for investigations of very weakly bound clusters (in particular helium dimers). The diffraction of heavier particles such as Ne, Ar, Kr, and Na atoms and Na_2 molecules has also been investigated (with grating constants of 100 nm) [Schöllkopf and Toennies (1997), Chapman et al. (1995)]. The most massive and complex object for which diffraction has been observed so far is the C_{60} molecule [Arndt et al. (1999)].

The nondestructive nature and the influence of the finite size of molecules upon diffraction from a transmission grating has recently been studied theoretically for the He_2 dimer and compared with the atomic diffraction [Hegerfeldt and Köhler (1998)]. He_2 is the largest diatomic molecule (its bond length is of the order of 60 Å) and it has an extremely low binding energy E (E/k ≈ -1.3 mK). While for the available nanofabricated gratings (grating constant 100 nm) atomic and molecular diffraction are practically identical, for smaller grating constants (50 and 25 nm) considerable differences are predicted. The intensity of the diffraction peaks is reduced due to the onset of molecular dissociation and in the even orders (for $\sigma = 0.5$, see (5.53)) an increasing intensity (with decreasing grating constant) is expected due to the finite size of the molecules, which acts like a diminishing of the slit width.

The latter effect has also been experimentally observed in the diffraction pattern of heavy rare gas atoms. The measurements shown in Fig. 5.25 have been performed with a grating having equal widths of slits and bars ($\sigma = 1/2$ within the accuracy of fabrication). Hence, the even orders are vastly suppressed. Diffraction patterns for Ne, Ar, and Kr beams using the same grating show that with increasing atom mass (increasing atom polarizability) the suppression of the even order peaks is more and more canceled. This empirical finding may be explained by the long-range van der Waals interaction between the atoms and the grating bars, which also acts like a reduction of σ. The relative intensities of the diffraction peaks can also be used to determine the long-range van der Waals interaction of the particles with the walls of the grating bars [Grisenti et al. (1999)].

The demands on the angular resolution of a diffraction apparatus are considerably reduced if slowed atoms are used. These beams offer a large spectrum of wavelengths, and grating constants in the μm range can be used to study diffraction phenomena. As an example, we mention here the diffraction of slow, metastable Ne atoms from a double slit (distance between the slits 6 μm) [Shimizu et al. (1992)]. The slow atoms are produced by free fall from a magnetooptical trap, hit

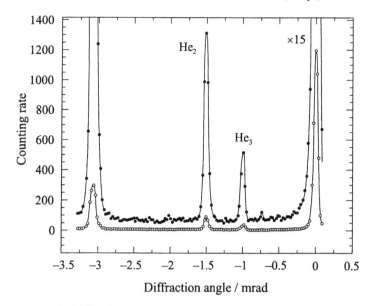

Fig. 5.26. Diffraction pattern of a low-temperature He beam (stagnation temperature $T = 5.2$ K). Between the zeroth and first diffraction peaks of the atoms the diffraction peaks of He dimers and trimers are clearly resolved

the double slit at a distance of 76 mm, and are detected by a microchannel plate 113 mm remote from the double slit. The de Broglie wavelength is of the order of 100 Å, resulting in a distance between successive diffraction orders of about 200 μm in the plane of the detector. The experiment is of fundamental importance, since it may be used to test the equality of the gravitational and the inertial mass. The spacing of the interference stripes is determined by the de Broglie wavelength and thus by the inertial mass, whereas the velocity of the particles gained in the free fall is determined by the gravitational mass.

Free-standing, microfabricated gratings have the advantage of working for any atom or molecule independent of its internal quantum state. On the other hand, they absorb (or backscatter) atoms which hit the grating bars. Thus, their total transmission is limited to about 40% by the opacity of the grating and the requisite supporting structure. Since the phases of the transmitted atoms are not significantly altered, these gratings are amplitude gratings.

Gratings from Standing Light Waves. Another possibility to generate transmission gratings, which is based on the optical dipole force, is provided by standing waves of light [Gould et al. (1986)]. Slow atoms, incident on an intense, red-detuned standing light wave, are attracted and collimated by the antinodes of the wave. Hence, the antinodes act like slits of a transmission grating, while the nodes of the standing wave play the role of the grating bars. In the case of a blue-detuned wave, the effect of nodes and antinodes on the atoms interchanges. In any case, the standing wave makes a phase grating, which does not absorb any atoms, so that its transmission is 100% [Dalibard et al. (1987), Pritchard (1991)]. Examples of

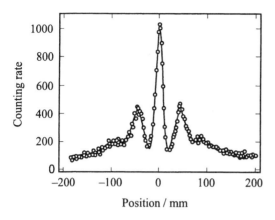

Fig. 5.27. Pattern of a metastable Ar beam diffracted from a resonant standing light wave [Abfalterer et al. (1997)]. The resolution is determined by the thermal speed distribution of the beam

applications of standing light wave transmission gratings are interference experiments, beam splitters, and atom lithography.

For metastable particles, an additional technique to generate optical transmission gratings exists. It is based on the quench effect caused by the standing wave and has been demonstrated for metastable Ar atoms by Abfalterer et al. (1997) (see Fig. 5.27): The metastable argon atoms ($1s_5$ state) cross a standing laser wave tuned to the transition $1s_5 \rightarrow 2p_8$ (801.7 nm). Only those atoms which hit the nodes of the standing wave (where the light intensity is zero) can pass through the "light grating" as metastable atoms. All other atoms are quenched by the resonant laser light into the $2p_8$ state, which decays spontaneously into the ground state. Hence, the standing wave acts as an amplitude grating. Since the detector (channeltron) is only sensitive to metastable atoms, the diffraction pattern of the metastable beam can be observed in the detector plane (1 m remote from the standing light wave). Figure 5.27 shows the diffraction pattern as a function of the detector position (detector width 10 μm). The thermal speed distribution of the metastable beam, which is produced in a gas discharge source, is responsible for the low resolution of the diffraction pattern.

Reflection Gratings by Light. The mirror action of an evanescent wave and the diffractive action of a standing wave can be combined to build a reflection grating, realized by a standing evanescent wave. This is achieved by reflecting the laser beam, which generates the evanescent wave by total internal reflection, by a mirror, as is shown in Fig. 5.28. [Hajnal and Opat (1989), Hajnal et al. (1989), Brouri et al. (1996), Deutschmann et al. (1993), Ertmer et al. (1996)]. According to (5.49), the electric field **E** of the standing evanescent wave is given by

$$\mathbf{E} = 2\mathbf{E}_0 \exp(-z/z_0)\exp(i\omega t)\cos(qx), \tag{5.62}$$

if the two superimposed waves have equal amplitudes. The periodic intensity

$$I(x) \propto E^2(x) \propto \cos^2(qx) \tag{5.63}$$

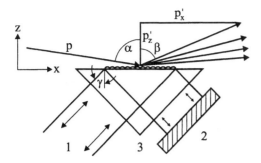

Fig. 5.28. Reflection grating with a standing evanescent wave. (1) laser beam, (2) mirror, (3) glass prism [Deutschmann et al. (1993)]

yields a reflection grating with a grating constant $d = \pi/q$. The elementary treatment of diffraction from a reflection grating yields for the angle of reflection β, at which the diffraction peak of n_0-th order is observed, the relationship

$$\sin\beta = \sin\alpha \pm n_0 \lambda/d, \qquad (5.64)$$

where λ is the wavelength and d the grating constant. α is the angle of incidence. $\sin\alpha$ and $\sin\beta$ may be expressed by the tangential components of the momentum of the incident p_x and reflected atoms p'_x and the total momentum p:

$$\sin\alpha = p_x/p; \quad \sin\beta = p'_x/p. \qquad (5.65)$$

With $\lambda = h/p$, $d = \pi/q$ the x component of the atom momentum in the n_0-th order follows from (5.64) as

$$p_x = p'_x \pm 2n_0 \hbar q. \qquad (5.66)$$

Thus, each diffraction order corresponds to a momentum transfer which is an integer multiple of $\hbar q$.

5.4.6 Prisms

Similar to a monochromatic light beam which can be deflected by a prism into any desired direction, an atom beam of uniform speed can be deflected by several techniques. One possibility is provided by the spontaneous or stimulated recoil force, as described in Sect. 5.1.6. Another method for particles with a magnetic dipole moment utilizes inhomogeneous magnetic two-pole fields, as described in detail in Sects. 3.4.1 and 4.2.2. Beam deflection, as in the optical case, can also be achieved by atom mirrors [Johnson et al. (1998)]. While the attainable deflections are small for thermal energy beams, arbitrary large deflections may be obtained with slow atoms [Nellessen et al. (1989a), Rowlands et al. (1996)]. The deflection by an inhomogeneous magnetic field and a magnetic mirror is state selective; the population of a given magnetic substate and thus the number of deflected atoms can be increased by optical pumping.

5.5 Atom Interferometry

The development of optical interferometers in the 19-th century began with the famous double slit experiment by Young (1802) and led to some of the most fundamental principles of modern physics. They are based on a simple basic principle. A plane light wave is split into two or more coherent waves which traverse different optical path lengths inside the interferometer and are then recombined by coherent superposition. An interference pattern develops caused by the phase difference between the recombined partial beams due to the differences in optical path length along their respective paths [see e.g. Tolansky (1973)].

In the area of interferometry with electron and neutron beams, this basic principle has been transferred to matter waves. The result was an interesting new field of physics [Marton et al. (1954), Mollenstedt and Duker (1956), Maier-Leibnitz and Springer (1962), Rauch et al. (1974)]. Numerous experiments to verify the foundations of quantum mechanics and particle optics have been performed with these interferometers [Rauch (1985), (1994), Klein and Werner (1993)]. They contributed decisively to gain new insights into the theoretical fundamentals and the intricacies of experimental techniques.

Microscopic atomic and molecular beam interferometry was carried out under a different name for many years in the field of intermolecular force investigations by molecular beam methods. Numerous studies of the total and differential scattering cross section were carried out, and it was realized that glory and rainbow scattering, carried out extensively in the 1960s and 1070s, are due to microscopic interferometric effects on an atomic scale [Pauly (1979), and references therein].

Recently, macroscopic interferometers have been realized with atoms and molecules [Adams et al. (1994a), (1994b), Berman (1997), Schmiedmayer et al. (1997), Batelaan et al. (1997)]. Compared to the interferometers with de Broglie waves known thus far they have several advantages. Contrary to electrons, atoms and molecules do not experience the disturbing influence of the Coulomb interaction, and contrary to a reactor as a neutron source, molecular beams of high intensity can be produced easily in rather compact devices. The complex inner structure of atoms and molecules offers additional possibilities for the investigations of their mutual interactions or their interaction with external fields.

Soon after the classical Young's double slit experiment with neutral atoms [Carnal and Mlynek (1991b)] and the first successful experiments with a Mach–Zehnder interferometer employing a transmission grating [Keith et al. (1991)], more matter wave interferometers were built and employed in a variety of experiments [Carnal and Mlynek (1991a), (1996), Kasevich and Chu (1991), (1992), Shimizu et al. (1992), Weiss et al. (1993a), (1993b), Schmiedmayer et al. (1995), Clauser and Li (1994a), (1994b), Müller et al. (1995), Weitz et al. (1996), Hinderthur et al. (1997), Gustavson et al. (1997), Peters et al. (1999)]. We restrict ourselves to briefly explain a three grating Mach–Zehnder interferometer as described in detail by Schmiedmayer et al. (1997). It utilizes microfabricated transmission gratings and has been used with thermal energy Na atomic and Na_2 molecular beams. A similar device utilizing diffraction at standing light waves acting as phase gratings, has been described by Rasel et al. (1995),

5.5.1 A Mach–Zehnder Interferometer

Figure 5.29 shows a schematic drawing of the interferometer. A Na atomic or Na_2 molecular beam is produced in the source (1) by nozzle jet expansion with argon as carrier gas. After two collimating apertures (2) the beam passes a first grating (3), serving as beam splitter. The beam is diffracted mainly into the orders −1, 0, and +1. After a distance of 60 cm a second grating follows which recombines the beams split up by the first (zeroth and first order of the first grating) by diffraction (orders +1 and −1 of the second grating) so that the two particle beams interfere with each other. Yet another grating serves as a mask and determines, by its transversal position in relation to the interference pattern, the particle flux passing through, which in turn is detected and measured by a Langmuir–Taylor detector (11) (wire of 50 µm ∅). The gratings are free-standing transmission gratings with a grating constant of 200 nm. By moving the third grating perpendicularly to the axis of the interferometer, the interference pattern can be determined. Figure 5.30 shows an example. The position of the gratings relative to each other is controlled by an optical interferometer. A similar interferometer where standing light waves are employed as gratings has been described by Batelaan et al. (1997) [Rasel et al. (1995)].

Behind the second grating a 10 cm long interaction zone (9) can be inserted, separating the two partial beams by a 10 µm thick copper foil which is suspended symmetrically between two metal electrodes. Thus, the two partial beams traverse

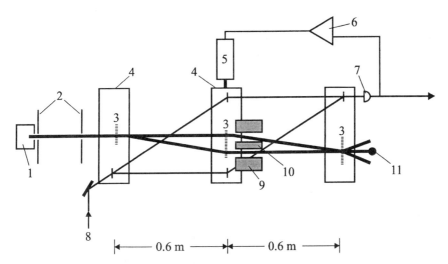

Fig. 5.29. Schematic (not to scale) of an atom interferometer [Schmiedmayer et al. (1997)]. (1) Na source (carrier gas argon), (2) apertures, (3) grating, (4) positioning devices, (5) active position stabilization, (6) reference signal, (7) photodiode, (8) He–Ne laser, (9) interaction zone, separated by a copper foil (10), (11) Langmuir–Taylor detector

two separate chambers and can be subjected to different fields, or one of them can interact with a gas while the other traverses through vacuum. A critical point in the design and operation of the interferometer is the adjustment of the gratings and stability from external vibrations. The axes of the three gratings must be parallel to each other with a precision better than half the grating constant over the height of the molecular beam. In this case it means that parallelism of all gratings has to be fulfilled to better than 20 seconds of arc (10^{-4} rad).

Vibration of the gratings results in a time dependence of the phases of the partial beams which, in the time average, leads to a smudging of the interference pattern. The best way to reduce this disturbance is to place the whole setup on a stable and heavy platform which is isolated against vibrations (e.g. from the room's floor or through vacuum hoses). Thermal isolation is advisable too. Despite all the isolation efforts, some drifts in the adjustment are unavoidable. An active feedback servomechanism, acting on the middle grating and controlled by an optical interferometer consisting of three phase gratings (3μm) rigidly connected to the atom-optics gratings, compensates these drifts. Thus, long term stability is ensured.

5.5.2 Some Experimental Results

Electric Polarizabilities. In order to measure the electric polarizability of sodium atoms one of the partial beams of the interferometer passes through a homogeneous electric field E which shifts its energy by the Stark potential

$$V = -\alpha E^2/2. \tag{5.67}$$

For the additional phase shift due to a potential V(x) along the x coordinate (interferometer path) while the other beam passes through field-free space, the semiclassical approximation yields

$$\Delta\varphi(k) = \int(k(x) - k_0(x))dx, \tag{5.68}$$

where k(x) and k_0(x) are the perturbed and unperturbed wave numbers in each branch of the interferometer

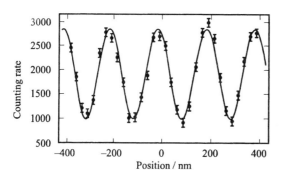

Fig. 5.30. Interference pattern as function of the position of the third grating (accumulation time 1s per point)

$$k(x) = \frac{1}{\hbar}\sqrt{2m(E - V(x))}; \quad k_0(x) = \frac{1}{\hbar}\sqrt{2mE}. \tag{5.69}$$

Integration in (5.68) is along the classical particle trajectory. With (5.67) for the Stark potential and assuming $V/E \ll 1$ one obtains the phase difference

$$\Delta\varphi = \frac{\alpha}{\hbar v}\int_{x_1}^{x_2} E^2(x)dx = \frac{\alpha}{2\hbar v}\left(\frac{U}{d}\right)^2 L_{\text{eff}}, \tag{5.70}$$

where v is the average velocity in the beam, U the applied voltage, and d the electrode distance. L_{eff} is the effective length of the field, taking into account the stray field on both ends. The average velocity v in the beam and the width of the distribution is determined from the diffraction by the grating with an accuracy of 0.15%. The phase shift $\Delta\varphi$ can be determined to an accuracy of 10 mrad. This allows an accuracy for the polarizability α of 0.2% [Ekstrom et al. (1995)]. Figure 5.31 shows the measured phase shift as a function of the applied voltage. For these measurements the voltage was applied alternately to the one (open points) or the other branch (solid points) of the interferometer. In between each changeover a zero check was performed.

Refractive Index of Matter Waves in Gases. Just as for light waves one can determine the refractive index n of a gas for a given matter wave by passing one of the partial beams of the interferometer through a gas-filled chamber with particle density N, gas particle mass m_k and length L, while the other partial beam passes through vacuum. The real part Re(n) of the refractive index follows immediately from the phase shift $\Delta\varphi$ relative to the unperturbed partial beam, as determined from the interference structure:

$$\text{Re}(n) = 1 + \frac{\Delta\varphi}{k_L L}. \tag{5.71}$$

k_L is the wave number in the laboratory system, given by

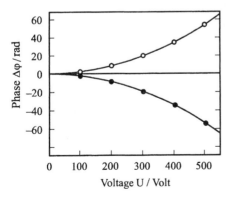

Fig. 5.31. Measured phase shift as a function of the voltage applied alternately to the partial beams along the two branches of the interferometer. The solid lines are a quadratic approximation

$$k_L = \frac{m_i v_i}{\hbar}. \tag{5.72}$$

m_i is the mass of the beam particles and v_i is their velocity (assumed as monochromatic). The physical meaning of the phase shift and the index of refraction follows from the wave function whose spatial dependence for the partial beam in the gas is represented by

$$\Psi_1 = A_1 \exp(ink_L z) \quad \text{or}$$
$$\Psi_1 = A_1 \exp(ik_L z) \exp\left(i \frac{2\pi}{k} Nz \, \text{Re}(f(k,0))\right) \exp\left(i \frac{2\pi}{k} Nz \, \text{Im}(f(k,0))\right). \tag{5.73}$$

For the wave through vacuum we have instead

$$\Psi_2 = A_2 \exp(ik_L z). \tag{5.74}$$

z is the coordinate along the interferometer axis, $k = \mu g/\hbar$ is the wave vector of the relative motion ($\mu = m_i m_k/(m_i + m_k)$ is the reduced mass and g is the relative velocity of the collision partners). We assume for the moment that there is a common center-of-mass system, i.e. that the target particles all have the same velocity (by value and direction). This can be realized experimentally by using a monochromatic target beam. The scattering amplitude in the forward the direction is designated by f(k,0). Equation (5.73) immediately yields for the complex index of refraction

$$n = 1 + \frac{2\pi}{k_L k} Nf(k,0) = 1 + \frac{2\pi}{k^2} Nf(k,0) \frac{g}{v_i} \frac{m_k}{m_i + m_k}. \tag{5.75}$$

Whenever the velocity of the beam particles is large compared to the velocity of the target particles (e.g. for thermal neutrons in a neutron interferometer), then $g/v_i \approx 1$ and this term in (5.75) has no influence. With this tacit assumption made, the factor g/v_i is often omitted [Vigué (1995), Audouard et al. (1995)].

In a scattering experiment one can only measure the beam intensity $I \propto \psi^*\psi$, i.e for the passage of a beam through a gas of particle density N follows Beer's law from (5.73)

$$I(N) = I(0) \exp\left(-\frac{4\pi}{k} N \, \text{Im}(f(k,0))\right) = I(0) \exp(-N\sigma_t(k)) \tag{5.76}$$

where the optical theorem has been utilized in the right part of the equation. σ_t is the total cross section.

In the interferometer, the superposition of the two partial waves ψ_1 and ψ_2 according to (5.73) and (5.74) leads to the interference term

$$\Im = A \exp\left(-\frac{2\pi}{k} NL \operatorname{Im}(f(k,0))\right) \cos\left(\frac{2\pi}{k} NL \operatorname{Re}(f(k,0))\right), \tag{5.77}$$

which yields for the phase shift due to the interaction with the gas

$$\Delta\varphi = \frac{2\pi}{k} N \operatorname{Re}(f(k,0)), \tag{5.78}$$

and for the amplitude A

$$\ln\frac{A(0)}{A(N)} = \frac{2\pi}{k} NL \operatorname{Im}(f(k,0)), \tag{5.79}$$

which is again Beer's law in another form. Having measured both quantities, one can calculate the ratio

$$\frac{\Delta\varphi(N)}{\ln(A(0)/A(N))} = \frac{\operatorname{Re}(f(k,0))}{\operatorname{Im}(f(k,0))} \tag{5.80}$$

independently of the product NL whose experimental determination is always prone to considerable errors.

As an example, Fig. 5.32 shows the ratio $\operatorname{Re}(f(0))/\operatorname{Im}(f(0))$ as a function of the reduced wave number kr_m (r_m is the equilibrium distance of the interatomic potential), calculated for a Lennard–Jones (12,6) potential with parameters corresponding to Na–Kr collisions. The ratio (as also separately the numerator and denomina-

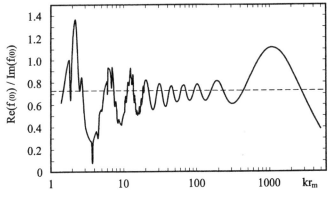

Fig. 5.32. Ratio of real and imaginary part of the scattering amplitude in the forward direction as a function of the reduced wave number kr_m, calculated for a Lennard–Jones (12,6) potential. For further details see text

tor) show explicit glory oscillations [Pauly (1979)] about an average value which the semiclassical approximation calculates as $\mathrm{tg}(\pi/5)$ (horizontal, dashed line in Fig. 5.32). At small values of kr_m orbiting resonances are superimposed on the glory oscillations.

In practice, the thermal motion of the target particles makes an interpretation of the measured data more difficult whenever the velocity of the beam particles is not large compared to the most probable velocity of the target particles. The wave number of the relative motion becomes a distribution even with a monochromatic beam, so that the measured quantities defined by (5.77) and (5.79), as well as the experimentally determined refractive index from (5.75), are values averaged over this distribution. In order to compare experimental data with theoretical calculations of the scattering amplitude, the equations (5.75), (5.78) and (5.79) as well as the measured quantities of the scattering experiment (see Sect. I.2.2.4) must be averaged over the distribution of the relative velocities (see (I.2.55)). Thus, similar to other scattering processes [Toennies et al. (1974), Köhler et al. (1977)], sharp resonances in the refractive index can be observed experimentally only with great effort and expense. The first measurements of the refractive index of Na and Na_2 beams have been carried out for a number of gases (He, Ne, Ar, Kr, Xe, N_2, CO_2, NH_3, and H_2O) [Schmiedmayer et al. (1995), (1997)].

Other interferometric measurements concern the determination of the Earth's gravitational acceleration with an accuracy of 3×10^{-9}, the gradient of the gravitational field of the Earth, the ratio h/m with an accuracy of 10^{-7} from precise measurements of the recoil which an absorbed photon exerts upon an atom, or fundamental quantum theoretical experiments [Peters et al. (1999), Kasevich and Chu (1991), Weiss et al. (1993a), (1993b), Clauser and Li (1994b), Lee et al. (1995), Courtois and Grynberg (1996), Batelaan et al. (1997), Snadden et al. (1998)].

References

Chapter 1

Abel, G., B. Stansfield, D. Michaud, G.G. Ross, J.L. Lachambre, and J.L. Gauvreau (1997): Rev. Sci. Instrum. **68**, 994
Aberle, W., J. Grosser, and W. Krüger (1980): J. Phys. B **13**, 2083
Abshagen, M., J. Kowalski, M. Meyberg, G. zu Putlitz, J. Slaby, and F. Träger (1991): Z. Phys. D **19**, 199
Almulhem, A.A. (1997): Physica Status Solidi B **201**, 105
Alvarino, J.M., C. Hepp, M. Kreiensen, B. Staudenmeyer, F. Vecchiocativi, and V. Kempter (1984): J. Chem. Phys. **80**, 765
Amdur, I. and H. Pearlman (1939): Rev. Sci. Instrum. **10**, 174
Amdur, I., C.F. Glick, and H. Pearlman (1948): Proc. Am. Acad. Arts Sci. **76**, 101
Amdur, I. (1968): in *Methods of Experimental Physics*, Vol. 7A, ed. by B. Bederson and W.L. Fite, p. 341 (Academic Press, New York)
Amirav, A. (1990): Comments At. Mol. Phys. **24**, 187
Amirav, A. (1991): Org. Mass Spectrom. **26**, 1
Amme, R.C. and P.O. Hangsjaa (1969): Phys. Rev. **177**, 230
Anderson, J.B., R.P. Andres, and J.B. Fenn (1965): Adv. At. Mol. Phys. **1**, 345
Anderson, R.W., V. Aquilanti, and D.R. Herschbach (1969): Chem. Phys. Lett. **4**, 5
Andersson, P.U. and J.B.C. Pettersson (1997): Z. Phys. D **41**, 57
Andresen, P. (1978): Max-Planck Institut für Strömungsforschung, Report **3**, Göttingen
Andresen, P., R. Düren, and H. Joswig (1983): J. Phys. B **16**, 3423
Anheier, N.C., M.J. Labiche, and M.M. Graff (1989): Rev. Sci. Instrum. **60**, 3080
Argoustoglou, E., W.R. Johnson, Z.W. Liu, and J. Sapirstein (1995): Phys. Rev. A **51**,1196
Aseyev, S.A., Y.A. Kudryavtsek, D.V. Laryushin, V.S. Letokhov, and V.V. Petrunin (1995): JETP Lett. **61**, 373
Aseyev, S.A., V.V. Petrunin, and V.S. Letokhov (1996): Comments At. Mol. Phys. **33**, 41
Aten, J.A. and J. Los (1975): J. Phys. E **8**, 408
Aten, J.A., G.E.H. Lanting, and J. Los (1977): Chem. Phys. **19**, 241
Auerbach, D., E.E.A. Bromberg, and L. Wharton (1966): J. Chem. Phys. **45**, 2160
Bacal, M. and W. Reichelt (1974): Rev. Sci. Instrum. **45**, 769
Baede, A.P.M., W.F. Jungman, and J. Los (1971): Physica **54**, 459
Bates, D.R., H.S.W. Massey, and A.L. Stewart (1953): Proc. Roy. Soc. A **216**, 437
Bayfield, J.E. and P.M. Koch (1974): Phys. Rev. Lett. **33**, 258
Bayfield, J.E., L.D. Gardner, and P.M. Koch (1977): Phys. Rev. Lett. **39**, 76
Becker, E.W., I. Bier, and H. Burghoff (1955): Z. Nat.forsch. A **10**, 7
Becker, E.W. and W. Henkes (1956): Z. Phys. **146**, 320
Behrisch, R. (1964): Ergeb. Exakt. Naturw. **35**, 295

Bell, E.W., X.Q. Guo, J.L. Forand, K. Rinn, D.R. Swenson, J.S. Thompson, and G.H. Dunn (1994): Phys. Rev. A **49**, 4585
Benoit, C. and Gauyacq, J.P. (1976): J. Phys. B **9**, L391
Beulens, J.J., C. Gastineau, N. Guerrassimov, J. Konlidiati, and D.J. Schram (1994): Plasma Chem. Plasma Proc. **14**, 15
Bickes, R.W. Jr., K.R. Newton, J.M. Herrmann, and R.B. Bernstein (1976): J. Chem. Phys. **64**, 3648
Bliek, F.W., R. Hoekstra, M.E. Bannister, and C.C. Havener (1997): Phys. Rev. A **56**, 426
Boffard, J.B., M.E. Lagus, L.W. Anderson, and C.C. Lin (1996): Rev. Sci. Instrum. **67**, 2738
Boileau, A., M. von Hellermann, M. Mandel, W. Summers, H.P. Weisen, and A. Zinoviev (1989): J. Phys. B **22**, L145
Bondybey, V.W., G.P. Schwarz, and J.H. English (1983): J. Chem. Phys. **78**, 11
Boring, J.W. and D.R. Humphries (1969): *Rarefied Gas Dynamics* **8**, Vol. 1, ed. by L. Trilling and H.Y. Wachmann, p. 1303 (Academic Press, New York)
Borisov, A.G., D. Teillet-Billy, and J.P. Gauyacq (1996): Phys. Rev. B **54**, 17166
Boudjarane, J., J. Lacoursière, and M. Lacoursière (1994): J. Chem. Phys. **101**, 10274
Brako, R. and D.M. Newns (1981): Surf. Sci. **108**, 253
Bransden, B.H. and M.R.C. McDowell (1992): *Charge Exchange and the Theory of Ion–Atom Collisions* (Clarendon Press, Oxford)
Brencher, L., B. Nawracala, and H. Pauly (1988): Z. Phys. D **10**, 211
Brisson, D., F.W. Baity, B.H. Quon, J.A. Ray, and C.F. Barnett (1980): Rev. Sci. Instrum. **51**, 511
Bruckmeier, R., Ch. Wanderlich, and H. Figger (1995): Phys. Rev. A **52**, 334
Buchinger, F., H.A. Schuessler, E.C. Benk, H. Imura, Y.F. Li, C. Bingham, and H.K. Carter (1992): Hyperfine Interactions **75**, 367
Buck, U., E. Leßner, and D. Pust (1980): J. Phys. B **13**, L125
Budrevich, A., B. Tsipinyuk, and E. Kolodney (1995): Chem. Phys. Lett. **234**, 253
Budrevich, A., B. Tsipinyuk, A. Bekkerman, and E. Kolodney (1997): J. Chem. Phys. **106**, 5771
Buelow, S., G. Radhakrishnan, J. Catanzarite, and C. Wittig (1985): J. Chem. Phys. **83**, 444
Buelow, S., G. Radhakrishnan, and C. Wittig (1987): J. Phys. Chem. **91**, 5409
Bull, T.H. and P.B. Moon (1954): Discuss. Faraday Soc. **17**, 54
Buntin, S.A., C.F. Giese, and W.R. Gentry (1987): J. Chem. Phys. **87**, 1443
Buntin, S.A. (1996): J. Chem. Phys. **105**, 2066
Caledonia, G.E., R.H. Krech, and B.D. Green (1987): AIAA J. **25**, 59
Campargue, R. (1984): J. Phys. Chem. **88**, 4466
Campos, F.X., C.J. Waltman, and S.R. Leone (1993): Chem. Phys. Lett. **201**, 399
Carter, G. and J.S. Colligon (1968): *Ion Bombardment of Solids* (Elsevier, New York)
Chen, M.H., K.T. Cheng, W.R. Johnson, and J. Sapirstein (1995): Phys. Rev. A **52**, 266
Ching, C.H., J.E. Bailey, P.W. Lake, A.B. Filuk, A.G. Adams, and J. McKenney (1997): Rev. Sci. Instrum. **68**, 354
Chitnis, C.E., R.S. Gao, J. Pinedo, K.A. Smith, and R.F. Stebbings (1988): Phys. Rev. A **37**, 687
Cho, C.C., J.C. Polanyi, and C.D. Stanners (1988): J. Phys. Chem. **92**, 6859
Christen, W., K.-L. Kompa, H. Schröder, and H. Stülpnagel (1992): Ber. Bunsenges. Phys. Chem. **96**, 1197
Churkin, I.N., V.I. Volosov, and A.G. Steshov (1998): Rev. Sci. Instrum. **69**, 856
Chutjian, A. and O.J. Orient (1996): Exp. Methods Phys. Sci. B **29**, 49
Clausnitzer, G. (1961): Helv. Phys. Acta (Suppl.) **6**, 35

Claytor, N.E., E.A. Hessels, and S.R. Landeen (1995): Phys. Rev. A **52**, 165
Coleman, M.L., R. Hammond, and J.W. Dubrin (1973): Chem. Phys. Lett. **10**, 271
Continetti, R.E., D.R. Cyr, D.L. Osborne, D.J. Leahy, and D.M. Neumark (1993): J. Chem. Phys. **99**, 2616
Cosby, P.C. and H. Helm (1989): J. Chem. Phys. **90**, 1434
Cosby, P.C. (1993): J. Chem. Phys. **98**, 7804, 9544 and 9560
Cousins, L.M. and S.R. Leone (1988): J. Mater. Res. **3**, 1158
Cousins, L.M. and S.R. Leone (1989): Chem. Phys. Lett. **155**, 162
Cousins, L.M., R.J. Levis, and S.R. Leone (1989): J. Chem. Phys. **91**, 5731
Croft H. and A.S. Dickinson (1996): J. Phys. B **29**, 57
Cross, J.B. and D.A. Cremers (1986): Nucl. Instrum. Methods **13**, 658
Cuthbertson, J.W., R.W. Motley, and W.D. Langer (1992): Rev. Sci. Instrum. **53**, 5279
Cyr, D.R., R.E. Continetti, R.B. Metz, D.L. Osborne, and D.M. Neumark (1992): J. Chem. Phys. **97**, 4937
Cyr, D.R., D.J. Leahy, D.L. Osborne, R.E. Continetti, and D.M. Neumark (1993): J. Chem. Phys. **99**, 8751
Dagan, S., A. Danon, and A. Amirav (1992): Int. J. Mass Spectrom. Ion Process. **113**, 157
Dagan, S., A. Amirav, and T. Fujii (1995): Int. J. Mass Spectrom. Ion Process. **151**, 159
Danon, A. and A. Amirav (1987): Rev. Sci. Instrum. **58**, 1724 and J. Chem. Phys. **86**, 4708
Danon, A., E. Kolodney, and A. Amirav (1988): Surf. Sci. **193**, 132
Danon, A. and A. Amirav (1988): Phys. Rev. Lett. **61**, 2961
Danon, A. and A. Amirav (1989): J. Phys. Chem. **93**, 5549
Danon, A. and A. Amirav (1990): J. Chem. Phys. **92**, 6968 and Int. J. Mass Spectrom. Ion Process. **96**, 139
Danon, A., A. Vardi, and A. Amirav (1990): Phys. Rev. Lett. **65**, 2038
Danon, A. and A. Amirav (1994): Int. J. Mass Spectrom. Ion Process. **133**, 187
Davis, B. and A.H. Barnes (1929): Phys. Rev. **34**, 1229
Davis, B. and A.H. Barnes (1931): Phys. Rev. **37**, 1368
Davis, S.L., D. Mueller, and C.J. Keane (1983): Rev. Sci. Instrum. **54**, 315
Davison, S.G., Z.L. Miskovic, A.T. Amos, B.L. Burrows, F.O. Goodman, and K.W. Sulston (1995): Prog. Surf. Sci. **48**, 193
Deck, F.J., E.A. Hessel, and S.R. Landeen (1993): Phys. Rev. A **48**, 4400
Demtröder, W. (1992): in *Atomic and Molecular Beam Methods*, Vol. 2, ed. by G. Scoles, p. 213 (Oxford University Press, Oxford)
de Juan, J., S. Callister, H. Reisler, G.A. Segal, and C. Wittig (1988): J. Chem. Phys. **89**, 1977
Devienne, F.M., B. Crave, J. Souquet, and R. Clapier (1963): *Rarefied Gas Dynamics* **3**, Vol 1, ed. by J.A. Laurmann, p. 362 (Academic Press, New York)
Devienne, F.M., J.C. Rouston, and R. Clapier (1967): *Rarefied Gas Dynamics* **5**, Vol 1, ed. by C.L. Brundin, p. 269 (Academic Press, New York)
Dewangan, D.P. (1973): J. Phys. B **6**, L20
Domen, K. and T.J. Chuang (1989): J. Chem. Phys. **90**, 3318 and 3332
Donnally, B.L., T. Clapp, W. Sawyer, and M. Schultz (1964): Phys. Rev. Lett. **12**, 502
Dose, V., W. Hett, R.E. Olson, P. Pradel, F. Roussel, A.S. Schlachter, and G. Spiess (1975): Phys. Rev. A **12**, 1261
Doughty, R.O. and W.J. Schaetzle (1969): Entropie **30**, 47
Dunn, P. and D.A. Reay (1976): *Heat Pipes* (Pergamon Press, New York)
Düren, R., M. Kick, and H. Pauly (1974): Chem. Phys. Lett. **27**, 118
Düren, R., U. Krause, and G. Moritz (1980): J. Phys. B **13**, 503

Eccles, A.J., J.A. van den Berg, A. Brown, and J.S. Vickermann (1986): J. Vac. Sci. Technol. A **4**, 1888
Eckert, H.U., F.L. Kelly, and H.L. Olson (1968): J. Appl. Phys. **39**, 1846
Endo,T., H. Yan, K. Abe, S. Nagase, Y. Ishida, and H. Nishiki (1997): J. Vac. Sci. Technol. A **15**, 1990
Eriksen, F.J., S.M. Fernandez, A.B. Bray, and E. Pollack (1975): Phys. Rev. A **11**, 1239
Even, U., P.J. de Lange, H.Th. Jonkman, and J. Kommandeur (1986): Phys. Rev. Lett. **56**, 965
Fahey, D.W., L.D. Schearer, and W.F. Parks (1978): Rev. Sci. Instrum. **49**, 503
Fayet, P. and L. Wöste (1986): Z. Phys. D **3**, 177
Feltgen, R., H. Ferkel, R.K.B. Helbing, D. Pikorz, and H. Vehmeyer (1993): J. Chem. Phys. **98**, 2466
Fenn, J.B. (1967): Entropie **18**, 11
Ferkel, H., R. Feltgen, and D. Pikorz (1991): Rev. Sci. Instrum. **62**, 2626
Fiedler, S. (1995): Report IPPIII/209, MPI für Plasmaphysik, Garching
Finck, K., Y. Wang, Z. Roller and H.O. Lutz (1988): Phys. Rev. A **38**, 6115
Fisher, D.S., C.W. Fehrenbach, S.R. Landeen, E.A. Hessels, and B.D. DePaola (1997): Phys. Rev. A **56**, 4656
Fleischmann, H.H., C.F. Barnett, and J.A. Ray (1974): Phys. Rev. A **10**, 569
Fluendy, M.A.D. and K.P. Lawley (1973): *Chemical Application of Molecular Beam Scattering* (Chapman and Hall, London)
Fluendy, M.A.D., I.H. Kerr, K.P. Lawley, and J.M. McCall (1975): J. Phys. B **8**, L190
Fluendy, M.A.D. and S.L. Lunt (1983): Mol. Phys. **49**, 1007
Folkerts, L., M.A. Haque, C.C. Havener, N. Shimakura, and M. Kimura (1995): Phys. Rev. A **51**, 3685
Foreman, P.B., A.B. Lees, and P.K. Rol (1976): Chem. Phys. **12**, 213
Fortson, N. (1993): in *Atomic Physics*, Vol. 13, ed. by H. Walther, T.W. Hänsch, and B. Neizert, p. 62 (AIP, New York)
Frauenfelder, H. (1950): Helv. Phys. Acta **23**, 347
Friichtenicht, J.F. (1974): Rev. Sci. Instrum. **45**, 51
Fumelli, M. (1974): Nucl. Instrum. Methods **118**, 337
Gabovich, M.D., N.V. Pleshivtsev, and N.N. Semashko (1989): *Ion and Atom Beams for Controlled Fusion and Technology* (Consultants Bureau, New York)
Gao, R.S., L.K. Johnson, D.E. Nitz, K.A. Smith, and R.F. Stebbings (1987): Phys. Rev. A **36**, 3077
Gao, R.S., L.K. Johnson, K.A. Smith, and R.F. Stebbings (1989): Phys. Rev. A **40**, 4914
Gardner, W.L., G.C. Barber, C.W. Blue, W.K. Dagenhart, H.H. Haselton, J. Kim, M.M. Menon, N.S. Ponte, P.M. Ryan, D.E. Schechter, W.L. Stirling, C.C. Tsai, J.H. Whealton, and R.E. Wright (1982): Rev. Sci. Instrum. **53**, 424
Gaupp, A., P. Kuske, and H.J. Andrä (1982): Phys. Rev. A **26**, 3351
Gaus, A.D., W.T. Htwe, J.A. Brand, T.J. Gay, and M. Schulz (1994): Rev. Sci. Instrum. **65**, 3739
Gentry, W.R., D.J. McClure, and C.H. Douglass (1975): Rev. Sci. Instrum. **46**, 367
Gerlich, D. (1993): J. Chem. Soc. Faraday Trans. **89**, 2199
Gersing, E., H. Pauly, E. Schädlich, and M. Vonderschen (1973): Discuss. Faraday Soc. **55**, 211
Gersing, E., M. Vonderschen, and H. Pauly (1976): in *50 Jahre MPI für Strömungsforschung*, ed. by MPIFS, p. 392 (Hubert & Co, Göttingen)
Giapis, K.P., T.A. Moore, and T.K. Minton (1995): J. Vac. Sci. Technol. **13**, 959

Gieler, M., F. Aumayr, J. Schweinzer, W. Koppensteiner, W. Husinsky, H.P. Winter, K. Lozhkin, and J.P. Hansen (1993): J. Phys. B **26**, 2137
Girard, J.M., A. Lebéhot, and R. Campargue (1993): J. Phys. D **26**, 1382
Girard, J.M., A. Lebéhot, and R. Campargue (1994): J. Phys. D **27**, 253
Götting, R. (1985): Max-Planck Institut für Strömungsforschung, Report **13**, Göttingen
Götting, R., H.R. Mayne, and J.P. Toennies (1986): J. Chem. Phys. **85**, 6396
Greene, D.B., E.A. Halprin, and J.G. Skofronik (1975): Rev. Sci. Instrum. **46**, 136
Greene, J.E. and N.-E. Lee (1997): Nucl. Instrum. Methods B **121**, 58
Grisham, L., D. Post, B. Johnson, K. Jones, J. Barrette, T. Kruse, I. Tseruya, and D.H. Wang (1982): Rev. Sci. Instrum. **53**, 281
Grover, G.M., T.P. Cotter, and G.F. Erickson (1964): J. Appl. Phys. **35**, 1990
Gurnee, E.F. and J.L. Magee (1957): J. Chem. Phys. **26**, 1237
Hagstrum, H.D. (1977): in *Inelastic Ion–Surface Collisions*, ed. by N.H. Tolk, J.C. Tully, W. Heiland, and C.W. White, p. 1 (Academic Press, New York)
Harbich, W., S. Fedrigo, F. Meyer, D.M. Lindsay, J. Lignieres, J.C. Rivoal, and D. Kreisle (1990): J. Chem. Phys. **93**, 8535
Harrison, I, J.C. Polanyi, and P.A. Young (1988): J. Chem. Phys. **89**, 1499
Hasted, J.B. (1972): in *Physics of Atomic Collisions*, 2nd edn. (Butterworths, London)
Häusler, D., J. Rice, and C. Wittig (1987): J. Phys. Chem. **91**, 5413
Havener, C.C., M.S. Huq, P.A. Schulz, and R.A. Phaneuf (1989): Phys. Rev. A **39**, 1725
Havener, C.C., A. Mueller, P.A.Z. Emmichoven, and R.A. Phaneuf (1995): Phys. Rev. A **51**, 2982
Havener, C.C. (1997): in *Accelerator-Based Atomic Physics Techniques and Applications*, ed. by S.M. Shafroth and J.C. Austin, p. 117 (AIP, New York)
Helbing, R.K.B. and E.W. Rothe (1968): Rev. Sci. Instrum. **39**, 1948
Helbing, R.K.B. and E. Rothe (1969): J. Chem. Phys. **51**, 1607
Helm, H. and P.C. Cosby (1987): J. Chem. Phys. **86**, 6813
Hershcovitch, A.I., B.M. Johnson, V.J. Kovarik, M. Meron, K.W. Jones, K. Prelec, and L.R. Grisham (1984): Rev. Sci. Instrum. **55**, 1744
Hessels, E.A., P.W. Arcuni, F.J. Deck, and S.R. Landeen (1992): Phys. Rev. A **46**, 2622
Hintz, E. and B. Scheer (1995): Plasma Phys. Control. Fusion **37**, A 87
Hoflund, G.B. and J.F. Weaver (1994): Meas. Sci. Technol. **5**, 201
Hollstein, M. and H. Pauly (1966): Z. Phys. **196**, 353
Hollstein, M., J.R. Sheridan, J.R. Peterson, and D.C. Lorents (1969): Phys. Rev. **182**, 152
Hopkins, J.B., P.R.R. Langridge-Smith, M.D. Morse, and R.E. Smalley (1983): J. Chem. Phys. **78**, 1627
Hu, Z., B. Shen, Q. Zhou, S. Deosaran, J.R. Lombardi, and D.M. Lindsay (1991): J. Chem. Phys. **95**, 2206
Hubers, M.M., A.W. Kleyn, and J. Los (1976): Chem. Phys. **17**, 303
Hulpke, E. and V. Kempter (1966): Z. Phys. **197**, 41
Hunter, L.R., D. Krause, D.J. Berkeland, and M.G. Boshier (1991): Phys. Rev. A **44**, 6140
Huq, M.S., C.C. Havener, and R.A. Phaneuf (1989): Phys. Rev. A **40**, 1811
Husinsky, W., G. Betz, and I. Girgis (1983): J. Vac. Sci. Technol. A **2**, 1689
Hüwel, L., J. Maier, R.K.B. Helbing, and H. Pauly (1980): Chem. Phys. Lett. **74**, 459
Hüwel, L., J. Maier, and H. Pauly (1981): J. Chem. Phys. **74**, 5613
Hüwel, L., J. Maier, and H. Pauly (1982): J. Chem. Phys. **76**, 4961
Ivanov, A.A. and G.V. Rosyakov (1980): Sov. Phys.-Techn. Phys. **25**, 1346
Ivanovskii, M.N., V.P. Sorokin, and I.V. Yagodkin, translated by R. Berman (1982): *The Physical Principles of Heat Pipes* (Clarendon Press, Oxford)
Johnson, L.K., R.S. Gao, K.A. Smith, and R.F. Stebbings (1988): Phys. Rev. A **38**, 2794

Jones, M.E., S.E. Roadman, A.M. Lam, G. Eres, and J.R. Engstrom (1996): J. Chem. Phys. **105**, 7140
Jordan, J.E., E.A. Mason, and I. Amdur (1972): in *Physical Methods of Chemistry*, Vol. 1, part III, Chap. 6, ed. by A. Weissberger and B.W. Rossiter (Wiley, New York)
Kachru, R., T.F. Gallagher, F. Gounand, K.A. Safinya, and W. Sandner (1983): Phys. Rev. A **27**, 795
Kadar-Kallen, M.A. and K.D. Bonin (1989): Appl. Phys. Lett. **54**, 2296
Kadar-Kallen, M.A. and K.D. Bonin (1994): Appl. Phys. Lett. **64**, 1436 and Phys. Rev. Lett. **72**, 828
Kaivola, M., O. Poulsen, E. Riis, and S.A. Lee (1985): Phys. Rev. Lett. **54**, 255
Kaminsky, M. (1965): in *Atomic and Ionic Impact Phenomena on Metal Surfaces* (Springer, Berlin)
Kaufman, S.L. (1976): Opt. Commun. **17**, 309
Keim, M., E. Arnold, W. Borchers, U. Georg, A. Klein, R. Neugart, L. Vermeeren, R.E. Silverans, and P. Lievens (1995): Nucl. Phys. A **586**, 219
Kempter, V., B. Kübler, and W. Mecklenbrauck (1974): J. Phys. B **7**, 149
Kempter, V. (1975): Adv. Chem. Phys. **30**, 417
Kessler, R.W. and B. Koglin (1966): Rev. Sci. Instrum. **37**, 682
Kessler, K.G. and H.M. Crosswhite (1967): in *Methods of Experimental Physics*, Vol. 4B, ed. by V.W. Hughes and H.L. Schulz, p. 49 (Academic Press, New York)
Kick, M. (1974): Max-Planck-Institut für Strömungsforschung, Report **126**, Göttingen
Kimmel, G.A. and B.H. Cooper (1993): Phys. Rev. B **48**, 12164
Kinoshita, H., J. Ikeda, M. Tagawa, M. Umeno, and N. Ohmae (1998): Rev. Sci. Instrum. **69**, 2273
Kishi, H. and T. Fujii (1995): J. Phys. Chem. A **99**, 11153
Kishi, H. and T. Fujii (1997): J. Phys. Chem. B **101**, 3788
Kita, S., H. Hübner, W. Kracht, and R. Düren (1981): Rev. Sci. Instrum. **52**, 684
Kleinermanns, K. and J. Wolfrum (1984): J. Chem. Phys. **80**, 1446
Knowles, M.P. and S.R. Leone (1997): J. Vac. Sci. Technol. A **15**, 2709
Knuth, E.L. (1964): Appl. Mech. Rev. **17**, 751
Koch, P.M. and D.R. Mariani (1980): J. Phys. B **13**, L645
Kolodney, E. and A. Amirav (1983): J. Chem. Phys. **79**, 4648
Kolodney, E. and A. Amirav (1984): in *Dynamics on Surfaces*, ed. by B. Pullman, J. Jortner, A. Nitzan, and B. Gerber, p. 231 (Reidel Publishing Company, Dordrecht)
Kolodney, E., B. Tsipinyuk, and A. Budrevich (1994): J. Chem. Phys. **100**, 8542
Können, G.P., J. Grosser, A. Haring, A.E. De Vries, and J. Kistemaker (1974): Radiation Effects **21**, 269
Kuipers, E.W., A. Vardi, A. Danon, and A. Amirav (1991): Phys. Rev. Lett. **66**, 116
Kushawaha, V.S. (1983): Z. Phys. A **313**, 155
Kuwano, H. and K. Nagai (1985): J. Vac. Sci. Technol. A **3**, 1809
Kuzel, M., R.D. DuBois, R. Maier, O. Heil, D.H. Jakabassa-Amandsen, M.W. Lucas, and K.O. Groeneveld (1994): J. Phys. B **27**, 1993
Kwan, C.K. (1974): Max-Planck-Institut für Strömungsforschung, Report **129**, Göttingen
Kwan, J.W., O.A. Anderson, C.F. Chan, W.S. Cooper, G.J. De Vries, W.B. Kunkel, K.N. Leung, A.F. Lietzke, W.F. Steele, C.F.A. van Os, R.P. Wells, and M.D. Williams (1992): Rev. Sci. Instrum. **63**, 2705
Lackschewitz, U., J. Maier, and H. Pauly (1986): J. Chem. Phys. **84**, 181
Lacmann, K. and D.R. Herschbach (1970): Chem. Phys. Lett. **6**, 97
Lagus, M.E., J.B. Boffard, L.W. Anderson, and C.C. Lin (1996): Phys. Rev. A **53**, 1505
Larsen, R.A., S.K. Neoh, and D.R. Herschbach (1974): Rev. Sci. Instrum. **45**, 1511

Leahy, D.J., D.L. Osborne, D.R. Cyr, and D.M. Neumark (1995): J. Chem. Phys. **103**, 2495
Leasure, E.L., C.R. Mueller, and T.Y. Ridley (1975): Rev. Sci. Instrum. **46**, 635
Lebéhot, A. and R. Campargue (1996): Phys. Plasmas **3**, 2502
Leonas, V.B. (1972): Sov. Phys. USPEKHI **15**, 266
Levine, L.P., J.F. Ready, and E. Bernal (1968): J. Quantum Electron. **QE-4**, 18
Levis, R.J., C.J. Waltman, L.M. Cousins, R.G. Copeland, and S.R. Leone (1990): J. Vac. Sci. Technol. A **8**, 3118
Lievens, P., P. Thoen, S. Bouckaert, W. Bouwen, E. Vandeweert, F. Vanhoutte, H. Weidele and R. Solverans (1997): Z. Phys. D **42**, 231
Lindsay, D.M., F. Meyer, and W. Harbich (1989): Z. Phys. D **12**, 15
Liu, S.M., W.E. Rodgers, and E.L. Knuth (1974): J. Chem. Phys. **61**, 902
Lo, S.-Y., J.D. Lobo, S. Blumberg, T.S. Dibble, X. Zhang, C.-C. Tsao, and M. Okumura (1997): J. Appl. Phys. **81**, 5896
Lorente, N. and R. Monreal (1997): Surf. Sci. **370**, 324
Love, R.P., J.M. Herrmann, R.W. Bickes Jr., and R.B. Bernstein (1977): J. Am. Chem. Soc. **99**, 8316
Ludin, A.I. and B.E. Lehmann (1995): Appl. Phys. B **61**, 461
Luhmann, N.C. and W.A. Peebles (1984): Rev. Sci. Instrum. **55**, 279
Lundeen, S.R. (1991): in *Atomic Physics*, Vol. 12, ed. by J.C. Zorn and R.R. Lewis, p. 288 (AIP New York)
MacAdam, K.B., L.G. Gray, and R.G. Rolfes (1990): Phys. Rev. A **42**, 5269
Madeheim, H., R. Hippler, and H.O. Lutz (1990): Z. Phys. D **15**, 327
Madson, J.M. and E.A. Theby (1983): Rev. Sci. Instrum. **54**, 958
Mapleton, R.A. (1972): *Theory of Charge Exchange* (Wiley-Interscience, New York)
Marmar, E.S., J.L. Terry, W.L. Rowan, and A.J. Wootton (1997): Rev. Sci. Instrum. **68**, 265
Marshall, D.G., P.B. Moon, J.E.S. Robinson, and J.T. Stringer (1948): J. Sci. Instrum. **25**, 348
Marshall, D.G. and T.H. Bull (1951): Nature **167**, 478
Maruyama, S., L.R. Anderson, and R.E. Smalley (1990): Rev. Sci. Instrum. **61**, 3686
Massey, H.S.W. and R.A. Smith (1933): Proc. Roy. Soc. A **142**, 142
Massey, H.S.W. (1949): Rep. Progr. Phys. **12**, 248
Massey, H.S.W., E.H.S. Burhop, and H.B. Gilbody (1974): *Electronic and Ionic Impact Phenomena*, Vol. 4 (Clarendon Press, Oxford)
McCormick, K., S. Fiedler, G. Kocsis, J. Schweinzer, and S. Zoletnik (1996): Report IPPIII/211, MPI für Plasmaphysik, Garching
McCullough, R.W. (1995): Nucl. Instrum. Methods B **98**, 170
McDaniel, E.W., J.B.A. Mitchell, and M.E. Rudd (1993): *Atomic Collisions* (Wiley, New York)
McGowan, R.W., D.M. Giltner, S.J. Sternberg, and S.A. Lee (1993): Phys. Rev. Lett. **70**, 251
Minton, T.K., K.P. Giapis, and T. Moore (1997): J. Phys. Chem. A **101**, 6549
Mochizuki, T. and K. Lacmann (1977): Chem. Phys. Lett. **49**, 604
Moon, P.B. (1953): J. Appl. Phys. **4**, 97
Moon, P.B., M.P. Ralls, J.B. Saul, and J.H. Broadhurst (1974): Phys. Bull. **25**, 511
Morgenstern, R., D.C. Lorents, J.R. Peterson, and R.E. Olson (1973): Phys. Rev. A **8**, 2372
Moritz, G. (1978): Max-Planck-Institut für Strömungsforschung, Report **24**, Göttingen
Morse, D.M. and R.E. Smalley (1984): Ber. Bunsenges. Phys. Chem. **88**, 228
Motley, R.W., S. Bernabei, and W.M. Hooke (1979): Rev. Sci. Instrum. **50**, 1586
Müller, U., U. Majer, R. Reichle, and M. Braun (1997): J. Chem. Phys. **106**, 7958

Murakami, K., Y. Fukuura, H. Mima, and I. Katsumata (1996): Rev. Sci. Instrum. **67**, 1196

Nagata, T., T. Kondow, K. Kuchitsu, K. Tabayashi, and K. Shobatake (1991): J. Chem. Phys. **95**, 1011

Natzle, W.C., D. Padowitz, and S.J. Sibener (1988): J. Chem. Phys. **88**, 7975

Nelson, R.S. (1968): in *The Observation of Atomic Collisons in Crystalline Solids* (North Holland, Amsterdam)

Neugart, R. (1987): in *Progress in Atomic Spectroscopy D*, ed. by H.J. Beyer and H. Kleinpoppen, p. 75 (Plenum Press, New York)

Neumann, W. and H. Pauly (1967): Entropie **18**, 93

Newman, J.H., J.D. Cogan, D.L. Zeigler, D.E. Nitz, R.D. Randel, K.A. Smith, and R.F. Stebbings (1982): Phys. Rev. A **25**, 2976

Newman, J.H., Y.S. Chen, K.A. Smith, and R.F. Stebbings (1986): J. Geophys. Res. **91**, 8947

Neynaber, R.H. (1968): *Methods of Experimental Physics*, Vol. 7A, ed. by B. Bederson and W.L. Fite, p. 476 (Academic Press, New York)

Neynaber, R.H. (1969): Adv. At. Mol. Phys. **5**, 57

Neynaber, R.H. and G.D. Magnuson (1975): Phys. Rev. A **11**, 865 and A **12**, 891

Neynaber, R.H. and G.D. Magnuson (1977): J. Chem. Phys. **67**, 430

Neynaber, R.H. and S.Y. Tang (1979): J. Chem. Phys. **70**, 4272

Nitz, D.E., R.S. Gao, L.K. Johnson, K.A. Smith, and R.F. Stebbings (1987): Phys. Rev. A **35**, 4541

Nordlander, P. (1990): Scanning Microsc. Suppl. **4**, 353

Nutt, C.W., T.J. Bale, P.J. Cosgrove, and M.J. Kirby (1977): VI. Int. Symp. Mol. Beams, Book of Abstracts, ed. by J. Kistemaker, J. Los, and A.E. De Vries, p. 256 (Noordwijkerhout)

Oakes, D.B., R.H. Krech, B.L. Upschulte, and G.E. Caledonia (1995): J. Appl. Phys. **77**, 2166

O'Brien, T.R. and J.E. Lawler (1996): Exp. Methods Phys. Sci. B **29**, 217

Ogawa, H., N. Sakamoto, I. Katayama, Y. Havyama, M. Saito, K. Yoshida, M. Tosaki, Y. Susuki, and K. Kimura (1996): Phys. Rev. A **54**, 5027

Ohara, Y., M. Akibya, M. Araki, H. Horiike, M. Kuriyama, S. Matsuda, M. Matsuoka, Y. Okumura, and S. Tanaka (1983): Rev. Sci. Instrum. **54**, 921

Ohara, Y. (1998): Rev. Sci. Instrum. **69**, 920

Oka, Y., Y. Tsumori, Y. Takeiri, O. Kaneko, M. Osakabe, E. Asano, T. Kawamoto, and R. Akiyama (1998): Rev. Sci. Instrum. **69**, 920

Okada, M. and Y. Murata (1997): J. Phys: Condensed Matter **9**, 1919

Okumura, Y., Y. Fujiwara, T. Inoue, K. Miyamoto, N. Miyamoto, A. Nagase, Y. Ohara, and K. Watanabe (1996): Rev. Sci. Instrum. **67**, 1092

Okumura, Y., Y. Fujiwara, M. Kashiwagi, T. Kitagawa, K. Miyamoto, T. Morshita, M. Takayanagi, M. Taniguchi, and K. Watanabe (2000): Rev. Sci. Instrum. **71**, 1219

Olamba, K., S. Szucs, J.P. Chenu, N. El Arbi, and F. Brouillard (1996): J. Phys. B. **29**, 2837

Oliphant, M.L.E. (1929): Proc. Roy. Soc. Lond. A **124**, 228

Oliphant, M.L.E. and P.B. Moon (1930): Proc. Roy. Soc. Lond. A **127**, 388

Olson, R.E. (1972): Phys. Rev. A **5**, 2094

Orient, O.J., A. Chutjian, and E. Murad (1990): Phys. Rev. **41**, 4106

Orient, O.J., K.E. Martus, A. Chutjian, and E. Murad (1992): Phys. Rev. A **45**, 2998

Orient, O.J., A. Chutjian, and E. Murad (1994): J. Chem. Phys. **101**, 8297

Orient, O.J., A. Chutjian, and E. Murad (1995): Phys. Rev. **51**, 2094

Osborne, D.L., D.J. Leahy, E.M. Ross, and D.M. Neumark (1995): Chem. Phys. Lett. **235**, 484
Osborne, D.L., H. Choi, D.H. Mordaunt, R.T. Bise, D.M. Neumark, and C.M. Rohlfing (1997): J. Chem. Phys. **106**, 3049
Palfrey, S.L. and S.R. Landeen (1984): Phys. Rev. Lett. **53**, 1141
Pang,H.M., C.H. Sin, D. Lubman, and J. Zorn (1986): Analyt. Chem. **58**, 1581
Parks, E.K., M. Inoue, and S. Wexler (1982): J. Chem. Phys. **76**, 1357
Parks, E.K., L.G. Pobo, and S. Wexler (1984): J. Chem. Phys. **80**, 5003
Parks, E.K., B.H. Weiller, P.S. Bechthold, W.F. Hoffmann, G.C. Nieman, L.G. Pobo, and S.J. Riley (1988): J. Chem. Phys. **88**, 1622
Parks, E.K. and S.J. Riley (1990): in *The Chemical Physics of Atomic and Molecular Clusters*, ed. by G. Scoles, p. 761 (North Holland, Amsterdam)
Parks, E.K., G.C. Niemann, and S.J. Riley (1996): J. Chem. Phys. **104**, 3531
Pauly, H. (1961): Fortschr. Phys. **9**, 613
Pauly, H. (1973): Faraday Discuss. Chem. Soc. **55**, 191
Pauly, H. and J. Schulz-Hennig (1975): in *50 Jahre MPI f. Strömungsforschung*, ed. by MPIFS, p. 468 (Hubert & Co, Göttingen)
Pauly, H. (1988): in *Atomic and Molecular Beam Methods*, ed. by G. Scoles, p. 124 (Oxford University Press, New York)
Pedersen, E.H. (1974): Phys. Rev. A **10**, 110
Pedersen, H.B., N. Djuric, M.J. Jensen, D. Kella, C.P. Safvan, H.T. Schmidt, L. Vejby-Christensen, and L.H. Andersen (1999): Phys. Rev. A **60**, 2882
Pellin, M.J., R.B. Wright, and D.M. Gruen (1981): J. Chem. Phys. **74**, 6448
Peng, T.C. and D.L. Liquornik (1967): Rev. Sci. Instrum. **30**, 989
Perel, J. and H.L. Daley (1971): Phys. Rev. A **6**, 1822 and VII. ICPEAC, Book of Abstracts, p. 603 (North Holland, Amsterdam)
Pieksma, M., M. Gargaud, R. McCarrol, and C.C. Havener (1996): Phys. Rev. A **54**, R13
Pieksma, M., M.E. Bannister, W. Wu, and C.C. Havener (1997): Phys. Rev. A **55**, 3526
Pieksma, M. and C.C. Havener (1998): Phys. Rev. A **57**, 1892
Pikorz, D. (1994): Max-Planck-Institut für Strömungsforschung, Report **23**, Göttingen
Piseri, P., A.L. Bassi, and P. Milani (1998): Rev. Sci. Instrum. **69**, 1647
Plastridge, B., K.A. Cowen, D.A. Wood, and J.V. Coe (1995): J. Phys. Chem. **99**, 118
Politiek, J., P.K. Rol, J. Los, and P.G. Ikelaar (1968): Rev. Sci. Instrum. **39**, 1147
Politiek, J. and J. Los (1969): Rev. Sci. Instrum. **40**, 1576
Poulsen, O. and N.I. Winstrup (1981): Phys. Rev. Lett. **47**, 1522
Powers, D.E., J.B. Hopkins, and R.E. Smalley (1981): J. Phys. Chem. **85**, 2711
Pradel, P., F. Roussel, A.S. Schlachter, G. Spiess, and A. Valance (1974): Phys. Rev. A **10**, 797
Quick, C.R. Jr. and J.J. Tiee (1983): Chem. Phys. Lett. **100**, 223
Quintana, E.J. and E. Pollak (1996): Phys. Rev. A **53**, 206 and A **54**, 4015
Radhakrishnan, G., S. Buelow, and C. Wittig (1986): J. Chem. Phys. **84**, 727
Rafac, R.J., C.E. Tanner, A.E. Livingston, K.W. Kukla, H.G. Berry, and C.A. Kurtz (1994): Phys. Rev. A **50**, R1976
Rafac, R.J., C.E. Tanner, A.E. Livingston, and H.G. Berry (1999): Phys. Rev. A **60**, 3648
Rapp, D. and W.E. Francis (1962): J. Chem. Phys. **37**, 2631
Reed, T.B. (1961): J. Appl. Phys. **32**, 821
Reid, S.A., F. Winterbottom, D.C. Scott, J. de Juan, and H. Reisler (1992): Chem. Phys. Lett. **189**, 430
Reynaud, C., J. Pommier, Vu Ngoc Tuan, and M. Barat (1979): Phys. Rev. Lett. **43**, 579
Rice, J., G. Hoffmann, and C. Wittig (1988): J. Chem. Phys. **88**, 2841

Riviere, A.C. (1968): in *Methods of Experimental Physics*, Vol. 7A, ed. by B. Bederson and W.L. Fite, p. 208 (Academic Press, New York)
Rouston, J.C. and F.M. Devienne (1967): Entropie **18**, 127
Royer, T., D. Dowek, J.C. Houver, J. Pommier, and N. Andersen (1988): Z. Phys. D **10**, 45
Rubin, K., P. Dittner, and B. Bederson (1964): Rev. Sci. Instrum. **35**, 1720
Ryan, S.G., T.C. Beers, C.P. Deliyannis, and J.A. Thorburn (1996): Astrophys. J. **458**, 543
Sakabe, S. and Y. Izawa (1991): At. Data Nucl. Data Tables **49**, 257
Sakabe, S. and Y. Izawa (1992): Phys. Rev. A **45**, 2086
Samano, E.C., W.E. Carr, M. Seidl, and B.S. Lee (1993): Rev. Sci. Instrum. **64**, 2746
Sarkissian, A.H., E. Charette, I. Coudrea, and B.C. Gregory (1997): Rev. Sci. Instrum. **68**, 289
Savola, W.J. Jr., F.J. Eriksen, and E. Pollack (1973): Phys. Rev. A **7**, 932
Schnieder, L., K. Seekamp-Rahn, F. Liedeker, H. Steuwe, and K.H. Welge (1991): Faraday Discuss. Chem. Soc. **91**, 259
Schorn, R.P., E. Wolfrum, F. Aumayr, E, Hintz, H. Rusbüldt, and H. Winter (1992): Nucl. Fusion **32**, 35
Schuessler, H.A., A. Alousi, M. Idrees, Y.F. Li, F. Buchinger, R.M. Evans, M. Brieger, and C.F. Fischer (1992): Phys. Rev. A **45**, 6459
Schweinzer, J., E. Wolfrum, F. Aumayr, M. Pöckl, H.P. Winter, R.P. Schorn, E. Hintz, and A. Unterreiter (1992): Plasma Phys. Control. Fusion **34**, 1173
Shao, H., P. Nordlander, and D.C. Langreth (1995): Phys. Rev. B **52**, 2988
Sheen, S.H., G. Dimoplon, E.K. Parks, and S. Wexler (1978): J. Chem. Phys. **68**, 4950
Sherman, J., A. Arvin, L. Hansborough, D. Hodgins, E. Meyer, J.D. Schneider, H.V. Smith Jr., M. Stettler, R.R. Stevens Jr., M. Thuot, T. Zaugg, and R. Ferdinand (1998): Rev. Sci. Instrum. **69**, 1003
Sherwood, C.R. and R.E. Continetti (1996): Chem. Phys. Lett. **258**, 171
Shimokawa, F. (1992): J. Vac. Sci. Technol. A **10**, 1352
Shimokawa, F. and H. Kuwano (1994): J. Vac. Sci. Technol. A **12**, 2739
Shobatake, K. and K. Tabayashi (1982): in *Atomic Collision Research in Japan*, Progress Report **8**, ed. by Y. Hatano, M. Matsuzawa, I. Shimamura, and S. Tsurubuchi, p. 157 (Society for Atomic Collision Research, Tokio)
Siegmann, B., G.G. Tepekan, R. Hippler, H. Madeheim, H. Kleinpoppen, and H.O. Lutz (1994): Z. Phys. D **30**, 223
Sigmund, P. (1969): Phys. Rev. **184**, 383
Sigmund, P. (1977): in *Inelastic Ion–Surface Collisions*, ed. by N.H. Tolk, J.C. Tully, W. Heiland, and C.W. White, p. 121 (Academic Press, New York)
Silver, J.A., A. Friedman, C.E. Kolb, A. Rahbee, and C.P. Dolan (1982): Rev. Sci. Instrum. **53**, 1714
Skinner, G.T. (1961): Phys. Fluids **4**, 1172
Skinner, G.T. and J. Moyzis (1965): Phys. Fluids **8**, 452
Smith, G.J., R.S. Gao, B.G. Lindsay, K.A. Smith, and R.F. Stebbings (1996): Phys. Rev. A **53**, 1581
Spiess, G., A. Valence, and P. Pradel (1972): Phys. Rev. A **6**, 746
Stacey, D.N. (1993): in *Atomic Physics*, Vol. 13, ed. by H. Walther, T.W. Hänsch, and B. Neizert, p. 46 (AIP, New York)
Staudenmaier, G. (1972): Rad. Eff. **13**, 87
Stebbings, R.F., C.J. Latimer, W.P. West, F.B. Dunning, and T.B. Cook (1975): Phys. Rev. A **12**, 1453
Stephen, T.M., B. van Zyl, and R.C. Amme (1996): Rev. Sci. Instrum. **67**, 1478
Stephen, T.M. and B.L. Peko (2000): Rev. Sci. Instrum. **71**, 1355

Stirling, W.L., C.C. Tsai, H.H. Haselton, D.E. Schechter, J.H. Whealton, W.K. Dagenhart, R.C. Davis, W.L. Gardner, J. Kim, M.M. Menon, and P.M. Ryan (1979): Rev. Sci. Instrum. **50**, 523

Sugai, T. and H. Shinohara (1997): Z. Phys. D **40**, 131

Sulston, K.W., A.T. Amos, and S.G. Davison (1988): Phys. Rev. B **37**, 9121

Summers, H.P. (1994): Adv. At. Mol. Opt. Phys. **33**, 275

Suzuki, S., T. Wakabayashi, H. Matsuma, H. Shirmaru, C. Kittaka, and Y. Achiba (1991): Chem. Phys. Lett. **182**, 12

Tabares, F.L., D. Tafalla, V. Herrero, and I. Tanarro (1997): J. Nucl. Mater. **241–243**, 1228

Tabayashi, K., S. Ohshima, and K. Shobatake (1984): J. Chem. Phys. **80**, 5335

Tanaka, M.Y., M. Bacal, M. Sasao, and T. Kuroda (1998): Rev. Sci. Instrum. **69**, 980

Tang, S.P., K.R. Chien, and M.J. Sabety-Dzvonik (1981): 8. Int. Symp. Mol. Beams, Book of Abstracts, ed. by F.M. Devienne, p. 71 (Cannes)

Tanner, C.E. (1995): in *Atomic Physics*, Vol. 14, ed. by D.J. Wineland, C.E. Wieman, and S.J. Smith, p. 130 (AIP, New York)

Tanner, C.E., A.E. Livingston, R.J. Rafac, K.W. Kukla, H.G. Berry, and C.A. Kurtz (1995): Nucl. Instrum Methods B **99**, 117

Tarnovsky, V. and K. Becker (1993): J. Chem. Phys. **98**, 7868

Theuws, P.G.A., H.C.W. Beijerinck, D.C. Schram, and N.F. Verster (1977): J. Appl. Phys. **48**, 2261

Theuws, P.G.A., H.C.W. Beijerinck, N.F. Verster, and D.C. Schram (1982): J. Phys. E **15**, 573

Thompson, M.W. (1968): Phil. Mag. **18**, 377

Treanor, C.E. and G.T. Skinner (1961): Plant. Space Sci. **3**, 253

Trujillo, S.M., R.H. Neynaber, and E.W. Rothe (1966): Rev. Sci. Instrum. **37**, 1655

Utterback, N.G. and G.H. Miller (1961): Rev. Sci. Instrum. **32**, 1101 and Phys. Rev. **124**, 1477

Utterback, N.G. (1966): J. Chem. Phys. **44**, 2540

Vályi, L. (1977): *Atom and Ion Sources* (Wiley, New York)

van Amersfoort, P.W., J.J.C. Geerlings, L.F.Tz. Kwakman, E.H.A. Granneman, and J. Los (1985): J. Appl. Phys. **58**, 2312

van de Runstaat, C.A., R.W. van Resandt, and J. Los (1970): J. Phys. E **3**, 575

van der Zande, W.J., W. Koot, J.R. Petersen, and J. Los (1988): Chem. Phys. **126**, 169

van der Zande, L.D.A. Siebbeles, J.M. Schins, and J. Los (1992): in *Electronic and Atomic Collisions*, ed. by MacGillivray, W.R., I.E. McCarthy, and M.C. Standage, p. 517 (Adam Hilger, Bristol)

van Zyl, B., N.G. Utterback, and R.C. Amme (1976): Rev. Sci. Instrum. **47**, 814

van Zyl, B., H. Neumann, T.Q. Le, and R.C. Amme (1978): Phys. Rev. A **18**, 506

Varentsov, V.L. and D.R. Hansevarov (1992): Nucl. Instrum. Methods A **317**, 1

Varentsov, V.L., D.R. Hansevarov, and D.V. Varentsov (1995): Nucl. Instrum. Methods A **352**, 542

Vidal, C.R. (1996): Exp. Methods Phys. Sci. B **29**, 67

Viehl, A., M. Kanyo, A. van der Hart, and J. Schelten (1993): Rev. Sci. Instrum. **64**, 732

Volosov, V.I. and I.N. Churkin (1997): Surf. Coatings Technol. **96**, 75

Vostrikov, A.A. and A.Y. Dubov (1991): Z. Phys. **20**, 61

Vu Ngoe Tuan, G. Gautherin, and A.S. Schlachter (1974): Phys. Rev. A **9**, 1242

Wakasugi, M., W.G. Jin, T.T. Inamura, T. Murayama, T. Wakui, T. Kashiwabara, H. Katsuragawa, T. Ariga, T. Ishizuka, M. Koizumi, and I. Sugai (1993): Rev. Sci. Instrum. **64**, 3487

Wakasugi, M., W.G. Jin, M.G. Hies, T.T. Inamura, T. Murayama, T. Ariga, T. Ishizuka, T. Wakui, H. Katsuragawa, J.Z. Ruan, I. Sugai, and A. Ikeda (1996): Phys. Rev. B **53**, 611

Wang, H., R. Craig, H. Haouari, J.-G. Dong, Z. Hu, A. Vivoni, J.R. Lombardi, and D.M. Lindsay (1995): J. Chem. Phys. **103**, 3289

Ward, R.F. Jr., W.G. Sturrus, and S.R. Landeen (1996): Phys. Rev. A **53**, 113

Warrington, R.B., D.M. Lucas, D.N. Stacey, and C.D. Thompson (1995): Physica Scripta T **59**, 424

Way, K.R., S.C. Yang, and W.C. Stwalley (1976): Rev. Sci. Instrum. **47**, 1049

Webster, H.C. (1930): Proc. Cambridge Phil. Soc. **27**, 116

Wehner, G.K. (1955): Adv. Electron. Electron Phys. **7**, 258

Wehner, G.K. (1958): Phys. Rev. **112**, 1120

Wehner, G.K. (1959): Phys. Rev. **114**, 1270

Wehner, G.K. (1960): J. Appl. Phys. **31**, 1392

Weiller, B.H., P.S. Bechthold, E.K. Parks, and S.J. Riley (1989): J. Chem. Phys. **91**, 4714

Weiner, J., E.B. Peatman, and R.S. Berry (1971): Phys. Rev. A **4**, 1824

Wells, S.B., Y. Takeiri, A.F. Newmann, R. McAdams, and A.J.T. Holmes (1992): Rev. Sci. Instrum. **63**, 2735

Wick, O. (1994): Thesis, University of Freiburg, Germany

Wien, W. (1919): Ann. Phys., **60**, 517

Wiesemann, K. (1981): 8th Int. Symp. Mol. Beams, Book of Abstracts, ed. by M.F. Devienne, p. 16 (Cannes)

Winicur, D.H. and E.L. Knuth (1967): J. Chem. Phys. **46**, 4318

Wittkower, A.B., P.H. Rose, R.P. Bastide, and N.B. Brooks (1964): Rev. Sci. Instrum. **35**, 1

Wolan, J.T., C.K. Mount, and G.B. Hofland (1997): J. Vac. Sci. Technol. A **15**, 2502

Wolfgang, R. (1965): in *Progress in Reaction Kinetics*, Vol. 3, ed. by G. Porter, p. 97 (Pergamon Press, Oxford)

Wolfgang, R. (1968): Sci. Am. **218**, No.4, 44

Wouters, E.R., L.D.A. Siebbeles, P.C. Schuddeboom, B.R. Chalamala, and W.J. van der Zande (1996): Phys. Rev. A **54**, 522

Xu N., Y. Du, Z. Ying, Z. Ren, and F. Li (1997): Rev. Sci. Instrum. **68**, 2994

Young, W.S., W.E. Rodgers, and E.L. Knuth (1969): Rev. Sci. Instrum. **40**, 1346

Young, C.E., C.M. Sholeen, A.F. Wagner, A.E. Proctor, L.G. Pobo, and S. Wexler (1981): J. Chem. Phys. **4**, 1770

Young, L., C.A. Kurtz, D.R. Beck, and D. Datta (1993): Phys. Rev. A **48**, 173

Young, L. (1996): Exp. Methods Phys. Sci. B **29**, 301

Zheng, L.S., P.J. Burcat, C.L. Pettiette, S. Yang, and R.E. Smalley (1985): J. Chem. Phys. **83**, 4273

Zinoviev, A.N. (1992): in *Electronic and Atomic Collisions*, ed. by MacGillivray, W.R., I.E. McCarthy, and M.C. Standage, p. 105 (Adam Hilger, Bristol)

Chapter 2

Abe, H., W.Schulze, and B. Tesche (1980): Chem. Phys. **47**, 95

Abraham, O., S.-S. Kim, and G.D. Stein (1981): J. Chem. Phys. **75**, 402

Abshagen, M., J. Kowalski, M. Meyberg, G. zu Putlitz, J. Slaby, and F. Träger (1991): Z. Phys. D **19**, 199

Adamczewski, J., D. Löffelmacher, J. Meijer, A. Stephan, H.H. Bukow, and C. Rolfs (1997): Nucl. Instrum. Methods B **130**, 57

Aitchison, J. and J.A.C. Brown (1973): *The Lognormal Distribution* (Cambridge University Press, Cambridge)
Alexander, M.L., N.E. Leviner, M.A. Johnson, D. Ray, and W.C. Lineberger (1988): J. Chem. Phys. **88**, 6200
Åman, C. and L. Holmlid (1991): J. Phys. D **24**, 1049
Ancilotto, F., P.B. Lerner, and M.W. Cole (1995): J. Low Temp. Phys. **101**, 1123
Andersson, M., J.L. Persson, and A. Rosén (1996): J. Chem. Phys. **100**, 12222
Ando, K., H. Akinaga, and T. Okuda (1994): Bull. Electrotech. Lab. **58**, 121
Aoki, T., J. Matsuo, Z. Insepov, and I. Yamada (1997): Nucl. Instrum. Methods B **121**, 49
Apsel, S.E., J.W. Emmert, J. Deng, and L.A. Bloomfield (1996): Phys. Rev. Lett. **76**, 1441
Arnold, M., J. Kowalski, G. zu Putlitz, T. Stehlin, and F. Träger (1985a): Surf. Sci. **156**, 149
Arnold, M., J. Kowalski, G. zu Putlitz, T. Stehlin, and F. Träger (1985b): Z. Phys. A **322**, 179
Baede, A.P.M., W.F. Jungman, and J. Los (1971): Physica **54**, 459
Bahat, D., O. Cheshnovsky, U. Even, N. Lavie, and Y. Magen (1987): J. Phys. Chem. **91**, 2460
Baker, S.H., S.C. Thornton, A.M. Keen, T.I. Preston, C. Norris, K.W. Edmonds, and C. Binns (1997): Rev. Sci. Instrum. **68**, 1853
Barr, D.L. (1987): J. Vac. Sci. Technol. B **5**, 184
Bartelt, A., J.D. Close, F. Federmann, N. Quaas, and J.P. Toennies (1996): Phys. Rev. Lett. **77**, 3525
Bartelt, A., J.D. Close, F. Federmann, K. Hoffmann, N. Quaas, and J.P. Toennies (1997): Z. Phys. D **39**, 1
Batson, P.E. and J.R. Heath (1993): Phys. Rev. Lett. **71**, 911
Baumert, T. and G. Gerber (1995): Adv. At. Mol. Opt. Phys. **35**, 163
Becker, E.W., K. Bier, and W. Henkes (1956): Z. Phys. **146**, 333
Becker, E.W., R. Klingelhöfer, and P. Lohse (1962): Z. Nat.forsch. A **17**, 432
Begemann, W., S. Dreihöfer, K.H. Meiwes-Broer, and H.O. Lutz (1986): Z. Phys. D **3**, 183
Benslimane, M., M. Châtelet, A. De Martino, F. Pradère, and H. Vach (1995): Chem. Phys. Lett. **237**, 223
Bentley, P.G. (1961): Nature **190**, 432
Bernstein, E.R. (ed.) (1990): *Atomic and Molecular Clusters* (Elsevier, Amsterdam)
Berkowitz, J. and W.A. Chupka (1964): J. Chem. Phys. **40**, 2735
Betz, G. and W. Husinsky (1997): Nucl. Instrum. Methods B **122**, 311
Bhaskar, N.D., R.F. Frueholz, C.M. Klimack, and R.A. Look (1987): Phys. Rev. B **36**, 4418
Billas, I.M.L., A. Châtelain, and W.A. de Heer (1994): Science **265**, 1682
Billas, I.M.L., A. Châtelain, and W.A. de Heer (1996): Surf. Rev. Lett. **3**, 429
Bischoff, L., J. Teichert, E. Hesse, P.D. Prewett, and J.G. Watson (1996): Microelectronic Engineering **30**, 245
Bjørnholm, S. (1990): Contemporary Physics **31**, 309
Bjørnholm, S., J. Borggreen, O. Echt, K. Hansen, J. Pederson, and H.D. Rasmussen (1991): Z. Phys. D **19**, 47
Blanckenhagen von, P., A. Gruber, and J. Gspann (1997): Nucl. Instrum. Methods B **122**, 322
Boesiger, J. and S. Leutwyler (1987): Z. Phys. Chem. **154**, 31
Bouwen, W., P. Thoen, F. Vanhoutte, S. Bouckaert, F. Despa, H. Weidele, R.E. Silverans, and P. Lievens (2000): Rev. Sci. Instrum. **71**, 54
Brack, M. (1993): Rev. Mod. Phys. **65**, 677

Brack, M. (1997): Sci. Am. **277**, 30

Bréchignac, C., Ph. Cahuzac, J. Leygnier, R. Pflaum, and J. Weiner (1988): Phys. Rev. Lett. **61**, 314

Bréchignac, C., Ph. Cahuzac, F. Carlier, J. Leygnier, and I.V. Hertel (1990): Z. Phys. D **17**, 61

Bréchignac, C., Ph. Cahuzac, F. Carlier, M. de Frutos, A. Masson, and J.Ph. Roux (1991): Z. Phys. D **19**, 195

Bromann, K., C. Félix, H. Brune, W. Harbish, R. Monot, J. Buttet, and K. Kern (1996): Science **274**, 956

Broyer,M., B. Cabaud, A. Hoareau, P. Mélinon, D. Rayane, and B. Tribollet (1987): Mol. Phys. **62**, 559

Broyer, M., M. Pellarin, B. Baguernard, J. Lerme, J.L. Vialle, P. Mélinon, J. Tuaillon, V. Dupuis, B. Prevel, and A. Perez (1996): Mat. Sci. Forum **232**, 27

Buck, U. and H. Pauly (1968): Z. Phys. **208**, 390

Buck, U. and H. Meyer (1984): Phys. Rev. Lett. **52**, 109

Buck, U., H. Meyer, and H. Pauly (1985): in *Flow of Real Fluids*, ed. by G.E.A. Meier and F. Obermeyer, p. 170 (Springer, Berlin)

Buck, U. and H. Meyer (1986): J. Chem. Phys. **84**, 4854

Buck, U. (1994): in *Clusters of Atoms and Molecules*, Vol. 1, ed. by H. Haberland, p. 232 (Springer, Heidelberg)

Buck, U. (1995): Adv. At. Mol. Opt. Phys. **35**, 121

Buck, U. and R. Krohne (1996): J. Chem. Phys. **105**, 5408

Campbell, E.E.B., H.G. Ulmer, B. Hasselberger, H.-G. Busmann, and I.V. Hertel (1990): J. Chem. Phys. **93**, 6900

Campbell, E.E.B., R. Ehlich, A. Hielscher, J.M.A. Frazao, and I.V. Hertel (1992): Z. Phys. D **23**, 1

Campbell, E.E.B., R. Tellgmann, N. Krawez, and I.V. Hertel (1997): J. Phys. Chem. Solids **58**, 1763

Cannon, W.R., S.C. Danforth, J.H. Flint, J.S. Haggerty, and R.A. Marra (1982): J. Am. Ceram. Soc. **65**, 324

Castleman, A.W. and R.G. Keesee (1988): Science **241**, 36

Cha, C.-Y., G. Ganteför, and W. Eberhardt (1992): Rev. Sci. Instrum. **63**, 5661

Chai, Y., T. Guo, C. Jin, R.E. Haufler, L.P.F. Chibante, J. Fure, L. Wang, J.L. Alford, and R.E. Smalley (1991): J. Phys. Chem. **95**, 7564

Châtelet, M., A. De Martino, J. Pettersson, F. Pradère, and H. Vach (1992): Chem. Phys. Lett. **196**, 563

Chibante, L.P.F. and R.E. Smalley (1993): in *On Clusters and Clustering*, ed. by P.J. Reynolds (North-Holland, Amsterdam)

Cho, M.-H., D.-H. Ko, K. Jeong, S.W. Whangbo, C.N. Whang, S.C. Choi, and S.J. Cho (1999): J. Appl. Phys. **85**, 2909

Cich, M., K. Kim, H. Choi, and S.T. Hwang (1998): Appl. Phys. Lett. **73**, 2116

Clampitt, R. and D.K. Jefferies (1978): Nucl. Instrum. Methods **149**, 739

Clark, W.M., R.L. Selliger, M.W. Utlaut, A.E. Bell, L.W. Swanson, G.A. Schwind, and J.B. Jergenson (1987): J. Vac. Sci. Technol. B **5**, 197

Coon, S.R., W.F. Calaway, J.W. Burnett, M.J. Pellin, D.M. Gruen, D.R. Spiegel, and J.M. White (1991): Surf. Sci. **259**, 275

Coon, S.R., W.F. Calaway, M.J. Pellin, and J.M. White (1993): Surf. Sci. **258**, 161

Cox, A.J., J.G. Louderback, and L.A. Bloomfield (1993): Phys. Rev. Lett. **71**, 923

Curl, R.F. (1997): Rev. Mod. Phys. **69**, 691

Cuvellier, J., P. Meynadier, P. de Pujo, O. Sublemontier, J.P. Visticot, J. Berlande, A. Lallement, and J.M. Mestdagh (1991): Z. Phys. D **21**, 265
DeBoer, B.G. and G.D. Stein (1981): Surf. Sci. **106**, 84
de Heer, W.A. and W.D. Knight (1988): in *Elemental and Molecular Clusters*, ed. by G. Benedek, T.P. Martin, and G. Pacchioni, p. 45 (Springer, Berlin)
de Heer, W.A. (1993): Rev. Modern Physics, **65**, 611
De Martino, A., M. Benslimane, M. Châtelet, C. Crozes, F. Pradère, and H. Vach (1993): Z. Phys. D **27**, 185
De Martino, A., M. Benslimane, M. Châtelet, F. Pradère, and H. Vach (1996): J. Chem. Phys. **105**, 7828
Delacrétaz, G. and L. Wöste (1985): Surf. Sci. **156**, 770
Denman, H.H., W. Heller, and J. Pangouis (1966): *Angular Scattering Functions for Spheres* (Wayne State University Press, Detroit)
Desai, S.R., C.S. Feigerle, and J.C. Miller (1994): J. Chem. Phys. **101**, 4526
Devienne, F.M. and J.C. Rouston (1982): Org. Mass Spectrometry **17**, 173
Diederich, L., E. Barborini, P. Piseri, A. Podestà, P. Milani, A. Schneuwly, and R. Gallay (1999): Appl. Phys. Lett. **75**, 2662
Dienes, J.K. and J.M. Walsh (1970): in *High Velocity Impact Phenomena*, ed. by R. Kinslow, p. 45 (Academic Press, New York)
Dietz, T.G., M.A. Duncan, D.E. Powers, and R.E. Smalley (1981): J. Chem. Phys. **74**, 6511
Ditmire, T., E. Springate, J.W.G. Tisch, Y.L. Shao, M.B. Mason, N. Hay, J.P. Marangos, and M.H.R. Hutchinson (1998a): Phys. Rev. A **57**, 369
Ditmire, T., P.K. Patel, R.A. Smith, J.S. Wark, S.J. Rose, D. Milathianaki, R.S. Marjorinanks, and M.H.R. Hutchinson (1998b): J. Phys. B **31**, 2825
Dombrowski, H., D. Grzouka, W. Hamsink, A. Khoukaz, T. Lister, and R. Santo (1996): Nucl. Phys. A **626**, 427
Dombrowski, H., A. Khoukaz, and R. Santo (1997): Nucl. Phys. A **619**, 97
Dornenburg, E. and H. Hintenberger (1959): Z. Nat.forsch. A **14**, 765
Dresselhaus, M.S., G. Dresselhaus, and P.C. Eklund (1996): *Science of Fullerenes and Carbon Nanotubes* (Academic Press, San Diego)
Echt, O. (1996): in *Large Clusters of Atoms and Molecules*, ed. by T.P. Martin, p. 443 (Kluwer, Doordrecht)
Ehbrecht, M., M. Färber, F. Rohmund, V.V. Smirnov, O.M. Stelmakh, and F. Huisken (1993a): Chem. Phys. Lett. **214**, 34
Ehbrecht, M., M. Stemmler, and F. Huisken (1993b): Int. J. Mass Spectrom. Ion. Process. **123**, R1
Ehbrecht, M., H. Ferkel, V.V. Smirnov, O.M. Stelmakh, W. Zhang, and F. Huisken (1995): Rev. Sci. Instrum. **66**, 3833
Ehbrecht, M., H. Ferkel, V.V. Smirnov, O.M. Stelmakh, W. Zhang, and F. Huisken (1996): Surf. Rev. Lett. **3**, 807
Ehbrecht, M., H. Ferkel, and F. Huisken (1997a): Z. Phys. D **40**, 88
Ehbrecht, M., B. Kohn, F. Huisken, M:A. Laguna, and V. Paillard (1997b): Phys. Rev. B **56**, 6958
Ehbrecht, M. and F. Huisken (1999): Phys. Rev. B **59**, 2975
Ehrenreich, H. and F. Spaepen (eds.) (1994): *Solid State Physics*, Vol. 48 (Academic Press, New York)
Engelking, P.C. (1987): J. Chem. Phys. **87**, 936
Fabre, C., M. Gross, J.M. Raimond, and S. Haroche (1983): J. Phys. B **16**, L671
Falter, H., O.F. Hagena, W. Henkes, and H.V. Wedel (1970): Int. J. Mass Spectrom. Ion Process. **4**, 145

Farges, J., B. Raoult, and G. Torchet (1973): J. Chem. Phys. **59**, 3454
Farges, J., M.F. de Feraudy, B. Raoult, and G. Torchet (1981): Surf. Sci. **106**, 95
Farges, J., M.F. de Feraudy, B. Raoult, and G. Torchet (1983): J. Chem. Phys. **78**, 5067
Farges, J., M.F. de Feraudy, B. Raoult, and G. Torchet (1986): J. Chem. Phys. **84**, 3491
Fárník, M., U. Henne, B. Samelin, and J.P. Toennies (1997): Z. Phys. D **40**, 93
Fayet, P. and L. Wöste (1985): Surf. Sci. **156**, 134
Fayet, P., J.P. Wolf, and L. Wöste (1986): Phys. Rev. B **33**, 6792
Feldman, Y., E. Wasserman, D.J. Srolovitz, and R. Tenne (1995): Science **267**, 222
Fort, E., F. Pradère, A. De Martino, H. Vach, and M. Châtelet (1998): Eur. Phys. J. D **1**, 79
Fortov, V.E., T.P. Novikova, A.N. Lebedev, G.S. Romanov, V. Skvortsov, and A.V. Teterev (1995): Int. J. Impact Engineering **17**, 323
Frank, F., W. Schulze, B. Tesche, J. Urban, and B. Winter (1985): Surf. Sci. **156**, 90
Ganteför, G., H.R. Siekmann, H.O. Lutz, and K.H. Meiwes-Broer (1990): Chem. Phys. Lett. **165**, 293
Gatz, P. and O.F. Hagena (1995): J. Vac. Sci. Technol. A **13**, 2128
Gingerich, K., I. Shim, S. Gupta, and J. Kingcade Jr. (1985): Surf. Sci. **156**, 495
Gehring, J.W. (1970): in *High-Velocity Impact Phenomena*, ed. by R. Kinslow, p. 463 (Academic Press, New York)
Geusic, M., M. Morse, S. O'Brian, and R. Smalley (1985): Rev. Sci. Instrum. **56**, 2123
Goldby, I.M., B. von Issendorff, L. Kuipers, and R.E. Palmer (1997): Rev. Sci. Instrum. **68**, 3327
Goto, M., J. Murakami, Y. Tai, K. Yoshimura, K. Igarashi, and S. Tanemura (1997): Z. Phys. D **40**, 115
Gough, T., D. Knight, and G. Scoles (1983): Chem. Phys. Lett. **97**, 155
Gough, T., M. Mengel, P. Rowntree, and G. Scoles (1985): J. Chem. Phys. **83**, 4958
Goyal, S., G.N. Robinson, D.L. Schutt, and G. Scoles (1991): J. Chem. Phys. **95**, 4186
Greene, F.T. and T.A. Milne (1963): J. Chem. Phys. **39**, 3150
Gromov, A., W. Krätschmer, N. Krawez, R. Tellgmann, and E.E.B. Campbell (1997): Chem. Commun. **20**, 2003
Gspann, J. (1982): in *Physics of Electronic and Atomic Collisions*, ed. by S. Datz, p. 79 (North-Holland, Amsterdam)
Gspann, J. and H. Vollmer (1979): *Rarefied Gas Dynamics* 11, Vol. 2, R. ed. by R. Campargue, p. 1193 (CEA, Paris)
Gspann, J. (1992): in *From Clusters to Crystals*, Vol. 2, ed. by P. Jena et al., p. 1115 (Kluwer, Doordrecht)
Gspann, J. (1995): Microelectronic Engineering **27**, 517
Gspann, J. (1996): in *Large Clusters of Atoms and Molecules*, ed. by T.P. Martin, p. 443 (Kluwer, Doordrecht)
Gu, X.J., D.J. Levandier, B. Zhang, G. Scoles, and D. Zhuang (1990): J. Chem. Phys. **93**, 4898
Haberland, H., M. Karrais, and M. Mall (1991): Z. Phys. D **20**, 413
Haberland, H., M. Karrais, M. Malland, and Y. Thurner (1992): J. Vac. Sci. Technol. A**10**, 3266
Haberland, H. (1994): in *Clusters of Atoms and Molecules*, Vol. 1, ed. by H. Haberland, p. 207 (Springer, Heidelberg)
Haberland, H., Z. Insepov, and M. Moseler (1995): Phys. Rev. B **51**, 11062
Haberland, H., M. Moseler, Y. Qiang, O. Rattunde, T. Reiners, and Y. Thurner (1996): Surf. Rev. Lett. **3**, 887
Hagena, O.F. and Henkes (1965): Z. Nat.forsch. A **20**, 1344
Hagena, O.F. and W. Obert (1972): J. Chem. Phys. **56**, 1793

Hagena, O.F. (1981): Surf. Sci. **106**, 101
Hagena, O.F. (1987): Z. Phys. D **4**, 291
Hagena, O.F. (1991): Z. Phys. D **20**, 425
Hagena, O.F. (1992): Rev. Sci. Instrum. **63**, 2374
Hanley, L. and S.L. Anderson (1985): Chem. Phys. Lett. **122**, 410
Hanley, L., S.A. Ruatta, and S.L. Anderson (1987): J. Chem. Phys. **87**, 260
Hartmann, M., R.E. Miller, J.P. Toennies, and A. Vilesov (1995): Phys. Rev. Lett. **75**, 1566
Hartmann, M., R.E. Miller, J.P. Toennies, and A. Vilesov (1996): Science **272**, 1631
Heath, J.R., Y. Liu, S.C. O'Brian, Q.-L. Zhang, R.F. Curl, F.K. Tittel, and R.E. Smalley (1985): J. Chem. Phys. **83**, 5520
Hegerfeldt, G.C. and T. Köhler (1998): Phys. Rev. A **57**, 2021
Henkes, P.R.W. (1961): Z. Nat.forsch. A **16**, 842
Henkes, P.R.W. (1962): Z. Nat.forsch. A **17**, 786
Henkes, P.R.W. and R. Klingelhöfer (1989): J. de Physique **50**, 159
Herrmann, A., S. Leutwyler, L. Wöste, and E. Schumacher (1979): Chem. Phys. Lett. **62**, 444
Hill, T.L. (1948): J. Chem. Phys. **16**, 181
Hintz, P.A. and K.M. Ervin (1994): J. Chem. Phys. **100**, 5715
Hirsch, A. (1998): *Fullerenes and Related Structures* (Springer, Heidelberg)
Hofmann, M., S. Leutwyler, and W. Schulze (1979): Chem. Phys. **40**, 145
Hofmeister, H., F. Huisken, and B. Kohn (1999): Eur. Phys. J. D **9**, 137
Hogg, E.O. and B.G. Silbernagel (1974): J. Appl. Phys. **45**, 593
Holmgren, L., H. Gronbeck, M. Anderson, and A. Rosén (1996): Phys. Rev. B **53**, 16644
Holmlid, L., J.B.C. Pettersson, C. Åman, and B. Lönn (1992): Rev. Sci. Instrum. **63**, 1966
Honig, R.E. (1963): *Advances in Mass Spectrometry*, Vol. 2, ed. by R.M. Elliott (Pergamon Press, London)
Hopkins, J.B., P.R.R. Langridge Smith, M.D. Morse, and R.E. Smalley (1983): J. Chem. Phys. **78**, 1627
Hu, C.W., A. Kasuya, A. Wawri, N. Horiguchi, R. Czajkam Y. Nishina, Y. Saito, and H. Fujita (1996): Mat. Sci. Engineering A **217–218**, 103
Huisken, F. and M. Stemmler (1993): J. Chem. Phys. **98**, 7680
Huisken, F., B. Kohn, R. Alexandrescu, S. Cojocaru, A. Crunteanu, C Reynaud, and G. Ledoux (1999a): J. Nanoparticle Research **1**, 293
Huisken, F., B. Kohn, and V. Paillard (1999b): Appl. Phys. Lett. **74**, 3776
Huisken, F., B. Kohn, R. Alexandrescu, and I. Morjan (1999c): Eur. Phys. J. D **9**, 141
Huisken, F., H. Hofmeister, B. Kohn, M.A. Laguna, and V. Paillard (2000): Appl. Surf. Sci. **154–155**, 305
Insepov, Z. and I. Yamada (1996): Nucl. Instrum. Methods B **112**, 16
Insepov, Z. and I. Yamada (1997): Nucl. Instrum. Methods B **121**, 44
Ishitani, T., K. Umemura, S. Hosoki, S, Takayama, and H. Tamura (1984): J. Vac. Sci. Technol. A **2**, 1365
Jena, P., S.N. Khanna, and B.K. Rao (1996): Surf. Rev. Lett. **3**, 993
Jortner, J. (1984): Ber. Bunsenges. Phys. Chem. **88**, 188
Jortner, J., D. Scharf, and U. Landman (1988): in *Elemental and Molecular Clusters*, ed. by G. Benedek, T.P. Martin, and G. Pacchioni, p. 148 (Springer, Berlin)
Jortner, J. (1995): J. Chem. Phys. **92**, 205
Kappes, M., R. Kunz, and E. Schumacher (1982): Chem. Phys. Lett. **91**, 413
Kappes, M. and E. Schumacher (1985): Surf. Sci. **156**, 1
Kappes, M., M. Schär, P. Radi, and E. Schumacher (1986): J. Chem. Phys. **84**, 1863
Kappes, M., M. Schär, E. Schumacher, and P. Vayloyan (1987): Z. Phys. D **5**, 359

Kappes, M. and S. Leutwyler (1988): in *Atomic and Molecular Beam Methods*, Vol. 1, ed. by G. Scoles, p. 380 (Oxford University Press, Oxford)
Kappes, M. (1988): Chem. Rev. **88**, 369
Karnbach, R., M. Joppien, J. Stapelfeldt, J. Wörmer, and T. Möller (1993): Rev. Sci. Instrum. **64**, 2838
Katakuse, I. (1987): *Microclusters*, ed. by S. Sugano, Y. Nishina, and S. Ohnishi, p. 10 (Springer, Heidelberg)
Kaya, K., K. Hoshino, T. Kurikawa, and H. Takada (1996): in *Structures and Dynamics of Clusters*, ed.by T. Kondow, K. Kaya, and A. Terasaki, p. 129 (Academy Press, Tokyo)
Keesee, R.G., R. Sievert, and A.W. Castleman (1984): Ber. Bunsenges. Phys. Chem. **88**, 906
Kelchner, C.L. and A.E. DePristo (1997): Nanostructured Materials **8**, 253
Kerker, M. (1969): *The Scattering of Light* (Academic Press, New York)
Khmel', S.Y. and R.G. Sharafutdinov (1997): Techn. Phys. **42**, 775
Khmel', S.Y. and R.G. Sharafutdinov (1998): Techn. Phys. **43**, 986
Kikushi, K., S. Suzuki, Y. Nakao, N. Nakahara, T. Wakabayashi, H. Shiromaru, Y. Saito, I. Ikemoto, and Y. Achiba (1993): Chem. Phys. Lett. **216**, 67
Kim, S.S. and G.D. Stein (1982): Rev. Sci. Instrum. **53**, 838
Kimoto, K. and I. Nishida (1977): J. Phys. Soc. Japan **42**, 2071
Kinne, M., T.M. Bernhardt, B. Kaiser, and K. Rademann (1997): Z. Phys. D **40**, 105
Kittel, C. (1976): *Introduction to Solid State Physics* (Wiley, New York)
Klabunde, K.J. (1994): *Free Atoms, Clusters, and Nanoscale Particles* (Academic Press, New York)
Klingelhöfer, R. and H. Moser (1972): J. Appl. Phys. **43**, 4575
Kojima, Y., S. Suzuki, and T. Wakabayashi (1996): in *Structures and Dynamics of Clusters*, ed. by T. Kondow, K. Kaya, and A. Terasaki, p. 505 (Universal Academy Press, Tokyo)
Kondow, T., K. Kaya, and A. Terasaki (eds.) (1996): *Structures and Dynamics of Clusters* (Universal Academy Press, Tokyo)
Kovalenko, S.I., D.D. Solnyshkin, E.A. Bondarenko, E.T. Verhovtseva, and V.V. Eremenko (1998): Fizika Niskikh Temperatur **24**, 481
Krätschmer, W., L.D. Lamb, K. Fostiropoulos, and D.R. Huffman (1990): Nature **347**, 354
Kresin, V.V., V. Kasperovich, G. Tikhonov, and K. Wong (1998): Phys. Rev. **57**, 383
Krohn, V.E. (1961): Prog. Astronaut. Rocketry **5**, 73
Kroto, H.W., J.R. Heath, S.C. O'Brien, R.F. Curl, and R.E. Smalley (1985): Nature **318**, 162
Kroto, H.W. (1997): Rev. Mod. Phys. **69**, 703
Kurikawa, T., M. Hirano, H. Takeda, K. Yagi, K. Hoshino, A. Nakajima, and K. Kaya (1995): J. Phys. Chem. **99**, 16248
Kurikawa, T., H. Takeda, A. Nakajima, and K. Kaya (1997): Z. Phys. D **40**, 65
Lallement, A., J. Cuvellier, J.M. Mestdagh, P. Meynadier, P. de Pujo, O. Sublemontier, J.P. Visticot, J. Berlande, and X. Biquard (1992): Chem. Phys. Lett. **189**, 182
Landman, U. and L. Charles (1992): Science **257**, 355
Lee, J.W. and G.D. Stein (1985): Surf. Sci. **156**, 112
Lee, J.W. and G.D. Stein (1987): J. Phys. Chem. **91**, 2450
Leutwyler, S. (1984): J. Chem. Phys. **81**, 5480
Levandier, D.J., S. Goyal, J. McCombie, B. Pate, and G. Scoles (1990): J. Chem. Soc. Faraday Trans. **86**, 2361
Lewerenz, M., B. Schilling, and J.P. Toennies (1993): Chem. Phys. Lett. **206**, 381
Lewerenz, M., B. Schilling, and J.P. Toennies (1995): J. Chem. Phys. **102**, 8191

Lievens, P., P. Thoen, S. Bouckaert, W. Bouwen, E. Vandeweert, F. Vanhoutte, and H. Weidele (1997): Z. Phys. D **42**, 231
Liu, X.H., X.G. Zhang, Y. Li, X.Y. Wang, and N.Q. Lou (1998): Chem. Phys Lett. **288**, 805
Lorents, D.C. (1997): Comments At. Mol. Phys. **33**, 125
Lu, W., R. Huang, J. Ding, and S. Yang (1996): J. Chem. Phys. **104**, 6577
Luo, F., C.F. Giese, and W.R. Gentry (1996): J. Chem. Phys. **104**, 1151
Ma, Z., S.R. Coon, W.F. Calaway, M.J. Pellin, D.M. Gruen, and E.I. von Nagy-Felsobuki (1994): J. Vac. Sci. Technol. A **12**, 2425
Macler, M. and Y.K. Bae (1997): J. Phys. Chem. A **101**, 145
Maguera, T.F., D.E. David, and J. Michel (1990): J. Chem. Soc. Faraday Trans. **86**, 2427
Mahoney, J.F., E.S. Parilis, and T.D. Lee (1994): Nucl. Instrum. Methods B **88**, 154
Mandich, M.L., W.D. Reents Jr., and V.E. Bondybey (1990): in *Atomic and Molecular Clusters*, ed. by E.R. Bernstein, p. 69 (Elsevier, Amsterdam)
Martin, T.P. (1984): J. Chem. Phys. **80**, 170
Martin, T.P. and A. Kakizaki (1984): J. Chem. Phys. **80**, 3956
Martin, T.P. (1986): Angew. Chem. Int. Edn. Engl. **25**, 197
Martin, T.P. (1996): Physics Reports **273**, 199
Maruyama, S., L.R. Anderson, and R.E. Smalley (1990): Rev. Sci. Instrum. **61**, 3686
Mélinon, P., V. Paillard, V. Dupuis, A. Perez, P. Jensen, A. Hoareau, J.P. Perez, J. Tuaillon, and M. Broyer (1995): Int. J. Mod. Phys. B **9**, 339
Mélinon, P., P. Kéghélian, B. Prével, A. Perez, G. Guiraud, J. Lebrusq, J. Lermé, M. Pellarin, and M. Broyer (1997): J. Chem. Phys. **107**, 10278
Messiah, A. (1969): *Quantum Mechanics*, Vol. 2 (North-Holland, Amsterdam)
Milani, P. and W.A. de Heer (1990): Rev. Sci. Instrum. **61**, 1835
Milani, P. and W.A. de Heer (1991): Phys. Rev. B **44**, 8346
Miller, D.R., M.A. Fineman, and H.R. Murphy (1985): *Rarefied Gas Dynamics* 13, Vol. 2, ed. by O.M. Belotserkovskii, M.N. Kogan, S.S. Kutateladse, and A.K. Rebrov, p. 923 (Plenum Press, New York)
Miller, D.R. (1988): in *Atomic and Molecular Beam Methods*, Vol. 1, ed. by G. Scoles, p. 14 (Oxford University Press, Oxford)
Möller, K. and L. Holmlid (1988): Surf. Sci. **204**, 98
Mühlbach, J., K. Sattler, and E. Recknagel (1981): Surf. Sci. **106**, 18
Murphy, H.R. and D.R. Miller (1984): J. Phys. Chem. **88**, 4474
Nakajima, A., T. Taguwa, K. Hortino, T. Sugoika, T. Naganuma, F. Ono, K. Watanabe, K. Nakao, Y. Konishi, R. Kishi, and K. Kaya (1993): Chem. Phys. Lett. **214**, 22
Nakajima, A., N. Zhang, H. Kawamata, T. Hayase, R. Kishi, and K. Kaya (1996): in *Structures and Dynamics of Clusters*, ed. by T. Kondow, K. Kaya, and A. Terasaki, p. 377 (Universal Academy Press, Tokyo)
Newton, R.G. (1966): *Scattering Theory of Waves and Particles* (McGraw-Hill, New York)
Papanikolas, J.M., J.R. Gord, N.E. Levinger, D. Ray, V. Vorsa, and W.C. Lineberger (1991): J. Chem. Phys. **95**, 8028
Parks, E.K., G.C. Niemann, and S.J. Riley (1996): J. Chem. Phys. **104**, 3531
Pauly, H. and J.P. Toennies (1965): Adv. At. Mol. Phys. **1**, 195
Pauly, H. and J.P. Toennies (1968): in *Methods of Experimental Physics*, Vol. 7A, ed. by B. Bederson and W.D. Fite, p. 227 (Academic Press, New York)
Pauly, H. (1979): in *Atom–Molecule Collision Theory*, ed. by R.B. Bernstein, p. 111 (Plenum Press, New York)
Pederson, J., S. Bjørnholm, J. Broggreen, K. Hansen, T.P. Martin, and H.D. Rasmussen (1991): Nature **353**, 733

Pedersen, D.B., J.M. Parnis, and D.M. Rayner (1998): J. Chem. Phys. **109**, 551
Penner, A. and A. Amirav (1993): J. Chem Phys. **99**, 9616
Perez, A., P. Mélinon, V. Dupuis, P. Jensen, B. Prevel, J. Tuaillon, L. Bardotti, C. Martet, and M. Treilleux (1997): J. Phys. D **30**, 709
Pfund, A.H. (1930): Rev. Sci. Instrum. **1**, 397
Piscoti, C., J. Yarger, and A. Zettle (1998): Nature **393**, 771
Powers, D.W., S.G. Hansen, M.E. Geusic, A.C. Pulu, J.B. Hopkins, T.G. Dietz, M.A. Duncan, P.R.R. Langridge-Schmith, and R.E. Smalley (1982): J. Phys. Chem. **86**, 2556
Pradère, F., M. Benslimane, M. Chateau, M. Bierry, M. Châtelet, D. Clement, A. Guilbaud, J.-C. Jeannot, A. De Martino, and H. Vach (1994): Rev. Sci. Instrum. **65**, 161
Pradère, F., M. Benslimane, M. Châtelet, A. De Martino, and H. Vach (1997): Surf. Sci. **375**, L375
Rademann, K. (1991): in *Intermolecular Forces*, ed. by P.L. Huyskens, W.A.P. Luck, and T. Zeegers-Huyskens, p. 297 (Springer, Berlin)
Raether, H. (1957): in *Handbuch der Physik*, Vol. 32, ed. by S. Flügge, p. 443 (Springer, Heidelberg)
Ramirez, A.P., R.C. Haddon, O. Zhou, R.M. Fleming, D.W. Murphy, J. Zhang, S.M. McClure, and R.E. Smalley (1994): Science, **265**, 84
Raoult, B. and J. Farges (1973): Rev. Sci. Instrum. **44**, 430
Rapoport, L., Y. Bilik, Y. Feldman, M. Homyonfer, S.R. Cohen, and R. Tenne (1997): Nature **387**, 791
Raz, T. and R.D. Levine (1996): Chem. Phys. **213**, 263
Recknagel, E. (1984): Ber. Bunsenges. Phys. Chem. **88**, 201
Reddic, J.E. and M.A. Duncan (1997): Chem. Phys. Lett. **264**, 157
Richter, C.E. and M. Trapp (1981): Int. J. Mass Spectrom. Ion Process. **38**, 21
Richtsmeier, S., E. Parks, K. Liu, L. Pobo, and S. Riley (1985): J. Chem. Phys. **82**, 3659
Riley, S., E.K. Parks, C. Mao, L.G. Pobo, and S. Wexler (1982): J. Phys. Chem. **86**, 3911
Riley, S., E.K. Parks, G.C. Nieman, L.G. Pobo, and S. Wexler (1984): J. Chem. Phys. **80**, 1360
Riley, S.J. (1994): in *Cluster of Atoms and Molecules*, Vol. 2, ed. by H.Haberland, p. 221 (Springer, Berlin)
Rohlfing, E., D.M. Cox, A. Kaldor, and K. Johnson (1984): J. Chem. Phys. **81**, 3322
Rongwu, L., P. Zhengying, and H. Yukun (1996): Phys. Rev. B **53**, 4156
Rosenberg, P. (1939): Phys. Rev. **55**, 1267
Rosenberg, P. (1940): Phys. Rev. **57**, 561
Roux,J.F., B. Cabaud, G. Fuchs, D. Guillot, A. Hoareau, and P. Mélinon (1994): Appl. Phys. Lett. **64**, 1212
Rutzen, M., S. Kakar, C. Rienecker, R. von Pietrowski, and T. Möller (1996): Z. Phys. D **38**, 89
Ryali, A. and J. Fenn (1984): Ber. Bunsenges. Phys. Chem. **88**, 245
Sattler, K., J. Mühlbach, and E. Recknagel (1980): Phys. Rev. Lett. **45**, 821
Sattler, K. (1985): Surf. Sci. **156**, 292
Sattler, K. (1993): Jpn. J. Appl. Phys. **32**, 1428
Saunders, W.A. and S. Fredrigo (1989): Chem. Phys. Lett. **156**, 14
Saunders, M. H.A. Jiménez-Vazquez, R.J. Cross, and R.J. Poreda (1993): Science **259**, 1428
Savas, T.A., S.N. Shah, M.L. Schattenburg, J.M. Carter, and H.I. Smith (1995): J. Vac. Sci. Technol. B **13**, 2732
Schaber, H. and T. Martin (1985): Surf. Sci. **156**, 64
Scheidemann, A., B. Schilling, and J.P. Toennies (1993): J. Phys. Chem. **97**, 2128

Schek, I. and J. Jortner (1996): J. Chem. Phys. **104**, 4337
Scheuring, T. and K. Weil (1985): Surf. Sci. **156**, 457
Schilling, B. (1993): Max-Planck-Institut für Strömungsforschung, Report **14**, Göttingen
Schlag, E.W. and H.L. Selzle (1990): J. Chem. Soc. Faraday Trans. **86**, 2511
Schmiedmayer, J., M.S. Chapman, C.R. Ekstrom, T.D. Hammond, D.A. Kokorowski, A. Lenef, R.A. Rubenstein, E.T. Smith, and D.E. Pritchard (1997): in *Atom Interferometry*, ed. by P.R. Berman, p. 1 (Academic Press, New York)
Schmid, G. (ed.) (1994): *Clusters and Colloids* (VCH, Weinheim)
Schöllkopf, W. and J.P. Toennies (1996): J. Chem. Phys. **104**, 1155
Schöllkopf, W., J.P. Toennies, T.A. Savas, and H.I. Smith (1998): J. Chem. Phys. **109**, 9252
Schulze Icking-Konert, G., H. Handschuh, G. Ganteför, and W. Eberhardt (1996): in *Structures and Dynamics of Clusters*, ed. by T. Kondow, K. Kaya, and A. Terasaki, p. 145 (Universal Academy Press, Tokyo)
Schütte, S. (1997): Max-Planck-Institut für Strömungsforschung, Report **7**, Göttingen
Schwarz, H., T. Weiske, D.K. Böhme, and J. Hrusák (1993): in *Buckminsterfullerenes*, Chap. 10, ed. by W.E. Billups and M.A. Ciufolini (VCH Publishers, New York)
Seeger, K., L. Köller, J. Tiggesbäumker, and K.-H. Meiwes-Broer (1998): Eur. Phys. J. D **3**, 179
Shinohara, H., M. Ohno, M. Kishida, S. Yamazaki, T. Hashizume, and T. Sakurai (1996): in *Structures and Dynamics of Clusters*, ed. by T. Kondow, K. Kaya, and A. Terasaki, p. 79 (Universal Academy Press, Tokyo)
Siekmann, H.R., Ch. Lüder, J. Faehrmann, H.O. Lutz, and K.H. Meiwes-Broer (1991): Z. Phys. D **20**, 417
Smalley, R.E. (1993): Mat. Sci. & Engineering **19**, 1
Smalley, R.E. (1997): Rev. Mod. Phys. **69**, 723
Smith, R.A., T. Ditmire, and J.W.G. Tisch (1998): Rev. Sci. Instrum. **69**, 3798
Stein, G.D. and P.P. Wegener (1967): J. Chem. Phys. **46**, 3685
Stienkemeier, F., J. Higgins, C. Callegari, S.I. Kanorsky, W.E. Ernst, and G. Scoles (1996): Z. Phys. D **38**, 253
Sugano, S. and H. Koizumi (1998): Microcluster Physics, 2nd edn. (Springer, Berlin)
Sugai, T. and H. Shinohara (1997): Z. Phys. D **40**, 131
Takagi, T. (1984): J. Vac. Sci. Technol. A **2**, 382
Tamir, S. and S. Berger (1996): Thin Solid Films **276**, 108
Tellgmann, R., N. Krawez, S.-H. Lin, I.V. Hertel, and E.E.B. Campbell (1996): Nature **382**, 407
Tenne, R., L. Margulis, M. Genut, and G. Hodes (1992): Nature **360**, 444
Thilgen, C., F. Diederich, and R.L. Whetten (1993): in *Buckminsterfullerenes*, ed. by W.E. Billups and M.A. Ciufolini, p. 59 (VHC, Weinheim)
Toennies, J.P. (1974): in *Physical Chemistry*, Vol. 6A, ed. by W. Jost, p. 227 (Academic Press, New York)
Tomaschko, Ch., R. Kügler, M. Schurr, and H. Voit (1996a): Nucl. Instrum. Methods Phys. Res. B **117**, 90
Tomaschko, Ch., Ch. Schoppmann, M. Kraus, K. Kragler, G. Kreiselmeyer, G. Saemann-Ischenko, and H. Voit (1996b): Nucl. Instrum. Methods Phys. Res. B **117**, 199
Tombrello, T.A. (1995): Nucl. Instrum. Methods B **99**, 225
Torchet, G., J. Farges, M.F. Feraudy, and B. Raoult (1990): in *The Chemical Physics of Atomic and Molecular Clusters*, ed. by G. Scoles, p. 413 (North-Holland, Amsterdam)
Ulmer, H.G., B. Hasselberger, H.-G. Busmann, and E.E.B. Campbell (1990): Appl. Surf. Science **46**, 272

Urban, J., H. Sack-Kongehl, and K. Weiss (1996): Z. Phys. D **36**, 73
Vach, H., A. De Martino, M. Benslimane, M. Châtelet, and F. Pradère (1994): J. Chem. Phys. **100**, 3526
Vach, H., M. Benslimane, M. Châtelet, A. De Martino, and F. Pradère (1995): J. Chem. Phys. **103**, 1972
van de Walle, J. and P. Joyes (1987): Phys. Rev. B **35**, 5509
van den Biesen, J.J.H. (1988): in *Atomic and Molecular Beam Methods*, Vol. 1., ed. by G. Scoles, p. 472 (Oxford University Press, Oxford)
Visticot, J.P., J. Berlande, J. Cuvellier, A. Lallement, J.M. Mestdagh, P. Meynadier, P. de Pujo, and O. Sublemontier (1992): Chem. Phys. Lett. **191**, 107
Wagner, R.L., W.D. Vann, and A.W. Castleman Jr. (1997): Rev. Sci. Instrum. **68**, 3010
Weiske, T., D.K. Böhme, J. Hrusak, W. Krätschmer, and H. Schwarz (1991): Angew. Chem. **103**, 898
Werwa, E. and K.D. Kolenbrander (1996): in *Advanced Laser Processing of Materials*, ed. by R. Singh, D. Norton, L. Laude, J. Navayan, and J. Cheung, p. 381 (Mater. Res. Soc., Pittsburgh)
Whetten, R., D. Cox, D. Frevor, and A. Kaldov (1985): Phys. Rev. Lett. **54**, 1494
Woenckhaus, J., R. Schäfer, and J.A. Becker (1996): Surf. Rev. Lett. **3**, 371
Wörmer, J., V. Guzielski, J. Stapelfeldt, and T. Möller (1989): Chem. Phys. Lett. **159**, 321
Wu, C. (1980): J. Phys. Chem. **87**, 1534
Wucher, A., M. Wahl, and H. Oechsner (1993): Nucl. Instrum. Methods B **82**, 337
Wucher, A., W.F. Calaway, and M.J. Pellin (1994): Surf. Sci. **304**, L439
Xenoulis, A.C., P. Trouposkiadis, C. Potiriadis, C. Papastaikoudis, A.A. Katsanos, and A. Clouvas (1996): Nanostructured Materials **7**, 473
Yamada, I. (1989): Appl. Surf. Sci. **43**, 23
Yamamura, Y. and T. Muramoto (1995): Nucl. Instrum. Methods B **102**, 322
Yannouleas, C. and U. Landman (1996): in *Large Clusters of Atoms and Molecules*, ed. by T.P. Martin, p. 131 (Kluwer, Doordrecht)
Yoshimura, S. and R.P.H. Chang (1998): *Supercarbon* (Springer, Heidelberg)
Zimmermann, U., N. Malinowski, U. Näher, S. Frank, and T.P. Martin (1994): Z. Phys. D **31**, 85

Chapter 3

Abfalterer, R., C. Keller, S. Bernet, M.K. Oberthaler, J. Schmiedmayer, and A. Zeilinger (1997): Phys. Rev. A **56**, R4365
Alcalay, J.A. and E.L. Knuth (1969): Rev. Sci. Instrum. **40**, 438 and 1652
Arndt, M., O. Nairz, J. Vos-Andreae, C. Keller, G. van der Zouw, and A. Zeilinger (1999): Nature **401**, 680
Arnold, M., J. Kowalski, G. zu Putlitz, T. Stehlin, and F. Träger (1985a): Surf. Sci. **156**, 149
Auerbach, D.J. (1988): in *Atomic and Molecular Beam Methods*, Vol. 1, ed. by G. Scoles, p. 362 (Oxford University Press, Oxford)
Bally, D., E. Tarina, and P. Pirlogea (1961): Rev. Sci. Instrum. **32**, 297
Bassi, D., M.G. Dondi, F. Tommasini, F. Torello, and U. Valbusa (1976): Phys. Rev. A **13**, 584
Baumfalk, R., U. Buck, C. Frischkorn, S.R. Ghandi, and C. Lauenstein (1997): Ber. Bunsenges. Phys. Chem. **101**, 606
Beijerinck, H.C.W., R.G.J.M. Moonen, and N.F. Verster (1974): J. Phys. E **7**, 31

Beitz, W. and K.-H. Küttner (1981): *Dubbels Taschenbuch für den Maschinenbau* (Springer, Heidelberg)
Bennett, A. (1954): Phys. Rev. **95**, 608
Bennewitz, H.G. (1956): Thesis, University of Bonn, Germany
Bennewitz, H.G., H. Busse, H.D. Dohmann, D.E. Oates, and W. Schrader (1972): Z. Phys. **253**, 435
Bergmann, K., W. Demtröder, and P. Hering (1975): Appl. Phys. **8**, 65
Bergmann, K., U. Hefter, and P. Hering (1978): Chem. Phys. **32**, 329
Bergmann, K. (1988): in *Atomic and Molecular Beam Methods*, Vol. 1, ed. by G. Scoles, p. 293 (Oxford University Press, Oxford)
Biraben, F., J.G. Garseau, L. Julien, and M. Allegrini (1990): Rev. Sci. Instrum. **61**, 1468
Boato, G. (1992): in *Atomic and Molecular Beam Methods*, Vol. 2, ed. by G. Scoles, p. 340 (Oxford University Press, Oxford)
Bolzinger, T. (1975): J. Phys. E **8**, 99
Bredewout, J.W. (1973): Thesis, University of Leiden, Nethderlands
Brouard, M., S.P. Duxon, and J.P. Simons (1994): Isr. J. Chem. **34**, 67
Buck, U. (1969): Thesis, University of Bonn, Germany
Buck, U. and H. Pauly (1968): Z. Nat.forsch. A **23**, 475
Buck, U. and H. Pauly (1971): J. Chem. Phys. **54**, 1929
Buck, U. and H. Meyer (1984): Phys. Rev. Lett. **52**, 209
Buck, U., H. Meyer, and H. Pauly (1985): in *Flow of Real Fluids*, ed. by G.E.A. Meier and F. Obermeier, p. 170 (Springer, Heidelberg)
Buckland, J.R., R.L. Folkerts, R.B. Balsod, and W. Allison (1997): Meas. Sci. Technol. **8**, 933
Caracciolo, G., S. Iannotta, G. Scoles, and U. Valbusa (1980): J. Chem. Phys. **72**, 4491
Cardillo, M.J., M.S. Chou, E.F. Greene, and D.B. Sheen (1971): J. Chem. Phys. **54**, 3054
Carter, J.M., D.B. Olster, M.L. Schattenburg, A. Yeu, and H.I. Smith (1992): J. Vac. Sci. Technol. B **10**, 2909
Chapman, M.S., C.R. Ekstrom, T.D. Hammond, R.A. Rubinstein, J. Schmiedmayer, S. Wehinger, and D.E. Pritchard (1995): Phys. Rev. Lett. **74**, 4783
Clampitt, R. and A.S. Newton (1969): J. Chem. Phys. **50** 1997
Cohen, V.W. and A. Ellett (1937): Phys. Rev. **52**, 502 and 509
Colgate, S.O. and T.C. Imeson (1965): Rev. Sci. Instrum. **36**, 932
Comsa, G., G. Mechtersheimer, B. Poelsema, and S. Tomoda (1979): Surf. Sci. **89**, 123
Comsa, G., R. David, and B.J. Schumacher (1981): Rev. Sci. Instrum. **52**, 789
Cowin, J.P., D.J. Auerbach, C. Becker, and L. Wharton (1978): Surf. Sci. **78**, 545
Cowly, L.T., M.A.D. Fluendy, and K.P. Lawley (1970): Rev. Sci. Instrum. **41**, 666
Dash, J.G. and H.S. Sommers Jr. (1953): Rev. Sci. Instrum. **24**, 91
David, R., K. Kern, P. Zeppenfeld, and G. Comsa (1986): Rev. Sci. Instrum. **57**, 2771
Dittner, P.F. and S. Datz (1971): J. Chem. Phys. **54**, 4228
Dohmann, H.-D. (1970): Thesis, University of Bonn, Germany
Düren, R., H.O. Hoppe, and H. Pauly (1975): in *50 Jahre MPI für Strömungsforschung*, ed. by MPIFS, p. 414 (Hubert & Co. Göttingen)
Düren, R., E. Hasselbrink, and G. Hillrichs (1984): Chem. Phys. Lett. **112**, 441
Düren, R., H. Hübner, S. Kita, and U. Krause (1986a): J. Chem. Phys. **85**, 2751
Düren, R., E. Gersing, S. Kita, W. Kracht, H. Pauly, E. Schädlich, and M. Vonderschen (1986b): J. Chem. Phys. **85**, 2738
Ebinghaus, H., G. Spindler, and E. Steffens (1974): Z. Phys. **267**, 15
Ehbrecht, M., H. Ferkel, and F. Huisken (1997): Z. Phys. D **40**, 88
Ekstrom, C.R., D.W. Keith, and D.E. Pritchard, Appl. Phys. B **54**, 369 (1992)

Eldridge, J.A. (1927): Phys. Rev. **30**, 931
Ellis, T.H., G. Scoles, U. Valbusa, H. Jónsson, and J.H. Weare (1985): Surf. Sci. **155**, 499
Este, G.O., B. Hilko, D. Sawyer, and G. Scoles (1975): Rev. Sci. Instrum. **46**, 223
Este, G.O., G. Knight, and G. Scoles (1978): Chem. Phys. **35**, 421
Este, G.O., G. Knight, G. Scoles, U. Valbusa, and F. Grein (1983): J. Phys. Chem. **87**, 2772
Estermann, I. and O. Stern (1930): Z. Phys. **61**, 95
Estermann, I., R. Frisch, and O. Stern (1931): Z. Phys. **73**, 348
Estermann, I., O.C. Simpson, and O. Stern (1947a): Phys. Rev. **71**, 238
Estermann, I., S.N. Foner, and O. Stern (1947b): Phys. Rev. **71**, 250
Fink, G.A. (1936): Phys. Rev. **50**, 738
Fischer, C. (1994): Max-Planck-Institut für Strömungsforschung, Report **7**, Göttingen
Fluendy, M.A.D. and K.P. Lawley (1973*): Chemical Application of Molecular Beam Scattering* (Wiley, New York)
Frankl, D.R. (1974): Rev. Sci. Instrum. **45**, 1375
Fremerey, J.K. (1985): J. Vac. Sci. Technol. A **3**, 1715
Gaily, T.D., S.D. Rosner, and R.A. Holt (1976): Rev. Sci. Instrum. **47**, 143
Gersing, E., H. Pauly, E. Schädlich, and M. Vonderschen (1973): Faraday Discuss. Chem. Soc. **55**, 211
Giles, R.A. and J.A. Logan (1972): J. Phys. B **5**, 2123
Greene, E.F., R.W. Roberts, and J. Ross (1960): J. Chem. Phys. **32**, 940
Grosser, A.E., R.P. Iczkowski, and J.L. Margrave (1963): Rev. Sci. Instrum. **34**, 116
Grosser, A.E. (1967): Rev. Sci. Instrum. **38**, 257
Hahn, E., H. Schief, V. Marsico, A. Fricke, and K. Kern (1994): Phys. Rev. Lett. **72**, 3378
Hammer, D., E. Benes, P. Blum, and W. Husinsky (1976): Rev. Sci. Instrum. **47**, 1178
Hasselbrink, E. (1985): Max-Planck-Institut für Strömungsforschung, Report 15, Göttingen
Hecht, E. (1987): *Optics*, 2nd edn. (Addison-Wesley, Reading, MA)
Hefter, U. and K. Bergmann (1988): in *Atomic and Molecular Beam Methods*, Vol. 1, ed. by G. Scoles, p. 193 (Oxford University Press, Oxford)
Heindorff, T. and B. Fischer (1984): Rev. Sci. Instrum. **55**, 347
Helbing, R. (1966): Thesis, University of Bonn, Germany
Höhne, E. (1961): Ann. Phys. **7**, 50
Hostettler, U. and R.B. Bernstein (1960): Rev. Sci. Instrum. **31**, 872
Huisken, F. and T. Pertsch (1987): Rev. Sci. Instrum. **58**, 1038
Husinsky, W., G. Betz, and I. Girgis (1983): Phys. Rev. Lett. **50**, 1689
Husinsky, W., G. Betz, and I. Girgis (1984): J. Vac. Sci. Technol. A **2**, 698
Jaduszliwer, B. and Y.C. Chan (1994): Rev. Sci. Instrum. **65**, 2928
Johnson, N.B., R.A. Heppner, B.M. Pridmore, and J.C. Zorn (1970): Rev. Sci. Instrum. **41**, 777
Johnston, G.W., S. Satyapal, R. Bersohn, and B. Katz (1990): J. Chem. Phys. **92**, 206
Kasevich, M., K. Moler, E. Riis, E. Sunderman, D. Weiss, and S. Chu (1991): *in Atomic Physics*, Vol. 12, ed. by J.C. Zorn and R.R. Lewis, p. 47 (AIP, New York)
Keith. D.W., M.L. Schattenburg, H.I. Smith, and D.E. Pritchard (1988): Phys. Rev. Lett. **61**, 1580
Keith, D.W., R.J. Soave, and M.J. Roots (1991a): J. Vac. Sci. Technol. B **9**, 2846
Keith, D.W., M.L. C.R. Ekstrom, Q.A. Turchette, and D.E. Pritchard (1991b): Phys. Rev. Lett. **66**, 2693
Kim, H.-J. (1991): Diplom thesis, University of Göttingen, Germany
Kinsey, J.L. (1966): Rev. Sci. Instrum. **37**, 61
Kinsey, J.L. (1977): J. Chem. Phys. **66**, 2560
Kita, S., H. Hübner, W. Kracht, and R. Düren (1981): Rev. Sci. Instrum. **52**, 684

Knauer, F. and O. Stern (1929): Z. Phys. **53**, 779
Kresin, V.V., V. Kasperovich, G. Tikhonov, and K. Wong (1998): Phys. Rev. A **57**, 383
Kristensen, A., R.L. Palmer, H. Saltsburg, and J.N. Smith, Jr. (1967): Rev. Sci. Instrum. **38**, 987
Kroon, J.P.C., H.C.W. Beijerinck, and N.F. Verster (1981): J. Chem. Phys. **74**, 6528
Kuhnke, K., K. Kern, R. David, B. Lindenau, and G. Comsa (1994): Rev. Sci. Instrum. **65**, 653
Kusch, P. and V.W. Hughes (1959): in *Handbuch der Physik*, Vol. 37/1, ed. by S. Flügge, p.1 (Springer, Heidelberg)
Laguna, M.A., V. Paillard, B. Kohn, M. Ehbrecht, F. Huisken, G. Ledoux, R. Papoular, and H. Hofmeister (1999): J. Luminescence **80**, 223
Lammert, B. (1929): Z. Phys. **56**, 244
Legere, R. and J. Gibble (1998): Phys. Rev. Lett. **81**, 5780
Levinstein, H. and H. Crane (1946): Phys. Rev. **69**, 679
Linse, C.A. (1977): Thesis, University of Leiden, Netherlands
Lohbrandt, P. (1993): Diplom thesis, University of Göttingen, Germany
Macler, M. and M.E. Fajardo (1994): Appl. Phys. Lett. **65**, 2275
Marcus, P.M. and J.H. McFee (1959): in *Recent Research in Molecular Beams*, ed. by I. Estermann, p. 43 (Academic Press, New York)
Maréchal, E., S. Guibal, J.-L. Bossennec, M.-P. Gorza, R. Barbé, J.-C. Keller, and O. Gorceix (1998): Eur. Phys. J. D **2**, 195
Martini, K.M., W. Franzen, and M. El-Batanouny (1987): Rev. Sci. Instrum. **58**, 1027
Miller, R.C. and P. Kusch (1955): Phys. Rev. **99**, 1314
Miers, R.E., R.W. York, P.T. Pickett, and W.L.Taylor (1988): Rev. Sci. Instrum. **59**, 1303
Molenaar, P.A., P. van der Straten, H.G.M. Heideman, and H. Metcalf (1998): Phys. Rev. A **55**, 605
Moon, P.B. (1953): Brit. J. Appl. Phys. **4**, 97
Moore, L. (1968): J. Phys. D **1**, 237
Moore, J.H., Jr. and C.B. Opal (1975): Space Sci. Instrum. **1**, 377
Morrison, J.D. (1963): J. Chem. Phys. **39**, 200
Moskowitz, W.B., B. Steward, R.M. Bilotta, J.L. Kinsey, and D.E. Pritchard (1984): J. Chem. Phys. **80**, 5496
Mull, T., B. Baumeister, M. Menges, H.-J. Freund, D. Weide, C. Fischer, and P. Andresen (1992): J. Chem. Phys. **96**, 7108
Park, J., N. Shafer, and R. Bersohn (1989): J. Chem. Phys. **91**, 7861
Pasternack, L. and P.J. Dagdigian (1977): Rev. Sci. Instrum. **48**, 226
Pauly, H. (1960): Z. Nat.forsch. A **15**, 277
Pauly, H. and J.P. Toennies (1968): in *Methods of Experimental Physics*, Vol. 7A, ed. by B. Bederson and W.L. Fite, p. 227 (Academic Press, New York)
Pellin, M.J., R.B. Wright, and D.M. Gruen (1981): J. Chem. Phys. **74**, 6448
Phillips, W.D., J.A. Serri, D.J. Ely, D.E. Pritchard, K.R. Way, and J.L. Kinsey (1978): Phys. Rev. Lett. **41**, 937
Poelsema, B., R.L. Palmer, G. Mechtersheimer, and G. Comsa (1982): Surf. Sci. **117**, 60
Politiek, J., P.K. Rol, J. Los, and P.G. Ikelaar (1968): Rev. Sci. Instrum. **39**, 1147
Porter, R.A.R. and A.E. Grosser (1980): Rev. Sci. Instrum. **51**, 140
Pritchard, D.E. (1993): in *Atomic Physics*, Vol. 13, ed. by H. Walther, T.W. Hänsch, and B. Neizert, p.185 (AIP, New York)
Ramsey, N.F. (1956): *Molecular Beams* (Clarendon Press, Oxford)
Reuss, J. (1988): in *Atomic and Molecular Beam Methods*, Vol. 1, ed. by G. Scoles, p. 276 (Oxford University Press, Oxford)

Roux, J.F., B. Cabaud, G. Fuchs, D. Guillot, A. Hoareau, and P. Mélinon (1994): Appl. Phys. Lett. **64**, 1212
Sandy, A.R., S.G.J. Mochrie, D.M. Zehner, G. Grübel, K.G. Huang, and D. Gibbs (1992): Phys. Rev. Lett. **68**, 2192
Schmiedmayer, J., M.S. Chapman, C.R. Ekstrom, T.D. Hammond, D.A. Kokorowski, A. Lenef, R.A. Rubenstein, E.T. Smith, and D.E. Pritchard (1997): in *Atom Interferometry*, ed. by P.R. Berman, p. 1 (Academic Press, San Diego)
Schief, H., V. Marsico, K. Kuhnke, and K. Kern (1996): Surf. Sci. **364**, L631
Schöllkopf, W. and J.P. Toennies (1994): Science **266**, 1345
Schöllkopf, W. and J.P. Toennies (1996): J. Chem. Phys. **104**, 1155
Schöttler, J. and J.P. Toennies (1968): Z. Phys. **214**, 472
Secrest, D. and H.D. Meyer (1972): Max-Planck-Institut für Strömungsforschung, Report **137**, Göttingen
Serri, J.A., J.L. Kinsey, and D.E. Pritchard (1981): J. Chem. Phys. **75**, 663
Shafer-Ray, N.E., H. Xu, R.P. Tuckett, M. Springer, and R.N. Zare (1994): J. Phys. Chem. **98**, 3369
Shafer-Ray, N.E., A.J. Orr-Ewing, and R.N. Zare (1995): J. Phys. Chem. **99**, 7591
Stark, J. (1905): Phys. Zeitschr. **6**, 893
Stern, O. (1920): Z. Phys. **2**, 49, **3**, 417 and 418
Stern, O. (1929): Naturwissensch. **17**, 391
Taatjes, C.A., J.I. Cline, and S.R. Leone (1990): J. Chem. Phys. **93**, 6554
Toschek, P. (1961): Thesis, University of Bonn, Germany
Trujillo, S.M., P.K. Rol, and E.W. Rothe (1962): Rev. Sci. Instrum. **33**, 841
Trujillo, S.M. (1975): *IX. Int. Conf. Electronic and Atomic Collisions, Book of Abstracts*, Vol. 2, ed. by J.S. Risley and R. Geballe, p. 953 (University of Washington Press, Seattle)
van den Meijdenberg, C.J.N. (1988): in *Atomic and Molecular Beam Methods*, Vol. 1, ed. by G. Scoles, p. 345 (Oxford University Press, Oxford)
van Steyn, R. and N.F. Verster (1972): J. Phys. E **5**, 691
Walther, H. (1994): Adv. At. Mol. Opt. Phys. **32**, 379
Weaver, G.C. and S.R. Leone (1995): Surf. Sci. **328**, 197
Weide, D., P. Andresen, and H.-J. Freund (1987): Chem. Phys. Lett. **136**, 106
Wertheim, G.K. (1975): J. Electron Spect. Relat. Phenom. **6**, 239
Wilhelmi, G. and F. Gompf (1970): Nucl. Instrum. Methods **81**, 36
Wykes, J. (1969): J. Phys. E **2**, 899
Xu, H., N.E. Shafer-Ray, F. Merkt, D.J. Hughes, M. Springer, R.P. Tuckett, and R.N. Zare (1995): J. Chem. Phys. **103**, 5157
Zacharias, J. (1954): Phys. Rev. **94**, 751
Zatselyapin, A.M. (1974): Instrum. Exp. Tech. (USSR) **17**, 1417
Zinn, W.H. (1947): Phys. Rev. **71**, 752

Chapter 4

Alagia, M., V.Aquilanti, D. Ascenzi, N. Balucani, D. Cappelletti, L. Cartechini, P. Cassevecchia, F. Pirani, G. Sanchini, and G. Volpi (1997): Isr. J. Chem. **37**, 329
Alexander, M.H. and P.J. Dagdigian (1985): J. Chem. Phys. **83**, 2191
Alguard, M.J., J.E. Clendenin, R.D. Ehrlich, V.W. Hughes, J.S. Ladish, M.S. Lubell, K.P. Schüler, G. Baum, W. Raith, R.H. Miller, and W. Lysenko (1979): Nucl. Instrum. Methods **163**, 29

References

Andersen, N., J.T. Broad, E.E.B. Campbell, J.W. Gallagher, and I.V. Hertel (1997): Phys. Rep. **278**, 107 and Phys. Rep. **279**, 251
Anderson, R.W. (1997): J. Phys. Chem. **101**, 7664
Antoine, R., D. Rayane, A.R. Allouche, M. Aubert-Frécon, E. Benichou, F.W. Dalby, Ph. Dugourd, M. Broyer, and C. Guet (1999): J. Chem. Phys. **110**, 5568
Apsel, S.E., J.W. Emmert, J. Deng, and L.A. Bloomfield (1996): Phys. Rev. Lett. **76**, 1441
Aquilanti, V., R. Candori, and F. Pirani (1988): J. Chem. Phys. **89**, 6157
Aquilanti, V., R. Candori, D. Cappelletti, V. Lorent, and F. Pirani (1992): Chem. Phys. Lett. **192**, 145
Aquilanti, V., D. Ascenzi, D. Cappelletti, and F. Pirani (1995): Int. J. Mass Spectrom. Ion Process. **149/150**, 355
Aquilanti, V., D. Ascenzi, E. Braca, D. Cappelletti, G. Liuti, E. Luzatti, and F. Pirani (1997): J. Phys. Chem. A **101**, 6523
Audoin, C., J.P. Schermann, and P. Grivet (1971): Adv. At. Mol. Phys. **7**, 2
Auerbach, D., E.E.A. Bromberg, and L. Wharton (1966): J. Chem. Phys. **45**, 2160
Bagnato, V.S., G.P. Lafyatis, A.G. Martin, E.L. Raab, R.N. Ahmad-Bitar, and D.E. Pritchard (1987): Phys. Rev. Lett. **58**, 2194
Bass, S.M., R.L. DeLeon, and J.S. Muenter (1987): J. Chem. Phys. **86**, 4305
Bass, S.M., R.L. DeLeon, and J.S. Muenter (1990): J. Chem. Phys. **92**, 71
Baum, G, C.D. Caldwell, and W. Schröder (1980): Appl. Phys. **21**, 126
Baum, G, W. Raith, and H. Steidl (1988): Z. Phys. D **10**, 171
Beck, D., U. Henkel, and A. Schultz (1968): Phys. Lett. **27A**, 272
Bederson, B., J. Eisinger, K. Rubin, and A. Salop (1960): Rev. Sci. Instrum. **31**, 852
Bederson, B. and J.E. Robinson (1966): in *Molecular Beams*, ed. by J. Ross, p. 1 (Wiley, New York)
Behrens, R., R. Herm, and C.M. Sholeen (1976): J. Chem. Phys. **65**, 4791
Belov, A.S., S.K. Esin, L.P. Netchaeva, Y.V. Plokhinski, G.A. Vasil'ev, V.S. Klenov, A.V. Turbabin, V.P. Yakushev, and V.G. Dudnikov (1996): Rev. Sci. Instrum. **67**, 1293
Bennewitz, H.G. and W. Paul (1954): Z. Phys. **139**, 489
Bennewitz, H.G., W. Paul, and Ch. Schlier (1955): Z. Phys. **141**, 6
Bennewitz, H.G., W. Paul, and P. Toschek (1956): Z. Nat.forsch. A **11**, 956
Bennewitz, H.G., K.H. Kramer, W. Paul, and J.P. Toennies (1964): Z. Phys. **177**, 84
Bennewitz, H.G. and R. Haerten (1969): Z. Phys. **227**, 399
Bennewitz, H.G., R. Haerten, and G. Müller (1971): Chem. Phys. Lett. **12**, 335
Bennewitz, H.G. and G. Buess (1978): Chem. Phys. **28**, 175
Berg, R.A., L. Wharton, W. Klemperer, A. Buchler, and J.L. Stauffer (1965): J. Chem. Phys. **43**, 2416
Bergeman, T., G. Erez, and H.J. Metcalf (1987): Phys. Rev. A **35**, 1535
Bergmann, K., U. Hefter, and J. Witt (1980): J. Chem. Phys. **72**, 4777
Bergmann, K. (1988): in *Atomic and Molecular Beam Methods*, ed. by G. Scoles, p. 293 (Oxford University Press, Oxford)
Bergmann, K., S. Schiemann, and A. Kuhn (1992): Phys. Bl. **48**, 907
Bergmann, K. and B.W. Shore (1995): in *Molecular Dynamics and Spectroscopy by Stimulated Emission Pumping*, ed. by H.L. Dai and R.W. Field, p. 315 (World Scientific, Singapore)
Bergmann, K., H. Theuer, and B.W. Shore (1998): Rev. Mod. Phys. **70**, 1003
Berkling, K., Ch. Schlier, and P. Toschek (1962): Z. Phys. **168**, 81
Bernheim, R.A. (1965): *Optical Pumping, an Introduction* (Benjamin, New York)
Bernstein, R.B. (1982): *Chemical Dynamics via Molecular Beam and Laser Techniques* (Clarendon Press, Oxford)

Bernstein, R.B., S.E. Choi, and S. Stolte (1989): J. Chem. Soc. Faraday Trans. II, **85**, 1097
Bersohn, R. and S.H. Lin (1969): Adv. Chem. Phys. **16**, 67
Bethlem, H.L., G. Berden, and G. Meijer (1999): Phys. Rev. Lett. **83**, 1558
Beuhler, R.J., R.B. Bernstein, and K.H. Kramer (1966): J. Am. Chem. Soc. **88**, 5331
Beuhler, R.J. and R.B. Bernstein (1969): J. Chem. Phys. **51**, 5305
Billas, I.M.L., A. Châtelain, and W.A. de Heer (1994): Science **265**, 1682
Billas, I.M.L., A. Châtelain, and W.A. de Heer (1996): Surf. Rev. Lett. **3**, 429
Bloomfield, L.A., J.P. Bucher, and D.C. Douglass (1993): in *On Clusters and Clustering*, ed. by P.J. Reynolds, p. 193 (North Holland, Amsterdam)
Borkenhagen, U., H. Malthan, and J.P. Toennies (1975): J. Chem. Phys. **63**, 3173
Borkenhagen, U., H. Malthan, and J.P. Toennies (1979): J. Chem. Phys. **71**, 1722
Borsella, E., R. Fantoni, A. Giardini-Guidoni, D. Masci, A. Palucci, and J. Reuss (1982): Chem. Phys. Lett. **93**, 523
Boughton, C.V., R.E. Miller, and R.O. Watts (1982): Austral. J. Phys. **35**, 611
Brandt, M., H. Müller, G. Zagatta, O. Wehmeyer, N. Böwering, and U. Heinzmann (1995): Surf. Sci. **331–333**, 30
Brandt, M., T. Greber, N. Böwering, and U. Heinzmann (1998): Phys. Rev. Lett. **81**, 2376
Braunstein, R. and J.W. Trischka (1955): Phys. Rev. **98**, 1092
Brencher, L., B. Nawracala, and H. Pauly (1988): Z. Phys. D **10**, 211
Brenner, T., S. Büttgenbach, J. Jacobs, M. Koster, H. Roeder, W. Rupprecht, and F. Träber (1988): Z. Phys. D **10**, 59
Bromberg, E.E.A., A.E. Proctor, and R.B. Bernstein (1975): J. Chem. Phys. **63**, 3287
Brooks, P.R. and E.M. Jones (1966): J. Chem. Phys. **45**, 3449
Brooks, P.R. (1976): Science **193**, 11
Brossel, J., A. Kastler, and J. Winter (1952): J. Phys. Radium **13**, 668
Bucher, J.P., D.C. Douglass, and L.A. Bloomfield (1991a): Phys. Rev. Lett. **66**, 3052
Bucher, J.P., D.C. Douglass, P. Xia, B. Haynes, and L.A. Bloomfield (1991b): Z. Phys. D **19**, 251
Budick, B. (1967): Adv. At. Mol. Phys. **3**, 73
Bullman, S.J. and P.J. Dagdigian (1984): Chem. Phys. **88**, 479 and J. Chem. Phys. **81**, 3347
Carter, G.M., D.E. Pritchard, M. Kaplan, and T.W. Ducas (1975): Phys. Rev. Lett. **35**, 1144
Celotta, R.J. and D.T. Pierce (1980): Adv. At. Mol. Phys. **16**, 102
Chakravorty, K.K., D.H. Parker, and R.B. Bernstein (1982): Chem. Phys. **68**, 1
Chamberlain, G.E. and J.C. Zorn (1963): Phys. Rev. **129**, 677
Chan, N., D.M. Crowe, M.S. Lubell, F.C. Tang, A. Vasilakis, F.J. Mulligan, and J. Slevin (1988): Z. Phys. D **10**, 393
Che, D.C., T. Kasai, H. Ohoyama, K. Ohashi, T. Fukawa, and K. Kuwata (1991): J. Phys. Chem. **95**, 8159
Che, D.C., T. Ogawa, H. Ohoyama, T. Kasai, and K. Kuwata (1994): in *Atomic Collision Research in Japan*, Nr. 20, ed. by I.H. Suzuki, p. 74 (The Society of Atomic Collision Research, Tokyo)
Childs, W.J. (1972): Case Stud. At. Phys. **3**, 215
Cho, D., K. Sangster, and E.A. Hinds (1989): Phys. Rev. Lett. **63**, 2559
Christensen, R.L., D.R. Hamilton, A. Lemonick, F.M. Pipkin, J.B. Reynolds, and H.H. Stroke (1956): Phys. Rev. **101**, 1389
Cohen-Tannoudji, C. (1975): in *Atomic Physics*, Vol. **4**, ed. by G. zu Putlitz, E.W. Weber, and A. Winnacker, p.589 (Plenum Press, New York)
Conaway, W.E., R.J.S. Morrison, and R.N. Zare (1985): Chem. Phys. Lett. **113**, 429
Coulter, D.R., F.R. Grabiner, L.M. Casson, G.W. Flynn, and R.B. Bernstein (1980): J. Chem. Phys. **73**, 281

Cox, D.M., D.J. Trevor, R.L. Whetten, E.A. Rohlfing, and A. Kaldor (1986): J. Chem. Phys. **84**, 4651
Cox, A.J., J.G. Louderback, S.E. Apsel, and L.A. Bloomfield (1994): Phys. Rev. B **49**, 12295
Crosby, D.A. and J.C. Zorn (1977): Phys. Rev. A **16**, 488
Crowe, D.M., X.Q. Guo, M.S. Lubell, J. Slevin, and M. Eminyan (1990): J. Phys. B **23**, L325
Cusma, J.T. and L.W. Anderson (1983): Phys. Rev. A **28**, 1195
Dagdigian, P.J., B.E. Wilcomb, and M.H. Alexander (1979): J. Chem. Phys. **71**, 1670
Dagdigian, P.J. and B.E. Wilcomb (1980): J. Chem. Phys. **72**, 6462
Davis, K.B., M.-O. Mewes, M.A. Joffe, M.R. Andrews, and W. Ketterle (1995a): Phys. Rev. Lett. **74**, 5202
Davis, K.B., M.-O. Mewes, M.R. Andrews, N.J. van Druten, D.S. Durfee, D.M. Kurn, and W. Ketterle (1995b): Phys. Rev. Lett. **75**, 3969
Dayton, I.E., F.C. Shoemaker, and R.F. Mozley (1954): Rev. Sci. Instrum. **25**, 485
de Heer, W.A. and W.D. Knight (1988): in *Elemental and Molecular Clusters*, ed. by G. Benedek, T.P. Martin, and G. Pachioni, p. 45 (Springer, Heidelberg)
de Heer, W.A., P. Milani, and A. Chatelain (1989): Phys. Rev. Lett. **63**, 2834
de Heer, W.A., P. Milani, and A. Chatelain (1990): Phys. Rev. Lett. **65**, 488
De Vries, M.S., G.W. Tyndall, C. Cobb, and R.M. Martin (1987): J. Chem. Phys. **86**, 2653
Deck, F.J., E.A. Hessels, and S.R. Lundeen (1993): Phys. Rev. A **48**, 4400
Dembczynski, J., W. Ertmer, V. Johann, S. Penselin, and P. Stiemer (1980): Z. Phys. A **294**, 313
Demtröder, W. (1976): Case Stud. At. Phys. **5**, 181
Denison, D.R. (1971): J. Vac. Sci. Technol. **8**, 266
DePristo, A.E., H. Rabitz, and R.B. Miles (1980): J. Chem. Phys. **73**, 4798
Desruelle, B., V. Boyer, P. Bouyer, G. Birkl, M. Lécrivain, F. Alves, C.I. Westbrook, and A. Aspect (1998): Eur. Phys. J. D **1**, 255
Dispert, H.H., M.W. Geis, and P.R. Brooks (1979): J. Chem. Phys. **70**, 5317
Dittmann, P., F.P. Pesl, J. Martin, G.W. Coulston, G.Z. He, and K. Bergmann (1992): J. Chem. Phys. **97**, 9472
Dohrmann, T. and B. Sonntag (1996): J. Electron Spectr. Rel. Phenom. **79**, 263
Douglass, D.C., A.J. Cox, J.P. Bucher, and L.A. Bloomfield (1993): Phys. Rev. B **47**, 12874
Dowek, D., N. Andersen, J.C. Houver, and C. Richter (1990): Phys. Rev. Lett. **64**, 1713
Doyle, J.M., B. Friedrich, J. Kim, and D. Patterson (1995): Phys. Rev. Lett. **52**, R2515
Drabbels, M., A.M. Wodtke, M. Yang, and M.H. Alexander (1997): J. Phys. Chem. A **101**, 6463
Drabbels, M. and A.M. Wodtke (1997): J. Chem. Phys. **106**, 3024
Drechsler, M. and E.W. Müller (1952): Z. Phys. **132**, 195
Dreves, W., W. Kamke, W. Broermann, and D. Fick (1981): Z. Phys. A **303**, 203
Dreves, W., H. Jänsch, E. Koch, and D. Fick (1983): Phys. Rev. Lett. **50**, 1759
Dunham, J.S., C.S. Galovich, H.F. Glavich, S.S. Hanna, D.G. Mavis, and S.W. Wissink (1984): Nucl. Instrum. Methods **219**, 46
Düren, R., E. Hasselbrinck, and G. Hillrichs (1984): Chem. Phys. Lett. **112**, 441
Düren, R. (1988): in *Atomic and Molecular Beam Methods*, Vol. 1, ed. by G. Scoles, p. 653 (Oxford University Press, Oxford)
Düren, R., M. Knepper, S. Mohr, S. te Lintel Hekkert. A.F. Linskens, and J. Reuss (1994): Chem. Phys. Lett. **228**, 41
Dyke, T.R. (1984): Topics. Current Chem. **120**, 86

Dymanus, A. (1992): in *Atomic and Molecular Beam Methods*, Vol. 2, ed. by G. Scoles, p. 58 (Oxford University Press, Oxford)
Ebenstein, W.L., C. Hanning, S. Shostak, and J.S. Muenter (19875): J. Chem. Phys. **87**, 1948
Eisel, D., W. Demtröder, W. Müller, and P. Botschwina (1983): Chem. Phys. **80**, 329
Ellenbroek, T., J.P. Toennies, M. Wilde, and J. Wanner (1981): J. Chem. Phys. **75**, 3414
English, T.C. and J.C. Zorn (1974): in *Methods of Experimental Physics*, Vol. 3 B, ed. by D. Williams, p. 669 (Academic Press, New York)
Eppink, A.T.J.B., D.H. Parker, M.H.M. Janssen, B. Buijsse, and W. van der Zande (1998): J. Chem. Phys. **108**, 1305
Ernst, W.E., S. Kindt, and T. Törring (1983): Phys. Rev. Lett. **51**, 979
Ernst, W.E. and S. Kindt (1983): Appl. Phys. B **31**, 79
Ertmer, W. and B. Hofer (1976): Z. Phys. **276**, 9
Essen, L. and V.L. Parry (1955): Nature **176**, 280
Everdij, J.J., A. Huijser, and N.F. Verster (1973): Rev. Sci. Instrum. **44**, 721
Fecher, G.H., N. Böering, M. Volkmer, B. Pawlitzky, and U. Heinzmann (1990): Surf. Sci. **230**, L169
Fehrenbach, C.W., S.R. Lundeen, and O.L. Weaver (1995): Phys. Rev. A **51**, R910
Feneuille, S. and P. Jacquinot (1981): Adv. At. Mol. Phys. **17**, 99
Fewell, M.P., B.W. Shore, and K. Bergmann (1997): Austral. J. Phys. **50**, 281
Fischer, E. (1959): Z. Phys. **156**, 1
Fisher, D.S., C.W. Fehrenbach, S.R. Lundeen, E.A. Hessels, and B.D. DePaola (1997): Phys. Rev. A **56**, 4656
Frerichs, V., W. Kaenders, and D. Meschede (1992): Appl. Phys. B **55**, 242
Freund, S.M., G.A. Fisk, D.R. Herschbach, and W. Klemperer (1971): J. Chem. Phys. **58**, 27
Fried, D.G., T.C. Killian, L. Willmann, D. Landhuis, S.C. Moss, D. Kleppner, and T.J. Greytak (1998): Phys. Rev. Lett. **81**, 3811
Friedburg, H. and W. Paul (1951): Naturwissensch. **38**, 159
Friedburg, H. (1952): Z. Phys. **130**, 493
Frisch, R. and E. Segré (1933): Z. Phys. **80**, 610
Fuchs, M. and J.P. Toennies (1986): J. Chem. Phys. **85**, 7062
Gandhi, S.R., T.J. Curtiss, Q.-X. Xu, S.E. Choi, and R. B. Bernstein (1986): Chem. Phys. Lett. **132**, 6
Gandhi, S.R., Q.-X. Xu, T.J. Curtiss, and R. B. Bernstein (1987): J. Phys. Chem. **91**, 5437
Gandhi, S.R. and R. B. Bernstein (1988): J. Chem. Phys. **88**, 1472
Gaubatz, U., P. Rudecki, S. Schiemann, and K. Bergmann (1990): J. Chem. Phys. **92**, 5363
Gerlach, W. and O. Stern (1924): Ann. Phys. **74**, 673
Gerlach, W. and O. Stern (1925: Ann. Phys. **76**, 163
Ghaffari, B., J.M. Gerton, W.I. McAlexander, K.E. Strecker, D.M. Homan, and R.G. Hulet (1999): Phys. Rev. A **60**, 3878
Goldenberg, H.M., D. Kleppner, and N.F. Ramsey (1960): Phys. Rev. Lett. **5**, 361
Goldner, L.S., C. Gerz, R.J.C. Spreeuw, S.L. Rolston, C.I. Westbrook, and W.D. Phillips (1994): Phys. Rev. Lett. **72**, 997
Gordon, J.P., H.J. Zeiger, and C.H. Townes (1954): Phys. Rev. **95**, 282
Gordon, J.P., H.J. Zeiger, and C.H. Townes (1955): Phys. Rev. **99**, 1264
Gott, Y.V., M.S. Ioffe, and V.G. Tel'kovskii (1962): Nucl. Fusion, Suppl., Pt. 3, 1045 and 1284
Gräff, G., F.G. Major, R.W.H. Röder, and G. Werth (1968): Phys. Rev. Lett. **21**, 340

Gräff, G., E. Klempt, and G. Werth (1969): Z. Phys. **222**, 201
Greytak, T.J. (1995): in *Bose–Einstein Condensation*, ed. by A. Griffin, D.W. Snoke, and S. Stringari, p. 131 (Cambridge University Press, Cambridge)
Grice, R., J.E. Mosch, S.A. Safron, and J.P. Toennies (1970): J. Chem. Phys. **53**, 3376
Grivet, P. and A. Septier (1960): Nucl. Instrum. Methods **6**, 126
Gupta, A., D.S. Perry, and R.N. Zare (1980): J. Chem. Phys. **72**, 6250
Haeberli, W., M.D. Barker, C.A. Gossett, D.G. Mavis, P.A. Quin, J. Sowinsky, T. Wise, and H.F. Glavish (1982): Nucl. Instrum. Methods **196**, 319
Hain, T.D., M.A. Weibel, K.M. Backstrand, P.E. Pope, and T.J. Curtiss (1997): J. Phys. Chem. A **101**, 7674
Halbach, K. (1990): Int. J. Mod. Phys. B **4**, 1201
Hall, W.D. and J.C. Zorn (1974): Phys. Rev. A **10**, 1141
Hamilton, D.R. (1939): Phys. Rev. **56**, 30
Hamilton, C.E., L.J. Kinsey, and R.W. Field (1986): Annu. Rev. Phys. Chem. **37**, 493
Hammond, M.S., F.B. Dunning, G.K. Walters, and G.A. Prinz (1992): Phys. Rev. B **45**, 3674
Han, D.J., R.H. Wynar, Ph. Courteille, and D.J. Heinzen (1998): Phys. Rev. A **57**, R4114
Hanne, G.F., J.J. McClelland, R.E. Scholten, and R.J. Celotta (1993): J. Phys. B **26**, L753
Happer, W. (1972): Rev. Mod. Phys. **44**, 169
Heer, C.V. (1963): Rev. Sci. Instrum. **34**, 532
Helmer, J.C., F.B. Jacobus, and P.A. Sturrock (1960): J. Appl. Phys. **31**, 458
Hermann, H. W., I.V. Hertel, W. Reiland, A. Stamatovic, and W. Stoll (1977): J. Phys. B **10**, 251
Herschbach, D.R. (1966): in *Molecular Beams*, ed. by J. Ross, p. 319 (Wiley, New York)
Hertel, I.V. and W. Stoll (1974): J. Phys. B **7**, 570 and 583
Hertel, I.V. (1975): in *Atomic Physics*, Vol.4, ed. by G. zu Putlitz, E.W. Weber, and A. Winnacker, p. 381 (Plenum Press, New York)
Hertel, I.V. and W. Stoll (1976): J. Appl. Phys. **47**, 214
Hertel, I.V., H. Hofmann, and K.J. Rost (1976): Phys. Rev. Lett. **36**, 861
Hertel, I.V. (1976): in *Electron and Photon Interactions with Atoms*, ed. by H. Kleinpoppen, and M.R.C. McDowell, p. 375 (Plenum Press, New York)
Hertel, I.V. and W. Stoll (1977): Adv. At. Mol. Phys. B **13**, 113
Hertel, I.V. (1982): Adv. Chem. Phys. **50**, 475
Hertenberger, R., K. El Albiary, P. Schiemenz, and G. Graw (1996): Rev. Sci. Instrum. **67**, 1354
Hertenberger, R., Y. Eisermann, A. Hofmann, A. Metz, P. Schiemenz, S. Trieb, and G. Graw (1998): Rev. Sci. Instrum. **69**, 750
Herzberg, G. (1968): *Molecular Spectra and Molecular Structure*, Vol. 2 (van Nostrand, Princeton)
Hishinuma, N. and O. Sueoka (1983): Chem. Phys. Lett. **98**, 414
Hishinuma, N. and O. Sueoka (1985): Chem. Phys. Lett. **121**, 293
Hishinuma, N. and C. Amano (1990): in *Atomic Collision Research in Japan*, No. 16, ed. by Y. Itihara, p. 115 (The Society of Atomic Collision Research, Tokyo)
Höpe, A., D. Haubrich, G. Müller, W.G. Kaenders, and D. Meschede (1993): Europhys. Lett. **22**, 669
Hoffmeister, M., L. Potthast, and H.J. Loesch (1983): Chem. Phys. **78**, 369
Hsu, D.S.Y., N.D. Weinstein, and D.R. Herschbach (1975): Mol. Phys. **29**, 237
Hughes, H.K. (1947): Phys. Rev. **76**, 1675
Hughes, V.W., R.L. Long Jr., M.S. Lubell, M. Posner, and W. Raith (1972): Phys. Rev. A **5**, 195

Hundhausen, E. and H. Pauly (1965): Z. Nat.forsch. A **20**, 625
Hüwel, L., J. Maier, and H. Pauly (1982): J. Chem. Phys. **76**, 4961
Jalink, H., D.H. Parker, K.H. Meiwes-Broer, and S. Stolte (1986a): J. Chem. Phys. **90**, 552
Jalink, H., F. Harren, D. van den Ende, and S. Stolte (1986b): Chem. Phys. **108**, 391
Jongma, R.T., G. Berden, Th. Rasing, H. Zacharias, and G. Meijer (1997): J. Chem. Phys. **107**, 252
Kaenders, W.G., F. Lison, I. Müller, A. Richter, R. Wynands, and D. Meschede (1996): Phys. Rev. A **54**, 5067
Kaesdorf, S., G. Schönhense, and U. Heinzmann (1985): Phys. Rev. Lett. **54**, 885
Kakati, D. and D.C. Lainé (1967): Phys. Lett. A **24**, 676
Kakati, D. and D.C. Lainé (1971): J. Phys. E **4**, 269
Karny, Z. and R.N. Zare (1978): J. Chem. Phys. **68**, 3360
Kartaschoff, P. (1978): *Frequency and Time* (Academic Press, New York)
Kasai, T., D.C. Che, K. Ohashi, H. Ohoyama, and K. Kuwata (1989): Chem. Phys. Lett. **163**, 246
Kasai, T., T. Fukawa, T. Matsunami, D.C. Che, K. Ohashi, Y. Fukunishi, H. Ohoyama, and K. Kuwata (1993): Rev. Sci. Instrum. **64**, 1150
Kasai, T., Y. Hanaki, F. Fukunishi, H. Ohoyama, and K. Kuwata (1995): in *Atomic Collision Research in Japan*, No. 21, ed. by S. Tsurubuchi, p. 74 (The Society of Atomic Collision Research, Tokyo)
Kastler, A. (1950): J. Phys. Radium **11**, 255
Kastler, A. (1954): J. Proc. Phys. Soc. **67** A, 853
Kastler, A. (1957): J. Opt. Soc. Am. **47**, 460
Katz, D.P. (1997): J. Chem. Phys. **107**, 8491
Keller, H.M., M. Külz, R. Setzkorn, G.Z. He, K. Bergmann, and H.-G. Rubahn (1992): J. Chem. Phys. **96**, 8819
Kessler, J. (1985): *Polarized Electrons*, 2nd edn. (Springer, Berlin)
Ketterle, W. and N.J. van Druten (1996): Adv. At. Mol. Opt. Phys. **37**, 181
Kim, J., B. Friedrich, D.P. Katz, D. Patterson, J.D. Weinstein, R. DeCarvalho, and J. M. Doyle al. (1997): Phys. Rev. Lett. **78**, 3665
King, J.G. and J.R. Zacharias (1956): Adv. Electron. Electron Phys. **8**, 1
Kircz, J.G., R. Morgenstern, and G. Nienhuis (1982): Phys. Rev. Lett. **48**, 610
Klucharev, A.N. and V. Vujnovic (1990): Phys. Rep. **185**, 55
Knight, W.D., K. Clemenger, W.A. de Heer, and A.W. Saunders (1985): Phys. Rev. B **31**, 2539
Korsuski, M.I. and J.M. Fogel (1951): Zurn. Eksper. Teor. Fiz. **21**, 25
Kramer, K.H. and R.B. Bernstein (1965): J. Chem. Phys. **40**, 200
Kügler, K.-J., W. Paul, and U. Trinks (1978): Phys. Lett. B **72**, 422
Kügler, K.-J., K. Moritz, W. Paul, and U. Trinks (1985): Nucl. Instrum. Methods **228**, 240
Kuhn, A., S. Steuerwald, and K. Bergmann (1998): Eur. Phys. J. D **1**, 57
Kulin, S., B. Saubamea, E. Peik, J. Lawall, T.W. Hijmans, M. Leduc, and C. Cohen-Tannoudji (1997): Phys. Rev. Lett. **78**, 4185
Külz, M., A. Kortyna, M. Keil, B. Schellhaass, and K. Bergmann (1995): Z. Phys., D **33**, 109
Külz, M., M. Keil, A. Kortyna, B. Schellhaass J. Hauck, and K. Bergmann (1996): Phys. Rev. A **53**, 3324
Kusch, P. and V.W. Hughes (1959): in *Handbuch der Physik*, Vol. 37/1, ed. by S. Flügge, p.1 (Springer, Heidelberg)
Lackschewitz, U., J. Maier, and H. Pauly (1986): J. Chem. Phys. **84**, 181

Lainé, D.C. (1992a,b): in *Atomic and Molecular Beam Methods*, Vol. **2**, ed. by G. Scoles, p. 103 and p. 132 (Oxford University Press, Oxford)
Lau, D.C., A.I. Sidorov, G.I. Opat, R.J. McLean, W.J. Rowlands, and P. Hannaford (1999): Eur. Phys. J. D **5**, 193
Legon, A.C. (1983): Annu. Rev. Phys. Chem. **34**, 275
Legon, A.C. and D.J. Millen (1986): Chem. Revs. **86**, 635
Lemonick, A. and F.M. Pipkin (1954): Phys. Rev. **95**, 1356
Lemonick, A., F.M. Pipkin, and D.R. Hamilton (1955): Rev. Sci. Instrum. **26**, 1112
Levy, C.D.P. and A.N. Zelenski (1998): Rev. Sci. Instrum. **69**,732
L'Hermite, J.-M., G. Rahmat, and R. Vetter (1991): J. Chem. Phys. **95**, 3347
Liedenbaum, C., S. Stolte, and J. Reuss (1988): Chem. Phys. **122**, 443
Liedenbaum, C., S. Stolte, and J. Reuss (1989): Phys. Rep. **178**, 1
Liepack, H. and M. Drechsler (1956): Naturwissensch. **43**, 52
Louderback, J.G., A.J. Cox, L.J. Lising, D.C. Douglass, and L.A. Bloomfield (1993): Z.Phys. D **26**, 310
Luijks, G., S. Stolte, and J. Reuss (1981): Chem. Phys. **62**, 217
Lynn, J.G., M.W. Hart, T.H. Jeys, and F.B. Dunning (1986): Appl. Opt. **25**, 2154
Lyons, H. (1952): N.Y. Acad. Sci. **55**, 831
Majorana, E. (1932): Nuovo Cimento **9**, 43
Malthan, H. and J.P. Toennies (1974): *Rarefied Gas Dynamics* **9**, ed. by M. Becker and M. Fiebig, Vol. 2, p. C.14 (DFVLR-Press, Porz-Wahn)
Mariella, R.P., D.R. Herschbach, and W. Klemperer (1973): J. Chem. Phys. **58**, 3785
Mariella, R.P., D.R. Herschbach, and W. Klemperer (1974): J. Chem. Phys. **61**, 4575
Martin, J., B.W. Shore, and K. Bergmann (1996): Phys. Rev. A **54**, 1556
Mastenbroek, J.W.G., C.A. Taatjes, K. Nanta, M.H.M. Janssen, and S. Stolte (1995): AIP Conf. Proc. **329**, 117
McClelland, J.J. (1996): Exp. Methods Phys. Sci. B **29**, 145
McColm, D. (1966): Rev. Sci. Instrum. **37**, 1115
McGeoch, M.W. and R.E. Schlier (1986): Phys. Rev. A **33**, 1708
Meijer, H.A.J. (1990): Z. Phys. D **17**, 257
Mestdagh, J.M., B.A. Balko, M.H. Covinsky, P.S. Weiss, M.F. Vernon, H. Schmidt, and Y.T. Lee (1987): Faraday Discuss. Chem. Soc. **84**, 145
Meyer, G. and J.P. Toennies (1980): Chem. Phys. **52**, 39
Meyer, G. and J.P. Toennies (1982): J. Chem. Phys. **77**, 798
Migdall, A.L., J.V. Prodan, W.D. Phillips, T.H. Bergeman, and H.J. Metcalf (1985): Phys. Rev. Lett. **54**, 2596
Miller, T.M. and B. Bederson (1977): Adv. At. Mol. Phys. **13**, 1
Miller, R.L., A.G. Suits, P.L. Houston, R. Toumi, J.A. Mack, and A.M. Wodtke (1994): Science **265**, 1831
Millman, S., I.I. Rabi, and J.R. Zacharias (1938): Phys. Rev. **53**, 384
Misewich, J., H. Zacharias, and M.M.T. Loy (1985): Phys. Rev. Lett. **55**, 1919
Misewich, J. and M.M.T. Loy (1986): J. Chem. Phys. **85**, 1939
Mo, Y., H. Katayanagi, M.C. Heaven, and T. Suzuki (1996): Phys. Rev. Lett. **77**, 830
Monroe, C., W. Swann, H. Robinson, and C. Wieman (1990): Phys. Rev. Lett. **65**, 1571
Mori, Y. (1996): Rev. Sci. Instrum. **67**, 1286
Mosch, J.E., S.A. Safron, and J.P. Toennies (1975): Chem. Phys. **8**, 304
Muenter, J.S. (1992): in *Atomic and Molecular Beam Methods*, Vol. **2**, ed. by G. Scoles, p. 15 (Oxford University Press, Oxford)
Müller, H., B. Dierks, F. Hamsa, G. Zagatta, G.H. Fecher, and U. Heinzmann (1992): Surf. Sci. **269/270**, 207

Müller, H., G. Zagatta, N. Böwering, and U. Heinzmann (1994a): Chem. Phys. Lett. **223**, 197
Müller, H., G. Zagatta, M. Brandt, O. Wehmeyer, N. Böwering, and U. Heinzmann (1994b): Surf. Sci. **307–309**, 159
Neumann, R., F. Träger, and G. zu Putlitz (1987): in *Progress in Atomic Spectroscopy D*, ed. by H.J. Beyer and H. Kleinpoppen, p. 75 (Plenum Press, New York)
Nierenberg, W.A. (1957): Annu. Rev. Nucl. Sci. **7**, 349
Ohoyama, H., T. Ogawa, H. Makito, T. Kasai, and K. Kuwata (1996): J. Chem. Phys. **100**, 4729
Opat, G.I., S.J. Wark, and A. Cimmino (1992): Appl. Phys. B **54**, 396
Oreg, J., F.T. Hioe, and J.H. Eberly (1984): Phys. Rev. A **29**, 690
Oró, D.M., Q. Lin, X. Zahng, F.B. Dunning, and G.K. Walters (1992): J. Chem Phys. **97**, 7743
Orr-Ewing, A.R., W.R. Simpson, T.P. Rakitzis, S.A. Kandel, and R.N. Zare (1997): J. Chem. Phys. **106**, 5961
Parker, D.H., K.K. Chakravorty, and R.B. Bernstein (1981): J. Phys. Chem. **85**, 466
Parker, D.H., K.K. Chakravorty, and R.B. Bernstein (1982): Chem. Phys. Lett. **86**, 113
Parker, D.H. and R.B. Bernstein (1989): Ann. Rev. Phys. Chem. **40**, 561
Paul, W., F. Anton, L. Paul, S. Paul, and W. Mampe (1989): Z. Phys. C **45**, 25
Paul, W. (1990): Rev. Mod. Phys. **62**, 531
Penselin, S. (1978): in *Progress in Atomic Spectroscopy*, ed. by W. Hanle and H. Kleinpoppen, p. 463 (Plenum Press, New York)
Pesnelle, A., J. Pascale, R. Trainham, and H.J. Andrä (1996): Phys. Rev. A **54**, 4051
Peter, M. and M.W.P. Strandberg (1957): J. Chem. Phys. **26**, 1657
Petrich, W., M.H. Anderson, J.R. Ensher, and E.A. Cornell (1995): Phys. Rev. Lett. **74**, 3352
Phipps, T.E. and O. Stern (1931): Z. Phys. **73**,185
Pinkse, P.W.H., A. Mosk, M. Weidemüller, M.W. Reynolds, T.W. Hijmans, and J.T.M. Walraven (1998): Phys. Rev. A **57**, 4747
Poelker, M., K.P. Coulter, R.J. Holt, C.E. Jones, R.S. Kowalczyk, L. Young, B. Zeidman, and D.K. Toporkov (1995): Nucl. Instrum. Methods **364**, 58
Pollack, E., E.J. Robinson, and B. Bederson (1964): Phys. Rev. **134**, A1210
Pritchard, D. E., G.M. Carter, F.Y. Chu, and D. Kleppner (1970): Phys. Rev. A **2**, 1922
Pritchard, D.E. (1983): Phys. Rev. Lett. **51**, 1336
Prodell, A.G. and P. Kusch (1952): Phys. Rev. **88**, 184
Rabi, I.I., J.M.B. Kellog, and J.R. Zacharias (1934): Phys. Rev. **46**, 157
Rabi, I.I., J.R. Zacharias, S. Millman, and P. Kusch (1938): Phys. Rev. **53**, 318
Rabi, I.I. (1945): Phys. Rev. **67**, 199
Radford, H.E. (1967): in *Methods of Experimental Physics*, Vol. 4 B, ed. by V. Hughes and H.L. Schultz, p. 105 (Academic Press, New York)
Ramsey, N.F. (1949): Phys. Rev. **76**, 966
Ramsey, N.F. (1956): *Molecular Beams* (Clarendon Press, Oxford)
Rettner, C.T. and R.N. Zare (1981): J. Chem. Phys. **75**, 3636
Rettner, C.T. and R.N. Zare (1982): J. Chem. Phys. **77**, 2416
Reuss, J. (1988): in *Atomic and Molecular Beam Methods*, Vol 1., ed. by G. Scoles, p.276 (Oxford University Press, Oxford)
Ricci, L., C. Zimmermann, V. Vuletic, and T.W. Hänsch (1994): Appl. Phys. B **59**, 195
Riddle, T. W., M. Onellian, F.B. Dunning, and G.K. Walters (1981): Rev. Sci. Instrum. **52**, 797
Rosenberg, P. (1939): Phys. Rev. **55**, 1267

Rosenberg, P. (1940): Phys. Rev. **57**, 561
Rosner, S.D., R.A. Holt, and T.D. Gaily (1975): Phys. Rev. Lett. **35**, 785
Rubahn, H.-G., J.P. Toennies, M. Wilde, and J. Wanner (1985): Chem. Phys. Lett. **120**, 11
Rubahn, H.-G. and J.P. Toennies (1988): J. Chem. Phys. **89**, 287
Rubahn, H.-G. and K. Bergmann (1990): Annu. Rev. Phys. Chem. **41**, 735
Rubahn, H.-G., E. Konz, S. Schiemann, and K. Bergmann (1991): Z. Phys. D **22**, 401
Rubahn, H.-G. and K. Bergmann (1993): Comments At. Mol. Phys. **28**, 211
Rubahn, H.-G., A. Slenczka, and J.P. Toennies (1994): J. Chem. Phys. **101**, 1262
Salgado, J., J.W. Thomson, N. Andersen, D. Dowek, A. Dubois, J.C. Houver, S.E. Nielsen, I. Reiser, and A. Svensson (1997): J. Phys. B **30**, 3059
Salop, A., E. Pollack, and B. Bederson (1961): Phys. Rev. **124**, 1431
Sarma, P.R. and R.K. Bhandari (1998): Rev. Sci. Instrum. **69**, 1293 and 2909
Sarma, P.R., S.K. Pattanayak, and R.K. Bhandari (1999): Rev. Sci. Instrum. **70**, 2655
Sathyamurty, N. (1983): Chem. Rev. **83**, 601
Schäfer, R., Woenckhaus, J., J.A. Becker, and F. Hensel (1995): Z. Nat.forsch. A **50**, 445
Scheffers, H. and J. Stark (1934): Phys. Zeitschr. **35**, 625
Scheffers, H. and J. Stark (1936): Phys. Zeitschr. **37**, 217
Schiemann, S., A. Kuhn, S. Steuerwald, and K. Bergmann (1993): Phys. Rev. Lett. **71**, 3637
Schiemenz, P. (1992): Rev. Sci. Instrum. **63**, 2519
Schlier, Ch. (1957): Fortschr. Phys. **5**, 378
Schmelzbach, P.A. (1994): Rev. Sci. Instrum. **65**, 1378
Schmiedmayer, J., M.S. Chapman, C.R. Ekstrom, T.D. Hammond, D.A. Kokorowski, A. Lenef, R.A. Rubenstein, E.T. Smith, and D.E. Pritchard (1997): in *Atom Interferometry*, ed. by P.R. Berman, p. 1 (Academic Press, New York)
Schreel, K., J. Schleipen, A. Eppink, and J.J. ter Meulen (1993): J. Chem. Phys. **99**, 8713
Schropp, D., D. Cho, T. Vold, and E.A. Hinds (1987): Phys. Rev. Lett. **59**, 991
Schulz, P.A., A.S. Sudbo, D.J. Krajnovich, H.S. Kwok, Y.R. Shen, and Y.T. Lee (1979): Annu. Rev. Chem. Phys. **30**, 379
Septier, A. (1961): Adv. Electronics Electron Phys, **14**, 85
Shimizu, F., K. Shimizu, and H. Takuma (1985): Phys. Rev. A **31**, 3132
Sholeen, C.M. and R. Herm (1976): J. Chem. Phys. **64**, 5261
Sholeen, C.M., L.A. Gundel, and R. Herm (1976): J. Chem. Phys. **65**, 3223
Shore, B.W. (1990): *The Theory of Coherent Atomic Excitation* (Wiley, New York)
Shostak, S. and J.S. Muenter (1991): J. Chem. Phys. **94**, 5883
Simpson, W.R., T.P. Rakitzis, S.A. Kandel, A.J. Orr-Ewing, and R.N. Zare (1995): J. Chem. Phys. **103**, 7313
Slobodrian, R.J., C. Rioux, J. Giroux, and R. Roy (1986): Nucl. Instrum. Methods A **244**, 127
Stenger, J., C. Grosshauser, W. Kilian, B. Ranzenberger, and N. Rith (1997): Phys. Rev. Lett. **78**, 4177
Stolte, S., J. Reuss, and H.L. Schwartz (1972): Physica **57**, 254
Stolte, S., A.E. Proctor, W.M. Pope, and R.B. Bernstein (1977): J. Chem. Phys. **66**, 3468
Stolte, S. (1988): in *Atomic and Molecular Beam Methods*, Vol. 1, ed. by G. Scoles, p. 631 (Oxford University Press, Oxford)
Suits, A.G., P. de Pujo, O. Sublemontier, J.-P. Visticot, J. Berlande, J. Cuvellier, T. Gustavsson, J.-M. Mestdagh, P. Meynadier, and Y.T. Lee (1992): J. Chem. Phys. **97**, 4094
Süptitz, W., B.C. Duncan, and P.L. Gould (1997): J. Opt. Soc. Am. B **14**, 1001
Sussmann, R., R. Neuhauser, and H.J. Neusser (1994): J. Chem. Phys. **100**, 4784
Sussmann, R., R. Neuhauser, and H.J. Neusser (1995): J. Chem. Phys. **103**, 3315

Suzuki, T., H. Katayanagi, Y. Mo, and K. Tonokura (1996): Chem. Phys. Lett. **256**, 90
Tan, W.S., Z. Shi, C.H. Ying, and L. Vuskovic (1996): Phys. Rev. A **54**, R3710
Tarnovsky, V., M. Bunimovicz, L. Vuškovic, B. Stumpf, and B. Bederson (1993): J. Chem. Phys. **98**, 3894
Theuer, H. and K. Bergmann (1998): Eur. Phys. J. D **2**, 279
Thorsheim, H.R., Y. Wang, and J. Weiner (1990): Phys. Rev. A **41**, 2873
Thuis, H., S. Stolte, and J. Reuss (1979): Chem. Phys. **43**, 351
Toennies, J.P. (1962): Faraday Discuss. Chem. Soc. **33**, 96
Toennies, J.P. (1965): Z. Phys. **182**, 257
Toennies, J.P. (1966): Z. Phys. **193**, 76
Tollett, J.J., C.C. Bradley, C.A. Sackett, and R.G. Hulet (1995): Phys. Rev. A **51**, R22
Trajmar, S., J.C. Nickel, and T. Antoni (1986): Phys. Rev. A **34**, 4154
Trischka, J.W. (1948): Phys. Rev. **74**, 718
Trischka, J.W. (1962): in *Methods of Experimental Physics*, Vol. 3, ed. by D. Williams, p. 589 (Academic Press, New York,)
Tsuo, L.Y., D.J. Auerbach, and L. Wharton (1977): Phys. Rev. Lett. **38**, 20
Tsuo, L.Y., D.J. Auerbach, and L. Wharton (1979): J. Chem. Phys. **70**, 5296
Urbain, X. and H. Pauly (1993): XVIII. Int. Conf. Phys. Electron. Atomic Collisions, Book of Abstracts, Vol. 2, ed. by T. Andersen, B. Fastrup, F. Folkmann, and H. Knudsen, p. 643 (Aarhus University, Aarhus)
Valentini, J.J. (1993): Proc. SPIE, **1858**, 150
van Brunt, R.J. and R.N. Zare (1968): J. Chem. Phys. **48**, 4304
van den Ende, D. and S. Stolte (1980): Chem. Phys. Lett. **76**, 13
van den Ende, D., S. Stolte, J.B. Cross, G.H. Kwei, and J.J. Valentini (1982): J. Chem. Phys. **77**, 2206
van Leuken, J.J., J. Bulthuis, S. Stolte, and J.G. Snijders (1996): Chem. Phys. Lett. **260**, 595
Vanier, J. and C. Audoin (1989): *The Quantum Physics of Atomic Frequency Standards* (Adam Hilger, Bristol)
Vauthier, C.R. (1949): Acad. Sci. Paris, **228**, 1113
Vohralik, P.F. and R.E. Miller (1985): J. Chem. Phys. **83**, 1609
von dem Borne, A., T. Dohrmann, A. Verweyen, B. Sonntag, K. Godehusen, and P. Zimmermann (1997): Phys. Rev. Lett. **78**, 4019
Vuletic, V., T.W. Hänsch, and C. Zimmermann (1996): Europhys. Lett. **36**, 349
Walker, T.G. and W. Happer (1997): Rev. Mod. Phys. **69**, 629
Walraven, J.T.M. and T.W. Hijmans (1994): Physica B **197**, 417
Weibel, M.A., T.D. Hain, and T.J. Curtiss (1997): J. Vac. Sci. Technol. A **15**, 2238
Weibel, M.A., T.D. Hain, and T.J. Curtiss (1998): J. Chem. Phys. **108**, 3134
Weiner, J., F. Masnou-Seeuws, and A. Giusti-Suzor (1990): Adv. At. Mol. Opt. Phys. **26**, 209
Weinstein, J.D., R. deCarvalho, J. Kim, D. Patterson, B. Friedrich, and J.M. Doyle (1998): Phys. Rev. A **57**, R3173
Weiser, C. and P.E. Siska (1987): Rev. Sci. Instrum. **58**, 2124
Weitz, M., T. Heupel, and T.W. Hänsch (1996): Phys. Rev. Lett. **77**, 2356
Willems, P.A. and K.G. Libbrecht (1995): Phys. Rev. A **51**, 1403
Wineland, D.J. and N.F. Ramsey (1972): Phys. Rev. A **5**, 821
Wing, W.H. (1980): Phys. Rev. Lett. **45**, 631
Wing, W.H. (1984): Prog. Quantum Electron. **8**, 181
Woenckhaus, J., R. Schäfer, and J.A. Becker (1996): Surf. Rev. Lett. **3**, 371
Wolfgang, R. (1968): Sci. Am. **218**, No.4, 44

Zacharias, J.R., J.G. Yates, and R.D. Haun (1955): Proc. IRE **43**, 364
Zandee, L. and J. Reuss (1977): Chem. Phys. **26**, 327
Zandee, L. and R.B. Bernstein (1978): J. Chem. Phys. **68**, 3760
Zare, R.N. (1972): Mol. Photochem. **4**, 1
Zhu, X. (1984): Acta Physica Sinica, **33**, 1605
Ziegler, G., S.V.K. Kumav, H.-G. Rubahn, A. Kuhn, B. Sun, and K. Bergmann (1991): J. Chem. Phys. **94**, 4252
Zorn, J.C. and T.C. English (1973): Adv. At. Mol. Phys. **9**, 244

Chapter 5

Aardema, T.G., R.M.S. Knops, S.P.L. Nijsten, K.A.H. van Leeuwen, J.P.J. Driessen, and H.C.W. Beijerinck (1996): Phys. Rev. Lett. **76**, 748
Abate, J.A. (1974): Opt. Commun. **10**, 269
Abfalterer, R., C. Keller, S. Bernet, M.K. Oberthaler, J. Schmiedmayer, and A. Zeilinger (1997): Phys. Rev. A **56**, R4365
Abraham, E.R.I., N.W.M. Ritchie, W.I. McAlexander, and R.G. Hulet (1995): J. Chem. Phys. **103**, 7773
Abraham, E.R.I., W.I. McAlexander, J.M. Gerton, R.G. Hulet, R. Côté, and A. Dalgarno (1997): Phys. Rev. A **55**, R3299
Adams, C.S. (1994): Contemporary Physics **35**, 1
Adams, C.S., M. Sigel, and J. Mlynek (1994a): Phys. Rep. **240**, 143
Adams, C.S., O. Carnal, and J. Mlynek (1994b): Adv. At. Mol. Opt. Phys. **34**, 1
Adams, C.S., H.J. Lee, N. Davidson, M. Kasevich, and S. Chu (1995): Phys. Rev. Lett. **74**, 3577
Adams, C.S. and E. Riis (1997): Prog. Quantum Electronics **21**, 1
Anderson, B.P. and M.A. Kasevich (1994): Phys. Rev. A **50**, R3581
Anderson, M.H., J.R. Ensher, M.R. Matthews, C.E. Wieman, and C.A. Cornell (1995): Science **269**, 198
Andrews, M.R., M.-O. Mewes, N.J. van Druten, D.S. Durfee, D.M. Kurn, and W. Ketterle (1996): Science **273**, 84
Andrews, M.R., C.G. Townsend, H.-J. Miesner, D.S. Durfee, D.M. Kurn, and W. Ketterle (1997): Science **275**, 637
Arndt, M., P. Szriftgiser, J. Dalibard, and A.M. Steane (1996): Phys. Rev. A **53**, 3369
Arndt, M., M. Ben Dahan, D. Guéry-Odelin, M.W. Reynolds, and J. Dalibard (1997): Phys. Rev. Lett. **79**, 625
Arndt, M., O. Nairz, J. Vos-Andreae, C. Keller, G. van der Zouw, and A. Zeilinger (1999): Nature **401**, 680
Ashkin, A. (1970): Phys. Rev. Lett. **25**, 1321
Ashkin, A. (1978): Phys. Rev. Lett. **40**, 729
Ashkin, A. and J.P. Gordon (1979): Opt. Lett. **4**, 161
Ashkin, A. and J.P. Gordon (1983): Opt. Lett. **8**, 511
Ashkin, A. (1984): Opt. Lett. **9**, 454
Ashkin, A., J.M. Dziedzic, and T. Yamane (1987): Nature **330**, 769
Aspect, A. (1997): Proc. SPIE, **2995**, 2
Audouard, E., P. Duplàa, and J. Vigué (1995): Europhys. Lett. **32**, 397
Auerbach, D., E.E.A. Bromberg, and L. Wharton (1966): J. Chem. Phys. **45**, 2160
Bahns, J.T., W.C. Stwalley, and P.L. Gould (1996): J. Chem. Phys. **104**, 9689
Baldwin, K.G.H. (1996): Australian J. Phys. **49**, 855
Balykin, V.I. and V.S. Letokhov (1987): Opt. Commun. **64**, 151

Balykin, V.I., V.S. Letokhov, B. Y.B. Ovchinnikov, and A.I. Sidorov (1988): J. Mod. Optics **35**, 17

Balykin, V.I. and V.S. Letokhov (1989): Phys. Today **42**, 23

Balykin, V.I. (1999): Adv. At. Mol. Opt. Phys. **41**, 182

Barrett, T.E., S.W. Dapore-Schwartz, M.D. Ray, and G.P. Lafyatis (1991): Phys. Rev. Lett. **67**, 3483

Basini, G.F., M. Inguscio, and T.W. Hänsch (eds.) (1989): *The Hydrogen Atom* (Springer, Heidelberg)

Batelaan, H., S. Bernet, M.K. Oberthaler, E.M. Rasel, J. Schmiedmayer, and A. Zeilinger (1997): in *Atom Interferometry*, ed. by P.R. Berman, p. 84 (Academic Press, New York)

Bauch, A. and R. Schröder (1997): Phys. Rev. Lett. **78**, 622

Baum, G., C.D. Caldwell, and W. Schröder (1980): Appl. Phys. **21**, 121

Bergeman, T., G. Erez, and H. Metcalf (1987): Phys. Rev. A **35**, 1535

Berggren, K.K., A. Bard, J.L. Wilbur, J.D. Gillaspy, A.G. Helg, J.J. McClelland, S.L. Rolston, W.D. Phillips, M. Prentiss, and G.M. Whitesides (1995): Science **269**, 1255

Berggren, K.K., R. Youngkin, E. Cheung, M. Prentiss, A. Black, G.M. Whitesides, D.C. Ralph, C.T. Black, and M. Tinkham (1997): Adv. Mat. **9**, 52

Berkhout, J.J., O.J. Luiten, I.D. Setija, T.W. Hijmans, T. Mizusaki, and J.T.M. Walraven (1989): Phys. Rev. Lett. **3**, 1689

Berkhout, J.J. and J.T.M. Walraven (1993): Phys. Rev. B **47**, 8886

Berman, P.R. (ed.) (1997): *Atom Interferometry* (Academic Press, New York)

Bernhardt, A.F., D.E. Duerre, J.R. Simpson, and L.L. Wood (1974): Appl. Phys. Lett. **25**, 617

Bernhardt, A.F., D.E. Duerre, J.R. Simpson, and L.L. Wood (1976): Opt. Commun. **16**, 166 and 169

Berthoud, P., A. Joyet, G. Dudle, N. Sagna, and P. Thomann (1998): Europhys. Lett. **41**, 141

Bethlem, H.L., G. Berden, and G. Meijer (1999): Phys. Rev. Lett. **83**, 1558

Bigelow, N.P. (1998): Low Temp. Phys. **24**, 106

Birkl, G., M. Gatzke, I.H. Deutsch, S.L. Rolston, and W.D. Phillips (1995): Phys. Rev. Lett. **75**, 2823

Bjorkholm, J.E., R.R. Freeman, A. Ashkin, and D.B. Pearson (1978): Phys. Rev. Lett. **41**, 1361

Bloch, I., T.W. Hänsch, and T. Esslinger (1999): Phys. Rev. Lett. **82**, 3008

Boesten, H.M.J.M., C.C. Tsai, J.R. Gardner, D.J. Heinzen, and B.J. Verhaar (1997): Phys. Rev. A **55**, 636

Boiron, D., A. Michaud, P. Lemonde, Y. Castin, C. Salomon, S. Weyers, K. Szymaniec, L. Cognet, and A. Clairon (1996): Phys. Rev. A **53**, R3734

Boiron, D., A. Michaud, J.M. Fourner, L. Simard, M. Sprenger, G. Grynberg, and C. Salomon (1998): Phys. Rev. A **57**, R4106

Bouchiat, M.A. (1991): in *Atomic Physics*, Vol. 12, ed. by J.C. Zorn and R.R. Lewis, p. 399 (AIP, New York)

Bradley, C.C., C.A. Sackett, J.J. Tollet, and R.G. Hulet (1995): Phys. Rev. Lett. **75**, 1687

Bradley, C.C. and R.G. Hulet (1996): Exp. Methods Phys. Sci. B **29**, 129

Brouri, R., R., Asimov, M. Gorlicki, S. Ferou, J. Reinhardt, V. Lorent, and H. Haberland (1996): Opt. Commun. **124**, 448

Burnett, K. (1996): Contemp. Phys. **1**, 1

Butts, D.A. and D.S. Rokhsar (1997): Phys. Rev. A **55**, 4346

Carnal, O. and J. Mlynek (1991a): Phys. Rev. Lett. **66**, 2689

Carnal, O. and J. Mlynek (1991b): Phys. Bl. **47**, 379
Carnal, O., M. Sigel, T. Sleator, H. Takuma, and J. Mlynek (1991a): Phys. Rev. Lett. **67**, 3231
Carnal, O., A. Faulstich, and J. Mlynek (1991b): Appl. Phys. B **53**, 88
Carnal, O. and J. Mlynek (1996): Exp. Methods Phys. Sci. B **29**, 341
Cataliotti, F.S., E.A. Cornell, C. Fort, M. Inguscio, F. Marin, M. Prevedelli, L. Ricci, and G.M. Tino (1998): Phys. Rev. A **57**, 1136
Celotta, R.J., R. Gupta, R.E. Scholten, and J.J. McClelland (1996): J. Appl. Phys. **79**, 6079
Chapman, M.S., C.R. Ekstrom, T.D. Hammond, R.A. Rubinstein, J. Schmiedmayer, S. Wehinger, and D. E. Pritchard (1995): Phys. Rev. Lett. **74**, 4783
Chu, S., L. Hollberg, J.E. Bjorkholm, A. Cable, and A. Ashkin (1985): Phys. Rev. Lett. **55**, 48
Chu, S., J.E. Bjorkholm, A. Ashkin, and A. Cable (1986): Phys. Rev. Lett. **57**, 314
Chu, S. (1991): Science **253**, 861
Chu, S. (1992): Sci. Am. **266**, 79
Chu, S. (1998): Rev. Mod. Phys. **70**, 685
Citron, M.L., H.R. Gray, C.W. Gabel, and C.R. Stroud Jr. (1977): Phys. Rev. A **16**, 1507
Clairon, A., C. Salomon, S. Guellati, and W. Phillips (1991): Europhys. Lett. **16**, 165
Clauser, J.F. and S. Li (1994a), Phys. Rev. A **49**, R2213
Clauser, J.F. and S. Li (1994b): Phys. Rev. A **50**, 2430
Cohen-Tannoudji, C. (1992): in *Laser Manipulation of Atoms and Ions*, ed. by A. Arimondo, W.D. Phillips, and F. Strumia, p. 99 (North-Holland, Amsterdam)
Cohen-Tannoudji, C. (1998): Rev. Mod. Phys. **70**, 707
Cook, R.J. (1979): Phys. Rev. A **20**, 224
Cook, R.J. and R.K. Hill (1982): Opt. Commun. **43**, 258
Courteille, Ph., R.S. Freeland, D.J. Heinzen, F.A. van Abeelen, and B.J. Verhaar (1998): Phys. Rev. Lett. **81**, 69
Courtois, J.-Y. and G. Grynberg (1996): Adv. At. Mol. Opt. Phys. **36**, 88
Dahan, M.B., E. Peik, J. Reichel, Y. Castin, and C. Salomon (1996): Phys. Rev. Lett. **76**, 4508
Dalfovo, F., S. Giorgini, L.P. Pitaevskii, and S. Stringari (1999): Rev. Mod. Phys. **71**, 463
Dalibard, J., C. Salomon, A. Aspect, H. Metcalf, A. Heidmann, and C. Cohen-Tannoudji (1987): in *Laser Spectroscopy* Vol. 8, ed. by S. Svanberg and W. Persson, p. 81 (Springer, Berlin)
Dalibard, J. and C. Cohen-Tannoudji (1985): J. Opt. Soc. Am. B **2**, 1707
Dalibard, J. and C. Cohen-Tannoudji (1989): J. Opt. Soc. Am. B **6**, 2023
Davis, K.B., M.O. Mewes, M.R. Andrews, N.J. van Druten, D.S. Durfee, D.M. Kurn, and W. Ketterle (1995): Phys. Rev. Lett. **75**, 3969
Demtröder, W. (1996): *Laser Spectroscopy*, 2nd edn. (Springer, Berlin)
Denschlag, J., D. Cassettari, and J. Schmiedmayer (1999): Phys. Rev. Lett. **82**, 2014
Deutschmann, R., W. Ertmer, and H. Wallis (1993): Phys. Rev. A **47**, 2169
Dieckmann, K., R.J.C. Spreeuw, M. Weidemüller, and J.T.M. Walraven (1998): Phys. Rev. A **58**, 3891
Doak, R.B., K. Kevern, A. Chizmeshya, R. David, and G. Comsa (1997): Proc. SPIE, **2995**, 146
Doak, R.B. (1989): Optical Society of America Technical Digest **15**, 250
Doak, R.B., R.E. Grisenti, S. Rehbein, G. Schmahl, J.P. Toennies, and Ch. Wöll (1999): Phys. Rev. Lett. **83**, 4229
Dowling, J.P. and J. Gea-Banacloche (1996): Adv. At. Mol. Opt. Phys. **37**, 1

Doyle, J.M., J.C. Sandberg, I.A. Yu, C.L. Cesar, D. Kleppner, and T.J. Greytak (1991): Phys. Rev. Lett. **67**, 603
Doyle, J.M., B. Friedrich, J. Kim, and D. Patterson (1995): Phys. Rev. Lett. **52**, R2515
Drndic, M., K.S. Johnson, J.H. Thywissen, M. Prentiss, and R.M. Westervelt (1998): Appl. Phys. Lett. **72**, 2906
Düren, R., H.O. Hoppe, and H. Pauly (1975): in *50 Jahre MPI für Strömungsforschung*, ed. by MPISF, p. 414 (Hubert & Co. Göttingen)
Ekstrom, C.R., D.W. Keith, and D.E. Pritchard (1992): Appl. Phys. B **54**, 369
Ekstrom, C.R., J. Schmiedmayer, M.S. Chapman, T.D. Hammond, and D.E. Pritchard (1995): Phys. Rev. A **51**, 3883
Ernst, U., A. Marte, F. Schreck, J. Schuster, and G. Rempe (1998): Europhys. Lett. **41**, 1
Ertmer, W., R. Blatt, and J.L. Hall (1985): Phys. Rev. Lett. **54**, 996
Ertmer, W., K. Sengstock, U. Sterr, J.H. Müller, D. Bettermann, V. Rieger, F. Ruschewitz, J.L. Peng, F. Dingler, A. Pabst, R. Strichirsch, M. Christ, A. Scholz, M. Schiffer, G. Wokurka, R. Deutschmann, H. Wallis, S. Friebel, and S. Penselin (1996): Laser Physics **6**, 278
Esslinger, T., M. Weidemüller, A. Hemmerich, and T.W. Hänsch (1993): Opt. Lett. **18**, 450
Faulstich, A. Schnetz, M. Sigel, T. Sleator, O. Carnal, V. Balykin, H. Takuma, and J. Mlynek (1992): Europhys. Lett. **77**, 393
Feron, S., J. Reinhardt, S. Le Boiteux, O. Gorceix, J. Baudon, M. Ducloy, J. Robert, C. Minatura, S. Chormaic, H. Haberland, and V. Lorent (1993): Opt. Commun. **102**, 83
Finer, J.T., R.M. Simmons, and J.A. Spudich, Nature **368**, 113 (1994)
Fioretti, A., D. Comparat, A. Crubellier, O. Dilieu, F. Masnou-Seeuws, and P. Pillet (1998): Phys. Rev. Lett. **80**, 4402
Fortagh, J., A. Grossmann, C. Zimmermann, and T.W. Hänsch (1998): Phys. Rev. Lett. **81**, 5310
Fried, D.G., T.C. Killian. L. Willmann, D. Landhuis, S.C. Moss, D. Kleppner, and T.J. Greytak (1998): Phys. Rev. Lett. **81**, 3811
Frisch, O.R. (1933): Z. Phys. **86**, 42
Gensemer, S.D., V. Sanchez-Villicana, K.Y.N. Tan, T.T. Grove, and P.L. Gould (1997): Phys. Rev. A **56**, 4055
Ghezali, S., P. Laurent, S.N. Lea, and A. Clairon (1996): Europhys. Lett. **36**, 25
Gibble, K. and S. Chu (1992): Metrologia **29**, 201
Gibble, K. and S. Chu (1993): Phys. Rev. Lett. **70**, 1771
Goepfert, A., I. Bloch, D. Haubrich, F. Lison, R. Schütze, R. Wynands, and D. Meschede (1997): Phys. Rev. A **56**, R3354
Gordon, J.P. and A. Ashkin (1980): Phys. Rev. A **21**, 1606
Gould, P.L., G.A. Ruff, and D.E. Pritchard (1986): Phys. Rev. Lett. **56**, 827
Grisenti, R.E., W. Schöllkopf, J.P. Toennies, G.C. Hegerfeldt, and T. Köhler (1999): Phys. Rev. Lett. **83**, 1755
Grynberg, G., B. Lounis, P. Verkerk, J.-Y. Courtois, and C. Salomon (1993): Phys. Rev. Lett. **70**, 2249
Guckert, R., X. Zhao, S.G. Crane, A. Hime, W.A. Taylor, D. Tupa, D.J. Vieira, and H. Wollnik (1998): Phys. Rev. A **58**, R1637
Gupta, R., J.J. McClelland, Z.J. Jabbour, and R.J. Celotta (1995): Appl. Phys. Lett. **67**, 1378
Gustavson, T.L., P. Bouyer, and M.A. Kasevich (1997): Phys. Rev. Lett. **78**, 2046
Hagley, E.W., L. Deng, M. Kozuma, J. Wen, K. Helmerson, S.L. Rolston, and W.D. Phillips (1999): Science **283**, 1706
Hänsch, T.W. and A.L. Schawlow (1975): Opt. Commun. **13**, 68

Hajnal, J.V. and G.I. Opat (1989): Opt. Commun. **71**, 119
Hajnal, J.V., K.G.H. Baldwin, P.T.H. Fisk, H.-A. Bachor, and G.I. Opat (1989): Opt. Commun. **73**, 331
Han, D.-J., R.H. Wynar, P. Courteille, and D.J. Heinzen (1998): Phys. Rev. A **57**, R4114
Hau, L.V., B.D. Busch, C. Liu, Z. Dutton, M.M. Burns, and J.A. Golovchenko (1998): Phys. Rev. A **58**, R54
Hecht, E. (1987): *Optics*, 2nd edn. (Addison-Wesley, Reading, MA)
Hegerfeldt, G.C. and T. Köhler (1998): Phys. Rev. A **57**, 2021
Heinzen, D.J. (1995): in *Atomic Physics*, Vol. 14, ed. by D.J. Wineland, C.E. Wieman, and S.J. Smith, p. 369 (AIP, New York)
Hemmerich, A. and T.W. Hänsch (1993): Phys. Rev. Lett. **70**, 410
Hemmerich, A., C. Zimmermann, and T.W. Hänsch (1994): Phys. Rev. Lett. **72**, 625
Henkel, C., A. Aspect, J.Y. Courtois, R. Kaiser, K. Mlmer, and C.I. Westbrook (1995): Annales de Physique **20**, 643
Henkel, C., K. Molmer, R. Kaiser, N. Vansteenkiste, C.I. Westbrook, and A. Aspect (1997): Phys. Rev. A **55**, 1160
Hess, H.F. (1986): Phys. Rev. B **34**, 3476
Higashi, G.S., R.S. Becker, Y.J. Chabal, and A. Becker (1991): Appl. Phys. Lett. **58**, 1656
Hinderthur, H., A. Pautz, V. Rieger, F. Ruschwewitz, J.L. Peng, K. Sengstock, and W. Ertmer (1997): Phys. Rev. A **56**, 2085
Holst, B. and W. Allison (1997): Nature **390**, 244
Hoffmann, D., S. Bali, and T. Walker (1996): Phys. Rev. A **54**, 1030
Hoffnagle, J. (1988): Opt. Lett. **13**, 102
Hughes, I.G., P.A. Barton, T.M. Roach, M.G. Boshier, and E.A. Hinds (1997a): J. Phys. B **30**, 647
Hughes, I.G., P.A. Barton, T.M. Roach, and E.A. Hinds (1997b): J. Phys. B **30**, 2119
Hundhausen, E. and H. Harrison (1966): Research Note 66-3, Boeing Scientific Research Laboratories, Seattle
Itano, W.M., L.L. Lewis, and D.J. Wineland (1982): Phys. Rev. A **25**, 1233
Ito, H., K. Sasaki, T. Nakata, W. Jhe, and M. Ohtsu (1995): Opt. Commun. **115**, 57
Ito, H., T. Nakata, K. Sasaki, M. Ohtsu, K.I. Lee, and W. Jhe (1996): Phys. Rev. Lett. **76**, 4500
Ito, H., K. Sasaki, W. Jhe, and M. Ohtsu (1997): Opt. Commun. **141**, 43
Ito, Y., A.L. Bleloch, and L.M. Brown (1998): Nature **394**, 49
Jacquinot, P., S. Liberman, J.L. Picqué, and J. Pinard (1973): Opt. Commun. **6**, 163
Jessen, P.S., C. Gerz, P.D. Lett, W.D. Phillips, S.L. Rolston, R.J.C. Spreeuw, and C.I. Westbrook (1992): Phys. Rev. Lett. **69**, 49
Jessen, P.S. and I.H. Deutsch (1996): Adv. At. Mol. Opt. Phys. **37**, 95
Johnson, K.S., M. Drndic, J.H. Thywissen, G. Zabow, R.M. Westervelt, and M. Prentiss (1998): Phys. Rev. Lett. **81**, 1137
Jones, K.M., P.S. Julienne, P.D. Lett, W.D. Phillips, E. Tiesinga, and C.J. Williams (1996): Europhys. Lett. **35**, 85 (1996)
Julienne, P.S., K. Burnett, Y.B. Band, and W.C. Stwalley (1998): Phys. Rev. A **58**, R796
Kaiser, R., Y. Lévy, N. Vansteenkiste, A. Aspect, W. Seifert, D. Leipold, and J. Mlynek (1994): Opt. Commun. **104**, 234
Kasevich, M., E. Riis, S. Chu, and R. DeVoe (1989): Phys. Rev. Lett. **63**, 612
Kasevich, M. and S. Chu (1991): Phys. Rev. Lett. **67**, 181
Kasevich, M., K. Moler, E. Riis, E. Sunderman, D. Weiss, and S. Chu (1991): in *Atomic Physics*, Vol. 12, ed. by J.C. Zorn and R.R. Lewis, p. 47 (AIP, New York)
Kasevich, M. and S. Chu (1992): Phys. Rev. Lett. **69**, 1741 and Appl. Phys. B **54**, 321

Ke, P.C., J. Szajman, X.S. Gau, and M. Gu (1997): Proc. SPIE **2984**, 42
Keith, D.W., M.L. Schattenburg, H.I. Smith, and D.E. Pritchard (1988): Phys. Rev. Lett. **61**, 1580
Keith, D.W., C.R. Ekstrom, Q.A. Turchette, and D.E. Pritchard (1991): Phys. Rev. Lett. **66**, 2693
Keith, D.W. and M.J. Rooks (1991): J. Vac. Sci. Technol. B **9**, 2846
Kelly, J.F. and A. Gallagher (1987): Rev. Sci. Instrum. **58**, 563
Ketterle, W., A. Martin, M.A. Joffe, and D.E. Pritchard (1992): Phys. Rev. Lett. **69**, 2483
Ketterle, W., K.B. Davis, M.A. Joffe, A. Martin, and D.E. Pritchard (1993): Phys. Rev. Lett. **70**, 2253
Ketterle, W. and N.J. van Druten (1996): Adv. At. Mol. Opt. Phys. **37**, 181
Kim, J.A., K.I. Lee, H.R. Noh, W. Jhe, and M. Ohtsu (1997a): Opt. Lett. **22**, 117
Kim, J., B. Friedrich, D.P. Katz, D. Patterson, J.D. Weinstein, R. DeCarvalho, and J.M. Doyle (1997b): Phys. Rev. Lett. **78**, 3665
King, J.G. (1959): Proc. 13th Annual Symp. on Frequency Control, Asbury Park, p. 603 (U.S. Army Signal Research and Development Laboratory, Fort Monmouth)
Klein, A.G. and S.A. Werner (1993): Rep. Prog. Phys. **46**, 259
Kleppner, D. (1997): Phys. Today **50**, 11
Köhler, K.A., R. Feltgen, and H. Pauly (1977): Phys. Rev. A **15**, 1407
Kokkelmans, S.J., B.J. Verhaar, K. Gibble, and D.J. Heinzen (1997): Phys. Rev. A **56**, R4389
Kreis, M., F. Lison, D. Haubrich, D. Meschede, S. Nowak, T. Pfau, and J. Mlynek (1996): Appl. Phys. B **63**, 649
Kuga, T., Y. Torii, N. Shiokawa, T. Hirano, Y. Shimizu, and H. Sasada (1997): Phys. Rev. Lett. **78**, 4713
Kuppens, S., M. Rauner, M. Schiffer, K. Sengstock, W. Ertmer, F.E. van Dorsselaer, and G. Nienhaus (1998): Phys. Rev. A **58**, 3068
Landragin, A., J.-Y. Courtois, G. Labeyrie, N. Vansteenkiste, C.-I. Westbrook, and A. Aspect (1996): Phys. Rev. Lett. **77**, 1464
Lau, D.C., A.I. Siderov, G.I. Opat, R.J. McLean, W.J. Roalands, and P. Hannaford (1999): Eur. Phys. J. D **5**, 193
Lawall, J., S. Kulin, F. Bardou, B. Saubamea, N. Bigelow, M. Leduc, and C. Cohen-Tannoudji (1995): Phys. Rev. Lett. **75**, 4194
Lee, H.J., C.S. Adams, N. Davidson, M. Weitz, B. Young, M. Kasevich, and S. Chu (1995): in *Atomic Physics*, Vol. 14, ed. by C.E. Wieman and D. Wineland, p. 258, (AIP, New York)
Lee, H.J., C.S. Adams, M. Kasevich, and S. Chu (1996): Phys. Rev. Lett. **76**, 2658
Legere, R. and J. Gibble (1998): Phys. Rev. Lett. **81**, 5780
Lepage, G.P. (1995): in *Atomic Physics*, Vol. 14, ed. by D.J. Wineland, C.E. Wieman, and S.J. Smith, p. 18 (AIP, New York)
Letokhov, V.S., V.G. Minogin, and B.D. Pavlik (1976): Opt. Commun. **19**, 72
Lett, P.D., R.N. Watts, C.I. Westbrook, W.D. Phillips, P.L. Gould, and H.J. Metcalf (1988): Phys. Rev. Lett. **61**, 169
Lett, P.D., P.S. Julienne, and W.D. Phillips (1995): Annu. Rev. Phys. Chem. **46**, 423
Lewenstein, M. and L. You (1996): Adv. At. Mol. Opt. Phys. **36**, 221
Liang, J. and C. Fabre (1986): Opt. Commun. **59**, 31
Lison, F., P. Schuh, D. Haubrich, and D. Meschede (2000): Phys. Rev. A **61**, 013405
Liu, Y., G.J. Sonek, M.W. Berns, and B.J. Tromberg (1996): Biophys. J. **71**, 2158
Lu, Z.T., K.L. Corwin, M.J. Renn, M.H. Anderson, E.A. Cornell, and C.E. Wieman (1996): Phys. Rev. Lett. **77**, 3331

Lu, Z.T., K.L. Corwin, K.R. Vogel, C.E. Wieman, T.P. Dinneen, J. Maddi, and H. Gould (1997): Phys. Rev. Lett. **79**, 994
Luppov, V.G., W.A. Kaufman, K.M. Hill, R.S. Raymond, and A.D. Krisch (1993): Phys. Rev. Lett. **71**, 2405
Maier-Leibnitz, H. and T. Springer (1962): Z. Phys. **167**, 368
Marcassa, L.G., V.S. Bagnato, Y. Wang, C. Tsao, J. Weiner, O. Dulieu, Y.B. Band, and P.S. Julienne (1993): Phys. Rev. A **47**, 4563
Marcassa, L.G., K. Helmerson, A.M. Tuboy, D.M.B.P. Milori, S.R. Muniz. J. Flemming, S.C. Zflio, and V.S. Bagnato (1996): J. Phys. B **29**, 3051
Marksteiner, S., C.M. Savage, P. Zoller, and S.L. Rolston (1994): Phys. Rev. A **50**, 2680
Marton, L., J.A. Simpson, and J.A. Suddeth (1954): Rev. Sci. Instrum. **25**, 1099
Marty, T. and D. Suter (1995): Helv. Physica Acta, **68**, 504
Mastwijk, H.C., J.W. Thomson, P. van der Straten, and A. Niehaus (1998): Phys. Rev. Lett. **80**, 5516
McClelland, J.J., R.E. Scholten, E.C. Palm, and R.J. Celotta (1993): Sience **262**, 877
McIntyre, D.H., S.K. Mayer, and N.J. Silva (1997): Proc. SPIE, **2995**, 68
Metcalf, H.J. (1986): in *Methods of Laserspectroscopy*, ed. by Y. Prior, A. Ben-Reuven, and M. Rosenbluth (Plenum Press, New York)
Metcalf, H.J. (1989): J. Opt. Soc. Am. B **6**, 2206
Metcalf, H.J. and P. van der Straten (1994): Phys. Rep. **244**, 203
Metcalf, H.J. and P. van der Straten (1999): *Laser Cooling and Trapping* (Springer, Berlin)
Mewes, M.-O., M.R. Andrews, D.M. Kurn, D.S. Durfee, C.G. Townsend, and W. Ketterle (1997): Phys. Rev. Lett., **78**, 582
Migdall, A.L., J.V. Prodan, W.D. Phillips, T.H. Bergeman, and H.J. Metcalf (1985): Phys. Rev. Lett. **54**, 2596
Miller, J.D., R.A. Cline, and D.J. Heinzen (1993): Phys. Rev. A **47**, 4567
Miret-Artés, S., W. Schöllkopf, and J.P. Toennies (1998): private communication
Moi, L. (1984): Opt. Commun. **50**, 349
Molenaar, P.A., P. van der Straten, H.G.M. Heideman, and H. Metcalf (1998): Phys. Rev. A **55**, 605
Mollenstedt, G. and H. Duker (1956): Z. Phys. **145**, 377
Monroe, C., W. Swann, H. Robinson, and C. Wieman (1990): Phys. Rev. Lett. **65**, 1571
Müller, J.H., D. Bettermann, V. Rieger, F. Ruschewitz, K. Sengstock, U. Sterr, M. Christ, M. Schiffer, A. Scholz and W. Ertmer (1995): in *Atomic Physics*, Vol. 14, ed. by D.J. Wineland, C.E. Wieman, and S.J. Smith, p. 240 (AIP, New York)
Nakagiri, K., J. Umezu, Y. Ohta, M. Kajita, N. Kotake, and T. Morikawa (1996): Proc. 5. Symp. Frequency Standards and Metrology, ed. by J.C. Bergquist, p. 417 (World Scientific, Singapore)
Nebenzahl, I. and A. Szöke (1974): Appl. Phys. Lett. **25**, 327
Nellessen, J., J.H. Müller, K. Sengstock, and W. Ertmer (1989a): J. Opt. Soc. Am. B **6**, 2149
Nellessen, J., K. Sengstock, J.H. Müller, and W. Ertmer (1989b): Europhys. Lett. **9**, 133
Nellessen, J., K. Sengstock, J.H. Müller, and W. Ertmer (1989c): in *Atomic Physics*, Vol. 11, ed. by S. Haroche, J.C. Gay, and G. Grynberg, p. 624, (World Scientific, Singapore)
Nellessen, J., J. Werner, and W. Ertmer (1990): Opt. Commun. **78**, 300
Nölle, B., H. Nölle, J. Schmand, and H.J. Andrä (1996): Europhys. Lett. **33**, 261
Nowak, S., T. Pfau, and J. Mlynek (1996): Appl. Phys. B **63**, 203
Nowak, S., T. Pfau, and J. Mlynek (1997): Microelectron. Eng. **35**, 427
Oates, C.W., K.R. Vogel, and J.L. Hall (1996): Phys. Rev. Lett. **76**, 2866

Ol'Shanii, M.A., Yu. Ovchinnikov, and V.S. Letokhov (1993): Opt. Commun. **98**, 77
Opat, G.I., S.J. Wark, and A. Cimmino (1992): Appl. Phys. B **54**, 396
Ovchinnikov, Y.B., I. Manek, and R. Grimm (1997): Phys. Rev. Lett. **79**, 2225
Pauly, H. (1979): in *Atom–Molecule Collision Theory, A Guide to the Experimentalist*, ed. by R.B. Bernstein, p. 111 (Plenum Publishing Company, New York)
Peters, A., K.Y. Chung, and S. Chu (1999): Nature **400**, 849
Phillips, W.D. and H. Metcalf (1982): Phys. Rev. Lett. **48**, 596
Phillips, W.D. (1992): in *Fundamental Systems in Quantum Optics*, ed. by J. Dalibard, J.M. Raimond, and J. Zinn-Justin, p. 165 (Elsevier, Amsterdam)
Phillips, W.D. (1995): in *Atomic Physics*, Vol. 14, ed. by D.J. Wineland, C.E. Wieman, and S.J. Smith, p. 211 (AIP, New York)
Phillips, W.D. (1998): Rev. Mod. Phys. **70**, 721
Pritchard, D.E., E.L. Raab, V.S. Bagnato, C.E. Wieman, and R.N. Watts (1986): Phys, Rev. Lett. **57**, 310
Pritchard, D.E. (1991): in *Atomic Physics*, Vol. 12, ed. by J.C. Zorn and R.R. Lewis, p.165 (AIP, New York)
Pritchard, D.E. (1993): in *Atomic Physics*, Vol. 13, ed. by H. Walther, T.W. Hänsch, and B. Neizert, p.185 (AIP, New York)
Prodan, J.V., A. Migdall, W.D. Phillips, I. So, H. Metcalf, and J. Dalibard (1985): Phys. Rev. Lett. **54**, 992
Raab, E., M. Prentiss, A. Cable, S. Chu, and D.E. Pritchard (1987): Phys, Rev. Lett. **59**, 2531
Raether, H. (1988): *Surface Plasmons* (Springer, Berlin)
Rasel, E.M., M.K. Oberthaler, H. Batelaan, J. Schmiedmayer, and A. Zeilinger (1995): Phys. Rev. Lett. **75**, 2633
Rauch, H., W. Treimer, and U. Bonse (1974): Phys. Lett. **47** A, 369
Rauch, H. (1985): Phys. Bl. **41**, 190
Rauch, H. (1994): Phys. Bl. **50**, 439
Rehbein, S., R.B. Doak, R.E. Grisenti, G. Schmahl, J.P. Toennies, and Ch. Wöll (1999): Proceedings of the Micro- and Nano-Engineering '99 Conference, Rome, Italy, ed. by M. Gentili, E. Di Fabrizio, and M. Meneghini (Elsevier, Amsterdam)
Renn, M.J., D. Montgomery, O. Vdovin, D.Z. Anderson, C.E. Wieman, and E. Cornell (1995): Phys. Rev. Lett. **75**, 3253
Renn, M.J., E.A. Donley, E.A. Cornell, C.E. Wieman, and D.Z. Anderson (1996): Phys. Rev. A **53**, R648
Riis, E., D.S. Weiss, K.A. Moler, and S. Chu (1990): Phys. Rev. Lett. **64**, 1658
Roach, T.M., H. Abele, M.G. Boshier, H.L. Grossmann, K.P. Zetie, and E.A. Hinds (1995): Phys. Rev. Lett. **75**, 629
Rooks, M.J., R.C. Tiberio, M. Chapman, T. Hammond, E. Smith, A. Lenef, R. Rubinstein, D. Pritchard, and S. Adams (1995): J. Vac. Sci. Technol. B **13**, 2745
Rowlands, W.J., D.C. Lau, G.I. Opat, A.I. Sidorov, R.J. Mclean, and P. Hannaford (1996): Austral. J. Phys. **49**, 577
Sakai, H., A. Tarasevich, J. Danilov, H. Stapelfeldt, R.W. Yip, C. Ellert, E. Constant, and P.B. Corkum (1998): Phys. Rev. A **57**, 2794
Salomon, C., J. Dalibard, W.D. Phillips, A. Clairon, and S. Guellati (1991): in *Atomic Physics*, Vol. 12, ed. by J.C. Zorn and R.R. Lewis, p. 73 (AIP, New York)
Santarelli, G., Ph. Laurent, P. Lemonde, A. Clairon, A.G. Mann, S. Chang. A.N. Luiten, and C. Salomon (1999): Phys. Rev. Lett. **82**, 4610
Sapirstein, J. (1995): in *Atomic Physics*, Vol. 14, ed. by D.J. Wineland, C.E. Wieman, and S.J. Smith, p. 45 (AIP, New York)

Savage C.M., S. Marksteiner, and P. Zoller (1993): in *Fundamentals of Quantum Optics* III, ed. by F. Ehlotzky, p. 60 (Springer, Berlin)
Savage, C. (1996): Austral. J. Phys. **49**, 745
Savas, T.A., S.N. Shah, M.L. Schattenburg, J.M. Carter, and H.I. Smith (1995): J. Vac. Sci. Technol. B **13**, 2732
Schieder R., H. Walther, and L. Wöste (1972): Opt. Commun. **5**, 337
Schiffer, M., M. Christ, G. Wokurka, and W. Ertmer (1997): Opt. Commun. **134**, 423
Schmiedmayer, J., M.S. Chapman, C.R. Ekstrom, T.D. Hammond, S. Wehinger, and D.E. Pritchard (1995): Phys. Rev. Lett. **74**, 1043
Schmiedmayer, J., M.S. Chapman, C.R. Ekstrom, T.D. Hammond, D.A. Kokorowski, A. Lenef, R.A. Rubenstein, E.T. Smith, and D.E. Pritchard (1997): in *Atom Interferometry*, ed. by P.R. Berman, p. 2 (Academic Press, New York)
Schöllkopf, W. and J.P. Toennies (1994): Science **266**, 1345
Schöllkopf, W. and J.P. Toennies (1996): J. Chem. Phys. **104**, 1155
Schöllkopf, W. and J.P. Toennies (1997): XVII. Int. Symp. on Molecular Beams, Book of Abstracts, p. 143 (Paris)
Schulze, Th., U. Drodofsky, B. Brezger, J. Stuhler, S. Nowak, T. Pfau, and J. Mlynek (1997): Proc. SPIE, **2995**, 80
Seideman, T. (1997): Phys. Rev. A **56**, 1217, J. Chem. Phys. **106**, 2881, and **107**, 10420
Seifert, W., R.Kaiser, A. Aspect, and J. Mlynek (1994a): Opt. Commun. **111**, 566
Seifert, W., C.S. Adams, V.I. Balykin, C. Heine, Yu. Ovchinnikov, and J. Mlynek (1994b): Phys. Rev. A **49**, 3814
Sengstock, K., U. Sterr, G. Hennig, D. Bettermann, J.H. Müller, and W. Ertmer (1993): Opt. Commun.**103**, 73
Sengstock, K., U. Sterr, J.H. Müller, V. Rieger, D. Bettermann, and W. Ertmer (1994): Appl. Phys. B **59**, 99
Sengstock, K. and W. Ertmer (1995): Adv. At. Mol. Opt. Phys. **35**, 1
Sheeby, B., S.Q. Shang, R. Watts, S. Hatamian, and H. Metcalf (1989): J. Opt. Soc. Am. B **6**, 2165
Shimizu, F., K. Shimizu, and H. Takuma (1992): Phys. Rev. A **46**, R17
Sidorov, A.I., R.J. McLean, W.J. Rowlands, D.C. Lau, J.E. Murphy, M. Walkiewicz, G.I. Opat, and P. Hannaford (1996): Quantum Semiclass. Opt. **8**, 713
Simon, E., P. Laurent, and A. Clairon (1998): Phys. Rev. A **57**, 436
Simsarian, J.E., G. Ghosh, G. Gwinner, L.A. Orozco, G.D. Sprouse, and P.A. Voytas (1996): Phys. Rev. Lett. **76**, 3522
Sleator, T., T. Pfau, V.I. Balykin, and J. Mlynek (1992): Appl. Phys. B **54**, 375
Snadden, M.J., J.M. McGuirk, P. Bouyer, K.G. Haritos, and M.A. Kasevich (1998): Phys. Rev. Lett. **81**, 971
Söding, J., R. Grimm, Y.B. Ovchinnikov, Ph. Bouyer, and Ch. Salomon (1997): Phys. Rev. Lett. **78**, 1420
Stapelfeldt, H., H. Sakai, E. Constant, and P.B. Corkum (1997): Phys. Rev. Lett. **79**, 2787
Sterr, U., K. Sengstock, J.H. Müller, D. Bettermann, and W. Ertmer (1992): Appl. Phys. B **54**, 341
Strohmeyer, P. (1990): Opt. Commun. **79**, 187
Suominen, K.-A. (1996): J. Phys. B **29**, 5981
Suominen, K.-A., Y.B. Band, I.Tuvi, K. Burnett, and P.S. Julienne (1998): Phys. Rev. A **57**, 3724
Swanson, T.B., N.J. Silva, S.K. Mayer, J.J. Maki, and D.H. McIntyre (1996): J. Opt. Soc. Am. B **13**, 1833
Taniguchi, N. (1996): *Nanotechnology* (Oxford University Press, Oxford)

Tanner, C.E., B.P. Masterson, and C.E. Wieman (1988): Opt. Lett. **13**, 357
Tennant, D.M., J.E. Bjorkholm, M.L. O'Malley, M.M. Becker, J.A. Gregus, and R.W. Epworth (1990): J. Vac. Sci. Technol. B **8**, 1975
Theuer, H. and K. Bergmann (1998): Eur. Phys. J. D **2**, 279
Thywissen, J.H., K.S. Johnson, R. Younkin, N.H. Dekker, K.K. Berggren, A.P. Chu, M. Prentiss, and S.A. Lee (1997): J. Vac. Sci. Technol. B **15**, 2093
Timp, G., R.E. Behringer, D.M. Tennant, and J.E. Cunningham (1992): Phys. Rev. Lett. **69**, 1636
Tkachuk, V.M. (1999): Phys. Rev. A **60**, 4715
Toennies, J.P., W. Welz, and G. Wolf (1974): J. Chem. Phys. **61**, 2461
Torii, Y., N. Shiokawa, T. Hirano, T. Kuga, Y. Shimizu, and H. Sasada (1998): Eur. Phys. J. D **1**, 239
Tolansky, S. (1973): *An Introduction to Interferometry* (Longman, London)
Ungar, J., D.S. Weiss, E. Riis, and S. Chu (1989): J. Opt. Soc. Am. B **6**, 2058
Vardi, A., D. Abrashkevich, E. Frishman, and M. Shapiro (1997): J. Chem. Phys. **107**, 6166
Vigué, J. (1995): Phys. Rev. A **52**, 3973
Vladimirskiî, V.V. (1961): Sov. Phys. JETP **12**, 740
Vuletic, V., T.W. Hänsch, and C. Zimmermann (1996): Europhys. Lett. **36**, 349
Walker, T. and P. Feng (1994): Adv. At. Mol. Opt. Phys. **34**, 125
Wallis, H., J. Werner, and W. Ertmer (1993): Comments At. Mol. Phys. **28**, 275
Wang, H., P.L. Gould, and W.C. Stwalley (1997a): J. Chem. Phys. **106**, 7899
Wang, H., J. Li, X.T. Wang, C.J. Williams, P.L. Gould, and W.C. Stwalley (1997b): Phys. Rev. A **55**, R1569
Wasik, G. and R. Grimm (1997): Opt. Commun. **137**, 406
Weidemüller, M., A. Hemmerich, A. Görlitz, T. Esslinger, and T.W. Hänsch (1995): Phys. Rev. Lett. **75**, 4583
Weiner, J. (1995): Adv. At. Mol. Opt. Phys. **35**, 45
Weiner, J., V.S. Bagnato, S. Zilio, and P.S. Julienne (1999): Rev. Mod. Phys. **71**, 1
Weinstein, J.D., R. deCarvalho, J. Kim, D. Patterson, B. Friedrich, and J.M. Doyle (1998): Phys. Rev. A **57**, R3173
Weiquan, C., C. Hongxin, L. Fusheng, S. Wei, S. Shundi, and W. Yuzku (1995): Chinese Phys. Lett. **12**, 269
Weiss, D.S., E. Riis, Y, Shevy, P.J. Ungar, and S. Chu (1989): J. Opt. Soc. Am. B **6**, 2072
Weiss, D.S., B.C. Young, and S. Chu (1993a): Phys. Rev. Lett. **70**, 2706
Weiss, D.S., M. Kasevich, B.C. Young, and S. Chu (1993b): in *Atomic Physics*, Vol. 13, ed. by H. Walther, T.W. Hänsch, and B. Neizert, p. 132 (AIP, New York)
Weitz, M., T. Heupel, and T.W. Hänsch (1996): Phys. Rev. Lett. **77**, 2356
Werner, J., H. Wallis, P. Hildebrandt, and A. Steane (1993): J. Phys. B **26**, 3063
Westphal, P., A. Horn, S. Koch, J. Schmand, and H.J. Andrä (1996): Phys. Rev. A **54**, 4577
Westphal, P., S. Koch, A. Horn, J. Schmand, and H.J. Andrä (1997): Phys. Rev. A **56**, 2784
Weyers, S., E. Aucouturier, C. Valentin, and N. Dimarcq (1997): Opt. Commun. **143**, 30
Wilson, R.J., B. Holst, and W. Allison (1999): Rev. Sci. Instrum. **70**, 2960
Xu, X., V.G. Minogin, K. Lee, Y. Wang, and W. Jhe (1999): Phys. Rev. A **60**,4796
Yen, A., M.L. Schattenburg, and H.I. Smith (1992): Appl. Opt. **31**, 2972
Yin, J., Y. Zhu, W. Jhe, and Z. Wang (1998): Phys. Rev. A **56**, 509
Young, Th. (1802): Phil. Trans. Roy. Soc. London XCII **12**, 387
Zhu, M., C.W. Oates, and J.S. Hall (1991): Phys. Rev. Lett. **67**, 46

Subject Index

Numbers in *italics* refer to pages where the subject is explained in more detail

Absorption spectroscopy 99
Acceleration
　aerodynamic 1, *25*
　mechanical 1, *51–52*
Adiabatic
　approximation 4, 8
　condition 234
　expansion 75, 76, 134
　passage *263*
Adiabaticity condition 233, 234
Aerodynamic acceleration 1, *25–29*
Alignment 252
Alternating gradient focusing 224
Approximation
　adiabatic 8
　Born 4
　Born–Oppenheimer 4
　semiclassical 268, 318
Arc-heated jet source *29–34*
Atom
　cooling *282–284*
　diffraction *120–122, 309–315*
　interferometry *316–322*
　laser 288, 293–295
　lenses *300–303*
　lithography 268, *297–298*, 314
　manipulation 268, 291, 299
　mirror 199, *202–204*, 251, 267, 286, 299, *303–308*, 315, 315
　optics 268, 289, *298–315*
　slowing *273–278*
　traps *284–296*, 304, 308
　waveguide *308*
Auger transition 22, 23

Atomic
　clock *242–244, 291*, 295
　collisions 2, *244–247*, 292
　fountain 192, 287, 291
　funnel *289*
　polarizability 248, 271, 312, 319
　trampoline 287, 306
Atoms, dissociated 48, 49
Autodetachment 2, 18, *21*

Back-illumination technique 40, 41
Beam
　arc-heated *29–34*
　compression *289*
　deflection 12, 98, 108, 197, 210, *278–281*, 315
　profile broadening *112–114*
　slowing *273–278*
Beer's law 108, 109, 321
Binary gas mixtures 26, 114
Binding energy 44, 45, 48, 71, 73, 75, 98, 123
Biot–Savart's law 200
Bolometer detector *67–68*, 69, 256
Born–Oppenheimer approximation 4
Bose–Einstein condensation 268, 288, *293–295*, 296
Buckminsterfullerene 72

Capture cross section *114–117*
Carrier gas *25–28*, 34, 41, 46, 50, 76, 78, 80–83, 87–91, 100, 101, 227, 229, 246, 248
Catalysis 71, 72

Subject Index

Cesium clock *242–244*, 291
Channeltron 41, 56, 61, 67, 68, 314
Charge exchange 2, 3, *4–18*, 20–23, 37, 53, 54, 56–58, 61, 63, 64, 72, 90, *127*
Chirped slowing *276–278*
Cluster
   aggregation 77, *99–100*
   applications *132–136*
   beam deposition 72, 95, 126, *132*
   beam formation *75–96*
   beam lithography 71, 72, *132–134*
   diagnostics *102–131*
   fragmentation 72, 77, 92, 99, 120, 123–125, 127, 133
   impact lithography *133–135*
   in high-energy physics 77, *136*
   magnetic properties 71, 96, 127, 198, 242, 245, *248*
   polarizability 248
   scaling law 71, 130
   size distribution 84, 87, 93, *102–131*
   sources *81–101*
   surface scattering *128–130*
   temperature 76, 77, 80, 85, 87, 88, 98, 99, 103, 104
   velocity selection *125–127*
   yield 84, 86, 87, 90
Coaxial laser spectroscopy 2, *53–55*
Cold atoms 278, 282, 283, 291, 292, 94, 295, 308
Collision
   cascade 43, 45
   cross section 107, 108–111, *175*, 244–248, 292
   experiment 2, 49, 193
   induced fluorescence *69–70*
   process 43, 175, 246–247, 261, 268, 289, *292*
   rate 289
Collisional detachment 2, 3, 18, *21*
Corona discharge 37
Cross-correlation method *161–164*
Cross section
   capture *114–115*
   charge exchange *4–7*, 72, 90, 127
   differential 2, 13, 103, 105, 106, 120–122, 175, 193, 245, 246, 292, 316
   gas kinetic 4, 111
   incomplete total 2
   Rayleigh scattering 105–107
   total *107–110*, 245, 292, 320

De Broglie wavelength 103, 120, 138, 193–195, 292, 293, 302, 308, 310, 312, 313
Debye–Scherrer rings 103
Deconvolution 179, 182, 183
Deflection of atoms by
   atom mirrors *303–307*
   gravity *191*
   inhomogeneous fields 137, *177–191*, 197, 198, *204–230*
   photon recoil *269*
Degree of
   dissociation 36
   freedom 99, 282, 288
   ionization *34*
   orientation 50, 277
   polarization 276, 277, 280
Detachment *18–21*, 63
Detector
   bolometer 67, 151
   calibration 20, 67–70
   channeltron 314
   cold surface ionization 12, 65–67
   collision-induced fluorescence *69*
   condensation target 41, 138
   Langmuir–Taylor 65–67, 98, 151, 191, 317
   laser-induced fluorescence 68, *172–175*
   microchannel plate 57
   pyroelectric 67
   secondary electron emission 20, 61, 65, 68, 69,
   universal beam 65, 120, 151, 157, 159, 169

Subject Index 371

Differential
  cross section 2, 103, 105, 106, 120–122, *175*, 193, 245, 246, 292, 316
  pumping 24, 26, 81–85, 93, 95, 96, 100, 136
Diffraction grating 195, *309–315*
Discharge sources
  arc discharge *29–34*
  corona discharge *37*
  glow discharge *37–38*
  hollow cathode discharge *37*
  laser-sustained discharge *34–36*
  radiofrequency discharge *34*
Dissociation
  degree of 36, 67
  efficiency 30
  molecular 2, 28, 48, 56, 57, 191, 312
Dissociative
  adsorption 50
  attachment 20
  ionization 66
Doped clusters *96–100*
Doppler
  broadening 46, 54, 254
  cooling 283
  effect 15, 17, 55, 280
  limit 268, 283
  profile 16, 176
  shift 15, 55, 58, *171–177*, 257, 274, 275, 276, 280, 291
  temperature 283
  tuning 17, 55
  width 54

Echelle grating *194*
Electric
  dipole moment 177, 218, 220–222, 230, 282, 300
  resonance method 198, 222, *240–242*
  polarizability 248, 259, 271, 312, 318, 319
  quadrupole moment 197, 241, 253

Electron
  affinity 20, 66
  detachment *18–22*, 56, 63
  diffraction *102–105*
  impact dissociation 57
  impact excitation 17, 56, 63
  impact ion source 7, 46, 135
  impact ionization 9, 20, 65, 83, 99, 120, 124, 125, 134, 135, 171, 255, 257, 262
  induced fluorescence *130*
  stimulated desorption 1, *50*
Electronic excitation 2, 27, 69, 70, 86, 265
Electrostatic trap *230–233*, 289
Energy
  defect 4, 5, 6, 8
  loss 62, 135
Equation
  Poisson 199, 200, 203, 232, 237
  Saha 33, 34
  Saha–Langmuir 66
Equilibrium fraction of dimers 75
Equivalent nozzle diameter 80
Evanescent wave 287, *305–307*, 308, 309, 314, 315
Excited state, lifetime 3, 12, 22, *57–59*, 157, 175, 256, 258, 259, 262, 270, 274, 275, 277, 291
Excited cluster 77, *101*

Fast beam
  applications *53–65*
  coaxial laser spectroscopy 2, *53–56*
  detection *65–70*
  lifetime measurements *57–59*
  of metastable particles 2, 8, *13-17*, 30, 33–35, 37, 59
  of radicals 9, 20, 37
  of Rydberg particles 2, 8, 16, *17–18*, 35
Fermi limit 22
Field ionization 17, 18, 101
Fizeau principle 159, *139*

Flow reactor 77, 92, *95–96*, 100, 101, 126
Fluorescence
　collision-induced 69, 70
　laser-induced 49, 50, 68, 166, 175, 246, 256, 259
Frequency standards 199, *242–244*, 268, 288
Fresnel zone plate *300–302*
Fullerenes 28, 67, 72, 73, 96, 101

Gas aggregation 76, 77, *82–85*, 86, 87, 88, 97, 125, 126
Gas discharge sources *29–38*
Gas–surface interaction *249*
Glow discharge *37*
Gravitational mass 191, 313
Gravito-optical trap 282, 286, 287

Heat pipe principle 11–13, *14*
Helium beams 15, 67, 89, 121, 138, 152, 160, 193, 194, 312
Hollow cathode discharge *37*

Inhomogeneous electric fields
　Rabi-field *222*, 236
　two-pole field *222*
　multipole fields *224–229*
Inhomogeneous magnetic fields
　two-wire field *206–210*, 236
　hexapole fields *214–219*, 236, 237
　quadrupole field *219–220*
　quadrupole sector field *211–213*
　Intensity of fast beams 7, 9, 10, 20, 24, 26, 34, 36, 45–47, 50, 54, 68
Internal energy 6, 9, 92
Ioffe trap *233*, 239
Ion source 3, 6, 7. 9, 13, 40, 41, 47, 53, 58, 63, 65, 77, 86, 91, 92, 123, 132, 135
　cluster 77, 90, 135
　liquid metal 77, 86, *91–92*
　pulsed arc 76, 77, *89–91*
Ion–surface neutralization *22–24*
Isotope
　radioactive 3, 53, 55, 220, 296

rare 286
separation 152, 280
shift 53, 55

Lambert's cosine law 129
Langmuir's space charge law 9, 10
Langmuir–Taylor detector *65–67*, 98, 151, 179, 191, 243, 317
Larmor frequency 233, 234, 235
Laser
　ablation 1, *39–42*, 46, 72, 76, 84, 85, *87–90*, *100–101*
　detonation source *35–36*
　induced fluorescence 49, 50, 68, 166, 175, 246, 256, 259
　induced pyrolysis 77, *92–96*,127
　sustained plasma *35–36*
Leidenfrost phenomenon 31, 128
Lennard–Jones potential 99, 121, 321
Light scattering *105–107*
Liquid metal ion source 77, *91–92*
Log-normal distribution *116–117*

Mach–Zehnder interferometer *317*
Magnetic
　dipole moment 137, 184, 197, *204*, 218, 219, 232, 241, 252, 268, 282, 304, 315
　mirror *303–304*
　resonance 198, 234, *240–244*
Magnetically suspended rotor 52, 152, 165
Magnetooptical trap 251, 282, 284–286, 287–290, 293, 295, 304, 308, 312
Magnetostatic trap *230–234*, 240
Majorana transitions *234*, 235
Massey criterion 5, 6
Matter wave
　coherent 288, 294
　interferometer *316–322*
　refractive index *319–321*
Maxwellian distribution 45, 46, 137, 150, 153, 179, 182, 268, 277, 285

drifting 31, 36, 39, 42
Mechanical acceleration 1, *51 - 52*
Merged beams 2, *59–62*
Mie theory 105
Mixed clusters *96–101*
Microchannel plate 56, 57, 69, 192, 313
Molecular beam epitaxy 47, 133
Molecular collisions 2, 50, 94, 241, *244–247*
Moving mirror *307*
Multipass cell 19, 255
Multiphoton dissociation *93–94*

Nanoparticles 74
Nanoscaled sieves *130–132*
Neutral beam injection 3, 21, *62*
Neutralization
  chamber 6–9, 11–13, *14*, 17
  efficiency 8, 19, 21, 24, 63
Nonresonant charge exchange 5, 8
Nozzle shape *78–80*
Nuclear
  fusion *62*
  moments 234
Nucleation rate 79

Optical
  dipole force *271–273*
  dipole trap 282, *286*
  lattice 268, *272–273*
  molasses 278, *282–283*, 284–286, 289, 290
  pumping 199, 250, *252*, 253–255, 260
Orbiting resonance 322
Oriented reactants *49–50*
Oscillator strength 57, 58

Parity nonconservation 57
Particle capture *97–99*
Penning discharge 38, 45
Photoassociative spectroscopy *295–296*
Photodetachment 2, *18–21*

Photodissociation 21, 50, 56, 57, *260*
Photofragment spectroscopy *56*
Photoionization 123–124, 250, 254, 262
Photolysis *48–50*, 96, 198
Photon recoil 267, *269–271*, 272, 273, 275, 278, 282
Photosensitization *93–95*
Pick-up source *97, 98,* 100, 114, 125
Plasma diagnostics 11, *63–64*
Plasma-heated source *29–32*
Polarizability, electric 198, 248, 271, 312, 318, 319
Potential
  atom–cluster *121*
  Lennard–Jones 99, 121, 321
Power broadening 254, 269, 277
Pseudorandom sequence *162, 164*
Pseudostatistical method *162*
Pulsed arc discharge 76, 77, *89–90*, 101
Pyroelectric detector *67, 68*

Quadrupole
  field *201*, 206, *219–220*, 225–227, 231–235, 238–242, 246, 285, 286, 309
  sector field *211–213*
Quartz microbalance 83
Quench lamp *15–16*

Rabi-field 197, 245, 249
Radiation pressure *268*
Radioactive isotopes 3, 47, 53, 55
Rainbow scattering 120, 177, 316
Rayleigh scattering *105–107*
Reactive collisions 2, 175, 246, 256
Recoil
  nuclei *52–53*
  temperature 283
Rydberg
  atoms *16–17*, 262

states 2, 8, 9, 17, 18, 35,157, 166, 259, 262, 263

Saha's equation 33, 34
Saha–Langmuir equation 66
Scaling
  law 77, 130
  parameter 78, 80, 81, 106, 107, 112, 130
Scattering amplitude 102, 103, 320, 321
Secondary electron emission 1, 20, 61, 65, *68–69*
Seeded beam technique *24,* 28, 29
Simulations
  molecular dynamics 104, 133, 134
  Monte Carlo 182
Slow atoms 137, 184, 192, 196, 267, 268, 275, 280, 282, 291, 292, 300, 312, 313, 315
Shock wave 1, *52*
Slotted cylinder velocity selector *152–155*
Slotted disk velocity selector 137, *139–153*, 154, 156, 164, 180, 188
Spin-flip, 233, 235
Sputtering 1, 2, 13, 24, *42–48*, 76, 84, 85, *86–87*
  threshold energy *44*
  yield 43–45
Stagnation pressure 76, 78, 79, 81, 89, 98, 104–107, 129, 130
Stark effect 199, *221–226*, 229, 230, 238, 253
Surface
  ionization 6, 11, 12, 13, *65–67*
  erosion 76, *85–92*
  neutralization 1, *22–24*
Supersonic jet 75, 76, *77–81*

Target beam 6, 12, 13, 16, 107, 109, 116, 152, 237, 245, 246, 250, 259, 262, 320
Thermal detector 65, *67*
Time-of-flight method *156–169*

calibration *169–171*
Total cross section 2, 107, *108–110*, 245
Tuning fork chopper 164
Two-photon ionization 124, 250

Universal detector 65, 120, 151, 157, 159, 169

Velocity distribution
  of sputtered particles *45–46*
  drifting Maxwellian 32, 37, 39, 43
Velocity loss *121–123*
Velocity resolution 160, *163*, 168, 174, 183, 188, 195, 202, 207, 209, 216
Velocity selection
  mechanical means *137–156*
  time-of-flight methods *156–169*
  Doppler shift *171–177*
  inhomogeneous magnetic fields *177–191*
  deflection by gravity *191–193*
  diffraction from gratings *193–195*
  deflection by photons *195–196*
Velocity selector
  calibration *150–152*
  sidebands *143–146*
  slotted cylinder *152–155*
  slotted disk *137–153*
  slotted plate *154*
  slotted ring *155*
  transmission *139–144*
Velocity slip 26, 130, 195

Wave
  evanescent *305–307*
  matter 288, 294, 316, 319
  plane 269, 305, 309
  plasmon wave 307
Wien filter 9, 18, 48, 56

Zeeman slowing 273, *274–275*, 277, 290

Conversion Factors for Pressure and Energy Units

Pressure Units

Unit	bar	mbar	Pa = N/m²	atm	Torr	mTorr
bar	1	10^3	10^5	0.9869	0.750×10^3	0.750×10^6
mbar	10^{-3}	1	10^2	0.9869×10^{-3}	0.750	0.750×10^3
Pa	10^{-5}	10^{-2}	1	0.9869×10^{-5}	0.750×10^{-2}	0.750×10^1
atm = 760 Torr	1.013	1.013×10^3	1.013×10^5	1	0.760×10^3	0.760×10^6
Torr	1.333×10^{-3}	1.333	1.333×10^2	1.316×10^{-3}	1	10^3
mTorr	1.333×10^{-6}	1.333×10^{-3}	1.333×10^{-1}	1.316×10^{-6}	10^{-3}	1

Energy Units

Unit	erg	K	eV	kcal/Mol	cm^{-1}	hartree	J
erg	1	7.2439×10^{15}	6.2418×10^{11}	1.439×10^{13}	5.03×10^{15}	6.2418×10^{11}	10^{-7}
K	1.3805×10^{-16}	1	8.6168×10^{-5}	1.9865×10^{-3}	6.89	8.6168×10^{-5}	1.3805×10^{-23}
eV	1.6021×10^{-12}	1.1605×10^4	1	2.3054×10^1	8.067×10^3	3.6752×10^{-2}	1.6022×10^{-19}
kcal/mol	6.9493×10^{-14}	5.0340×10^2	4.3377×10^{-2}	1	0.3498	4.3377×10^{-2}	6.9467×10^{-21}
cm^{-1}	1.9863×10^{-16}	1.4388	1.2391×10^{-4}	2.8591	1	1.2396×10^{-4}	1.986×10^{-23}
hartree	4.3593×10^{-11}	3.1577×10^5	2.72097×10^1	6.2728×10^2	2.195×10^5	1	4.359×10^{-18}
J	10^7	7.2439×10^{22}	6.2420×10^{18}	1.4395×10^{20}	5.035×10^{22}	2.294×10^{17}	1

Druck: Strauss Offsetdruck, Mörlenbach
Verarbeitung: Schäffer, Grünstadt

Springer Series on
ATOMIC, OPTICAL, AND PLASMA PHYSICS

Editors-in-Chief:

Professor G.F. Drake
Department of Physics, University of Windsor
401 Sunset, Windsor, Ontario N9B 3P4, Canada

Professor Dr. G. Ecker
Ruhr-Universität Bochum, Fakultät für Physik und Astronomie
Lehrstuhl Theoretische Physik I
Universitätsstrasse 150, 44801 Bochum, Germany

Editorial Board:

Professor W.E. Baylis
Department of Physics, University of Windsor
401 Sunset, Windsor, Ontario N9B 3P4, Canada

Professor R.N. Compton
Oak Ridge National Laboratory
Building 4500S MS6125, Oak Ridge, TN 37831, USA

Professor M.R. Flannery
School of Physics, Georgia Institute of Technology
Atlanta, GA 30332-0430, USA

Professor B.R. Judd
Department of Physics, The Johns Hopkins University
Baltimore, MD 21218, USA

Professor K.P. Kirby
Harvard-Smithsonian Center for Astrophysics
60 Garden Street, Cambridge, MA 02138, USA

Professor P. Lambropoulos, Ph.D.
Max-Planck-Institut für Quantenoptik, 85748 Garching, Germany, and
Foundation for Research and Technology – Hellas (F.O.R.T.H.),
Institute of Electronic Structure & Laser (IESL),
University of Crete, PO Box 1527, Heraklion, Crete 71110, Greece

Professor P. Meystre
Optical Sciences Center, The University of Arizona
Tucson, AZ 85721, USA

Professor J. Mlynek
Universität Konstanz
Universitätsstrasse 10, 78434 Konstanz, Germany

Professor Dr. H. Walther
Sektion Physik der Universität München
Am Coulombwall 1, 85748 Garching/München, Germany

Springer Series on
Atoms+Plasmas

Editors: G. Ecker   P. Lambropoulos   I. I. Sobel'man   H. Walther
Founding Editor: H. K. V. Lotsch

1 **Polarized Electrons**   2nd Edition
   By J. Kessler

2 **Multiphoton Processes**
   Editors: P. Lambropoulos and S. J. Smith

3 **Atomic Many-Body Theory**
   2nd Edition
   By I. Lindgren and J. Morrison

4 **Elementary Processes
   in Hydrogen-Helium Plasmas**
   Cross Sections
   and Reaction Rate Coefficients
   By R. K. Janev, W. D. Langer, K. Evans Jr.,
   and D. E. Post Jr.

5 **Pulsed Electrical Discharge
   in Vacuum**
   By G. A. Mesyats and D. I. Proskurovsky

6 **Atomic and Molecular Spectroscopy**
   3rd Edition
   Basic Aspects and Practical Applications
   By S. Svanberg

7 **Interference of Atomic States**
   By E. B. Alexandrov, M. P. Chaika
   and G. I. Khvostenko

8 **Plasma Physics**   3rd Edition
   Basic Theory
   with Fusion Applications
   By K. Nishikawa and M. Wakatani

9 **Plasma Spectroscopy**
   The Influence of Microwave
   and Laser Fields
   By E. Oks

10 **Film Deposition
   by Plasma Techniques**
   By M. Konuma

11 **Resonance Phenomena
   in Electron-Atom Collisions**
   By V. I. Lengyel, V. T. Navrotsky
   and E. P. Sabad

12 **Atomic Spectra
   and Radiative Transitions**   2nd Edition
   By I. I. Sobel'man

13 **Multiphoton Processes
   in Atoms**   2nd Edition
   By N. B. Delone and V. P. Krainov

14 **Atoms in Plasmas**
   By V. S. Lisitsa

15 **Excitation of Atoms
   and Broadening of Spectral Lines**
   2nd Edition
   By I. I. Sobel'man, L. Vainshtein,
   and E. Yukov

16 **Reference Data
   on Multicharged Ions**
   By V. G. Pal'chikov and V. P. Shevelko

17 **Lectures
   on Non-linear Plasma Kinetics**
   By V. N. Tsytovich

18 **Atoms and Their
   Spectroscopic Properties**
   By V. P. Shevelko

19 **X-Ray Radiation
   of Highly Charged Ions**
   By H. F. Beyer, H.-J. Kluge,
   and V. P. Shevelko

20 **Electron Emission
   in Heavy Ion–Atom Collision**
   By N. Stolterfoht, R. D. DuBois,
   and R. D. Rivarola

21 **Molecules and Their
   Spectroscopic Properties**
   By S. V. Khristenko, A. I. Maslov,
   and V. P. Shevelko

22 **Physics of Highly Excited Atoms
   and Ions**
   By V. S. Lebedev and I. L. Beigman

23 **Atomic Multielectron Processes**
   By V. P. Shevelko and H. Tawara

24 **Guided-Wave-Produced Plasmas**
   By Yu. M. Aliev, H. Schlüter, and
   A. Shivarova

25 **Quantum Statistics of Strongly
   Coupled Plasmas**
   By D. Kremp, W. Kraeft, and
   M. Schlanges

26 **Atomic Physics with Heavy Ions**
   By H. F. Beyer and V. P. Shevelko